MW01482453

Increasing Occupational Health and Safety in Workplaces

This book is dedicated to my co-editor, Ron Burke, who passed away suddenly, but peacefully, on 19 March 2019. Ron took me under his wing when I was a new graduate and we worked together ever since – for more than 30 years. We shared an interest in workplace health and well-being, and collaborated on many projects related to improving work environments. He was an extremely prolific researcher, a good colleague and scientific mentor, and not least, a good friend. I will miss him greatly.

Increasing Occupational Health and Safety in Workplaces

Individual, Work and Organizational Factors

Edited by

Ronald J. Burke

Emeritus Professor, Schulich School of Business, York University, Canada

Astrid M. Richardsen

Professor Emerita, BI Norwegian Business School, Norway

 Edward Elgar
PUBLISHING

Cheltenham, UK • Northampton, MA, USA

Published by
Edward Elgar Publishing Limited
The Lypiatts
15 Lansdown Road
Cheltenham
Glos GL50 2JA
UK

Edward Elgar Publishing, Inc.
William Pratt House
9 Dewey Court
Northampton
Massachusetts 01060
USA

A catalogue record for this book
is available from the British Library

Library of Congress Control Number: 2019946810

This book is available electronically in the **Elgar**online
Business subject collection
DOI 10.4337/9781788118095

Printed on elemental chlorine free (ECF)
recycled paper containing 30% Post-Consumer Waste

ISBN 978 1 78811 808 8 (cased)
ISBN 978 1 78811 809 5 (eBook)

Typeset by Servis Filmsetting Ltd, Stockport, Cheshire
Printed and bound in the USA

Contents

Figures

Tables and boxes

TABLES

BOXES

Contributors

Çakıl Agnew is Assistant Professor of Psychology at Heriot Watt University Dubai Campus. Çakıl received her PhD from the University of Aberdeen in 2010. She worked as a postdoctoral research fellow in the Industrial Psychology Research Centre at the University of Aberdeen and joined the Heriot Watt University in 2014. Her research focuses on the impact of safety culture and leadership on performance at work in high-risk industries. She has published and presented her work in international journals and conferences.

Christopher T. Austin is a PhD student at Washington State University Vancouver working in the Coalition for Healthy and Equitable Workplaces laboratory. His research centres on threats to occupational health and safety, in particular accident under-reporting, among vulnerable populations, with a primary focus on employees with stigmatized identities.

Alan Becker (PhD, MPH) is an Associate Professor of Environmental and Occupational Health in the Institute of Public Health in the College of Pharmacy and Pharmaceutical Sciences at Florida A&M University. He is an environmental toxicologist with emphasis in chemical and pesticide exposure.

Erica L. Bettac is a PhD student in the Industrial/Organizational Psychology programme at Washington State University Vancouver. A member of the Coalition for Healthy and Equitable Workplaces laboratory, she primarily examines work and family conflict, work–life balance, and several work- and health-related outcomes. Specifically, she is interested in the interaction between these constructs and cross-cultural research, most recently focusing on traditionally employed and self-employed individuals.

Gunhild Bjaalid is an organizational psychologist at the University of Stavanger with a PhD in leadership and organization. Gunhild specializes in human resource management (HRM), clinical psychology and organizational psychology, and has several years of both clinical and organizational work experience as a psychologist.

Carmel Bofinger is Associate Professor at the Minerals Industry Safety and Health Centre at The University of Queensland. Her role involves education and training, consultancies, project work and research in health and safety in the mining and minerals processing industry. She has previously worked independently and for mining companies and government agencies as a specialist occupational health and safety (OHS) and risk management consultant.

Ronald J. Burke passed away during the completion of this book. He was Emeritus Professor of Organizational Studies, Schulich School of Business, York University in Toronto. His most recent research interests included creating psychologically healthy workplaces, women in management, occupational health and safety, and work and health. The Founding Editor of the *Canadian Journal of Administrative Sciences*, he served on the editorial boards of over 25 journals. A Fellow of the Canadian Psychological Association, he was recently awarded a distinguished scholar practitioner award by the Academy of Management.

Sharon Clarke is Professor of Organizational Psychology at Alliance Manchester Business School, University of Manchester. She has research interests in safety culture, safety climate, leadership, well-being and health. Her work has been widely published in leading academic and practitioner journals, and co-authored books. She is currently Editor-in-Chief for the *Journal of Occupational and Organizational Psychology* and Associate Editor for the *Journal of Occupational Health Psychology*.

David Cliff is Professor of Occupational Health and Safety in Mining at The University of Queensland. His primary role is providing education, applied research and consulting across a wide spectrum of health and safety in the mining and minerals processing industry. He has provided expert assistance in the areas of health and safety to the mining industry for over 28 years.

David M. DeJoy is Professor Emeritus of Health Promotion and Behavior in the College of Public Health at the University of Georgia and founding director of the Workplace Health Group. He has been involved in workplace safety and health for over 35 years as a researcher, instructor and consultant. His areas of research include: climate and culture, work organization, safe work practices, integrated programming, and theory-based intervention design and intervention effectiveness.

Maureen F. Dollard is Professor of Work and Organizational Psychology and Director of the Asia Pacific Centre for Work Health and Safety at the University of South Australia, and Honorary Professor at Nottingham University. Her research concerns workplace psychosocial factors and she has published five edited books and 170 papers or book chapters. She is on the editorial board for *Work and Stress, Journal of Organizational Behavior* and *European Journal of Work & Organizational Psychology*.

Vinita Duraisingam is a registered psychologist and holds a Master's degree in organizational psychology. She has more than ten years of experience in drug and alcohol-related research and workforce development. Her research and practice interests are in workforce issues, including employee health and well-being, drug and alcohol use among workers, career development and workforce sustainability.

Mari-Amanda Dyal earned her PhD in health promotion and behavior from the University of Georgia. Her research interests are rooted in workplace health promotion in the areas of job demands and job resources. Her health education and promotion background spans many specialties, settings and populations. She is currently an Assistant Professor at Kennesaw State University in the Public Health Education programme.

Rhona Flin is Professor of Industrial Psychology, Aberdeen Business School, Robert Gordon University and Emeritus Professor of Applied Psychology, University of Aberdeen. Her work examines human performance in high-risk work settings, such as healthcare, aviation and the energy industries, with current studies focusing on safety culture and non-technical skills.

Laura Fruhen is a Lecturer in Applied Psychology at the University of Western Australia's School of Psychological Sciences. She received her PhD at the University of Aberdeen in Scotland in 2012 for her research on senior managers' influence on safety in organizations. Her main research interests concern the links of culture, leadership, work design, and teamwork with health, well-being and safety at work.

Joseph G. Grzywacz (PhD) is Chair and Norejane Hendrickson Professor of Family and Child Sciences at Florida State University. His research focuses on the health-related implications of everyday work and family life, and how these experiences contribute to health inequalities among those in poverty, and among racial and ethnic minorities. Much of this research has been supported by the National Institutes of Health and the Centers for Disease Control, and focuses on Latino workers in the agricultural and construction sectors.

Duygu Biricik Gulseren holds an MA degree in social and organizational psychology. She is currently a PhD candidate in industrial/organizational psychology at Saint Mary's University. Her research interests include leading healthy workplaces, chronic pain at work and work–family interface.

Oliver Hamlet is a PhD student at the University of Aberdeen. Having studied human factors since his undergraduate degree, he is now exploring non-technical skills in relation to high-risk environments. Currently his research is focused on exploring the non-technical skills of offshore transport and search and rescue helicopter pilots.

Amy Irwin attained a PhD in psychology from the University of Sheffield. She currently works as a Lecturer at the University of Aberdeen. She has studied non-technical skills since 2010 with a particular focus on agriculture, aviation and healthcare. She is currently the project lead for NTS Ag, a research group devoted to developing non-technical skills research in farming.

Maria Therese Jensen has a PhD in psychology and is a registered nurse. Currently she is a postdoctoral at the University of Stavanger working on research concerning emotional climate, burnout and performance. Previous research areas within work–family conflict, health and well-being.

Sabine Kaiser is a postdoctoral student in the Regional Center for Child and Adolescent Mental Health at UiT The Arctic University of Norway. Her PhD work was related to collaboration, burnout, engagement and service quality in the healthcare sector. Her research interests also include research methods, meta-analysis and mental health.

E. Kevin Kelloway is the Canada Research Chair in Occupational Health Psychology and Professor of Psychology at Saint Mary's University, Nova Scotia. His research focuses on organizational and occupational health psychology including the study of occupational stress, safety, leadership and well-being.

Danielle Lim is a research assistant at the Ontario Institute for Studies in Education and an undergraduate student in the Department of Psychology at the University of Toronto. She is broadly interested in the impacts of stress, trauma and violence on the mental health of children and families.

Helen Lingard is Professor and Director of the Centre for Construction Work Health and Safety Research at RMIT University, Melbourne. She undertakes applied research into occupational health and safety in the construction industry and has completed research projects examining organizational and workplace safety cultures, safety in design, client safety leadership, occupational health, work–life balance and wellbeing.

John S. Luque is Associate Professor in the Institute of Public Health in the College of Pharmacy and Pharmaceutical Sciences at Florida A&M University. He received his PhD from the University of South Florida. He has served as principal investigator (PI) on multiple research projects on cultural factors related to cancer screening and attitudes toward cancer prevention in Latino and African-American populations in the US South. He has published over 60 peer-reviewed articles in scholarly journals.

Jen MacGregor is Senior Research Associate at Western University and Community Research Associate at Western's Centre for Research & Education on Violence against Women & Children. She was a lead researcher on the first Canadian national study on the impacts of domestic violence on the workplace. Currently her work focuses on health and social service sector responses to family violence.

Barb MacQuarrie is Community Director, Centre for Research & Education on Violence against Women & Children, Western University. She coordinated the first Canadian national study on the impacts of domestic violence on the workplace and led the implementation of similar multilingual surveys in several other countries. Currently she is working to adapt new methodology to measure the costs of domestic violence to medium- and large-sized workplaces.

Monica Martinussen is Professor of Psychology in the Regional Center for Child and Adolescent Mental Health at UiT The Arctic University of Norway and Adjunct Professor at the Norwegian Defense University College. Her research interests include research methods and psychometrics, aviation psychology, mental health, and work- and organizational-psychology. She is currently Associate Editor of the *Journal of Aviation Psychology and Applied Human Factors*.

Katharine McMahon is a PhD student of Portland State University's Industrial and Organizational programme. Her research interests include workplace mistreatment, occupational stress, and motivation and affect. Recently she has been working on projects regarding the nursing population's exposure to aggression, self-regulation processes and workplace productivity.

Aslaug Mikkelsen is Professor in Management at the University of Stavanger Business School and Adjunct Professor at Stavanger University Hospital. She has conducted several research projects in occupational health, management, corporate social responsibility and strategic human resource management, with international publications from these projects. She was elected rector at the University of Stavanger for the period 2007–11.

Lee S. Newman is Professor of Environmental and Occupational Health and Professor of Epidemiology at University of Colorado, Anschutz Medical Campus, USA. He is the founding Director of the University of Colorado's Center for Health, Work & Environment. His research spans the field of occupational safety and health, most recently focusing on the adoption of Total Worker Health® concepts in small enterprises and in international agribusinesses.

Valerie O'Keeffe is Lecturer, Human Factors and Organizational Safety Management and a member of the Asia Pacific Centre for Work Health and Safety at the University of South Australia. Her research and consulting focus on work health and safety practice across diverse industries and in policy, and on understanding people and their interactions using a human factors systems perspective. She has a special interest in healthcare, psychosocial factors and qualitative research.

Espen Olsen is Professor of Organization and Leadership and Head of Department of Innovation, Leadership and Marketing at University of Stavanger Business School. His research areas are in job characteristics, culture and climate, occupational health and safety, work motivation, management and leadership. His research has involved organizations in healthcare and the petroleum industry.

Laura Olszowy is a PhD candidate in the Faculty of Education at Western University and research assistant for the Centre for Research & Education on Violence against Women & Children. Her research has focused on the impact of domestic violence on families and communities. More recently her work has focused on the child protection system response to domestic violence.

Lauren S. Park is a PhD student in industrial and organizational psychology at Portland State University. Her Master's thesis focused on incivility and emotional well-being in the field of nursing. Her research interests include workplace diversity and inclusion, health and well-being, and quantitative methods.

Ken Pidd is Associate Professor and Deputy Director of the National Centre for Education and Training on Addiction at Flinders University. He has extensive experience in the alcohol and drug field and has produced numerous publications concerning the workplace and employee drug use. He also provides consultancy on this issue to government and non-government bodies, employer and employee groups, and individual employers.

Silvia Pignata is a Senior Lecturer in Aviation within the School of Engineering at the University of South Australia. She has research expertise in workplace psychosocial factors and organizational interventions, particularly their potential to reduce psychological strain. She has published book chapters, refereed journal articles and international conference papers in the areas of work stress, interventions, employee awareness of organizational support and email load.

Tony Pooley is an adjunct Associate Professor and Risk Lecturer in the School of Engineering at the University of South Australia. His Associate Professor role is an honorary one in recognition of his achievements in risk management in Australia. He presents regularly at major industry conferences on risk management and process safety, and is co-author of the organizational risk management book, *Risk Bandits: Rescuing Risk Management from Tokenism*.

Tahira M. Probst is Professor of Psychology at Washington State University Vancouver, where she directs the Coalition for Healthy and Equitable Workplaces laboratory. Her research focuses on the relationship between workplace stressors and safety-related outcomes with a particular emphasis on accident under-reporting. She is currently Co-Editor of *Stress & Health* and sits on the editorial boards of several journals in the field of occupational health psychology.

Astrid M. Richardsen is Professor Emerita at BI Norwegian Business School and her research interests include occupational health, job stress and burnout, work engagement and various work outcomes. Her research has been published in international journals and edited books, and presented at international conferences. She was Department Head from 2007 to 2015.

Ann Roche is Professor and Director of the National Centre for Education and Training on Addiction at Flinders University. She has held academic posts at several universities and has worked in clinical, public health and community settings. Her professional activities have primarily focused on alcohol and other drug issues, particularly policy development, best practice, workforce development and research dissemination.

Michael D. Saxton is a PhD candidate in the Faculty of Education at Western University. His research has largely focused on the impact domestic violence has on victims and their families. Recently he has been working with the Canadian Domestic Homicide Prevention Initiative completing research on the police response to domestic violence.

Natalie V. Schwatka is Assistant Professor in the Department of Environmental and Occupational Health at University of Colorado, Anschutz Medical Campus. She is also a member of the Center for Health, Work & Environment, one of six centres of excellence for Total Worker Health® (TWH). She conducts research on TWH leadership and culture and climate assessment and interventions in high-risk work environments.

Katreena Scott is Associate Professor in the Department of Applied Psychology and Human Development at the University of Toronto and the Canada Research Chair in Family Violence Prevention and Intervention.

She leads an applied research programme aimed at reducing violence in family relationships, with specific expertise in addressing violence perpetration in men and fathers.

Todd D. Smith is Assistant Professor in the Department of Applied Health Science at the Indiana University – Bloomington School of Public Health. His research focuses on occupational safety and health and occupational health psychology, with a primary focus on the influence of psychosocial and organizational factors on worker safety, health and well-being.

Philip Matthew Stinson Sr is Professor of Criminal Justice at Bowling Green State University. His primary area of research is police behaviors, including police crime, police corruption and police misconduct. He is the principal investigator on a research project to study police crime across the United States. His research has been published in numerous peer-reviewed journals and has been featured in many news publications.

Liliana Tenney is the Associate Director for Outreach at the Center for Health, Work & Environment and a Senior Instructor at the Colorado School of Public Health, where she is currently completing her DrPH. Her work focuses on Total Worker Health® (TWH) research and public health practice, with expertise in research translation and dissemination, program planning, intervention science, and strategic partnerships.

Tabatha Thibault holds an MSc in applied psychology focusing on industrial/organizational psychology. She is currently a PhD candidate in industrial/organizational psychology at Saint Mary's University. Her research interests include the Dark Tetrad of personality, incivility and cyber deviance in the workplace, and leadership.

Nadine Wathen is Professor at Western University and a research scholar at Western's Centre for Research & Education on Violence against Women & Children. Her research examines the health and social service sector response to violence against women and children, interventions to reduce health inequities, and the science of knowledge mobilization, with a focus on partnerships to enhance the use of research in policy and practice.

Liu-Qin Yang is Associate Professor of Psychology at Portland State University. Her research interests include workplace mistreatment and occupational stress, affect and motivation, and quantitative methodologies. Recently she has been working on projects focused on the role of affective shift in influencing worker motivation and productivity, interventions to prevent workplace mistreatment toward workers in high-risk industries and reliability issues in experience sampling research.

PART I

Introduction

1. Increasing occupational health and safety in workplaces: why it matters*

Ronald J. Burke

There are at least two ways to emphasize the importance of the health and safety themes that form the basis of this chapter. The first is at the big-picture level. Biddle (2004), reported that over 100000 American workers died annually from occupational injuries. Her research developed a computer program estimating the national cost from these injuries, a figure of over US$100 million annually. The International Labour Organization (ILO) estimates that about 2.3 million workers die annually from occupational accidents and diseases. Also, 1 million workers will suffer a workplace accident every day. These accidents cost an estimated 4 percent of gross domestic product (US$2.8 trillion) owing to organizational and individual costs, for example, lost working time, worker's compensation, medical expenses and interruption to production. Finally, surveys of safety in a variety of industries, both small and large, indicate that many use outmoded methods to manage safety information, making little if any use of these data. This is particularly true of small business and businesses in general in the developing world. Finally, workers in many countries suffer exploitation at the hands of organizations in the developed world in terms of exposure to unsafe conditions and the handling of hazardous compounds.

The second way is to consider occupational health and safety at the micro level, the experiences every day of women and men at work. Here are some examples reported in the public domain, many on the same day.

Consider the following:

- Hurricane Harvey dumped over 50 inches of rain over parts of Texas in late August 2017 shutting down the majority of organizations, causing a few deaths and billions of dollars of destruction. You cannot fight Mother Nature but you can prepare for such disasters.
- A US military plane carrying 18 soldiers on route to Arizona for training, crashed on 11 July 2017 in Mississippi killing all 18. Witnesses said they had heard loud explosions when the plane was still high in the sky. Foul play has been ruled out.

- There has been a large increase in the manufacture of electric motor vehicles which is expected to accelerate. The Tesla battery contains about 150 pounds of graphite, a toxic product so dangerous it is mainly mined in China where little attention is paid to worker safety. China produces 66 percent of the world's graphite; India is second at 14 percent. Villages in China near where graphite is mined get enveloped in a grey shroud, covering homes, clothes and crops, and polluting drinking water. A Canadian company had planned to enter this lucrative market mining graphite in Canada but failed to meet safety standards. A related important issue is how do we dispose of these batteries.
- Worker death rates in China are 20 times higher than those in the UK.
- A study carried out in 20 countries by the Organisation for Economic Co-operation and Development (OECD) of the occurrence of physicians leaving objects in patients following surgery (for example, sponges, towels and knives) placed Canada in the top quarter of these mistakes.
- A US missile destroyer collided with a Japanese cargo ship killing seven US marine crew in the Sea of Japan in June 2017. The boat commander, two other senior officers and a few lower ranked seamen were relieved of duty but not fired as a result.
- France Telecom had a number of employee suicides in 2008 and 2009, and Renault, also in France, had a spate of suicides in both 2006 and 2008. Government studies attributed these in part to job demands and levels of stress, and large-scale organizational changes in strategy and size (downsizing).

ADDING SOME NEW ELEMENTS TO OCCUPATIONAL AND SAFETY RESEARCH AND INTERVENTION

This chapter has six important features that make it unique in the occupational health and safety field:

1. The chapter places a renewed and reinvigorating face on occupational health and safety issues. Accidents, injuries and deaths in the workplace continue on a daily basis worldwide. The challenges here, some ongoing and some new (for example, terrorism), have increased.
2. It examines traditional occupational health and safety content but also includes a wider range of health-related outcomes relevant to health

and safety (for example, incivility, toxic leadership, and disgruntled present and former employees).

3. It elevates the role of workplace stress in understanding and addressing occupational health and safety outcomes (accidents, errors, homicides and suicides).

4. The chapter documents the unique needs of small businesses (burglaries and robberies), which have typically been ignored in this area.

5. It includes a discussion of organizational responses to major occupational safety and health disasters.

6. It also examines a wider range of interventions to impact occupational health and safety outcomes including stress reduction initiatives (stress audits, mindfulness training, napping rooms for nurses working 12-hour shifts and de-escalation training for police officers).

The chapter gives health a higher priority, that is, the same standing as safety. It still considers the above listed more traditional content (for example, the handling of hazardous materials, and transformational leadership in improving safety performance), but expands the occupational health and safety experiences and outcomes to include more consideration of the emotional and psychological aspects of work experiences, and their effects on both health and safety (for example, bullying, sexual harassment, long work hours, workplace violence and active-shooter incidents).

Workplace stress affects workplace health and safety as well as employee well-being. An increasing number of specialists in workplace safety agree there is a direct association of increased levels of worker stress and increases in workplace accidents. This can be attributed to stress coping methods such as the use of alcohol or medications, preoccupation and distractions resulting from elevated stress levels, and higher levels of fatigue. Researchers with the National Institute for Occupational Safety and Health (NIOSH) have advocated accident reduction through stress management initiatives (Murphy et al 1986; Steffy et al. 1986). Sneddon et al. (2013) also advocate stress reduction as an intervention to reduce accident levels. They specifically highlight fatigue and sleep deprivation as reducing levels of alertness and attention, which increase the frequency of accidents.

INCREASING INTEREST IN OCCUPATIONAL HEALTH AND SAFETY CONCERNS

There is now a growing group of international researchers and practitioners interested in reducing occupational health and safety issues and threats in the workplace and increasing individual and organizational health and performance. More professionals and managers in all sectors have become active in human resource management, career and talent development, healthy work, healthy people and healthy organizations as these are being affected by the work experiences of men and women. More professionals are now employed in government agencies (NIOSH, OECD and ILO) who are responsible for work and health. This area has become more important to governments and organizations as the threats to and incidence of adverse occupational health and safety outcomes has increased.

Occupational Health and Safety: A Necessary Growth Industry

As I started gathering recent material to help me with this chapter I found several sites, usually occupational health and safety consultants or law firms in the US. I started receiving materials from them several times a week about seminars, webinars and meetings on specific aspects of this general topic. It is clear that the US is increasingly interested in improving occupational health and safety. My own country, Canada, seems to be less active here.

A sample of courses and webinars addressed the following topics:

- dealing with an active shooter;
- investigating an accident or near accident;
- building a safety culture;
- increasing manager and executive investment in safety;
- giving special attention to temporary workers;
- the onboarding of new employees;
- dealing with toxic employees;
- increasing employee commitment to safety;
- types of safety glasses and safety gloves;
- the importance of wearing safety glasses on some types of jobs; and
- understanding new OSHA safety guidelines.

In addition, organizations can purchase training programs from vendors in many of these areas. An example would be the availability of drug-testing kits and drug-testing training programs.

Why Investing in Occupational Health and Safety Makes Sense

Investments in occupational health and safety can be justified in terms of what accidents and injuries cost. It has been estimated that a fatal injury costs between US$1 billion and US$1.4 billion. Thus investing in health and safety will save money for organizations in the long run. However, a more important motivation is keeping workers safe. Organizations investing in occupational health and safety have more engaged employees who are also less likely to quit. Investing in occupational safety and health, then, is a business issue associated with future business success.

Senior executives often have difficulty communicating the value of such investments to others in their organization. Issues involve the difference between a cost and an investment and the failure to integrate occupational health and safety into their business strategy. Issues of employee safety, job performance and company reputation need to be put front and center. Occupational health and safety goals need to be closely tied to company business objectives, and the business value of successful occupational safety and health objectives more clearly identified.

STRESS AND EMPLOYEE HEALTH AND SAFETY

Murphy et al. (1986) reviewed the limited literature on the relationship of work stress and accidents or injury occurrence at work. Work stress is widespread and can influence individual feelings, perceptions, thought processes and performance, as well as physiological and biochemical responses. High levels of anxiety and depression and low self-worth have been found in workers prior to an accident. Pilots and crew members involved in aircraft accidents indicated more stressful life changes than others not involved in these accidents. Murphy et al. presented a model in which both work and non-work stressors increase anxiety, fatigue and alcohol use, and reduce motivation, which decrease capacities such as accuracy, reaction time, attention, reasoning, judgment and memory, which in turn increase risky work behaviors (for example, not wearing safety glasses) and improper use of equipment, thereby increasing workplace accidents and near misses. They reported the results of two studies of stress management training showing benefits in terms of accident claims and insurance losses. Employees also have a role to play here by following health and safety rules, wearing protective eye equipment, not texting while driving, following their training information and materials, and protecting co-workers and themselves from risk.

Sneddon et al. (2013) examined the effects of both work stress and fatigue on work situation awareness among drillers working off shore on the UK continental shelf. Drillers reporting high levels of stress, sleep disruption and fatigue indicated lower levels of work situation awareness, with stress being the strongest predictor.

Stress and Medical Errors

It is estimated that about 98 000 people die annually from medical errors in hospitals, more than deaths from motor vehicle accidents, breast cancer or AIDS (Committee on Quality of Health Care in America 2000). More people die from medical errors than from workplace accidents. Physicians suffer burnout, and physician burnout has been shown to increase medical errors (Shanafelt et al. 2010). Shanafelt et al. found that a stress reduction program that reduced burnout levels also reduced malpractice suits.

Medical errors by care providers lead to adverse health outcomes, as do to other factors such as improper use of medical equipment, poor or fatigued decision making, stress and cognitive overload (Peters and Peters 2007; Bogner 2010).

Stress and Medical Malpractice

Jones et al. (1988) carried out four studies on the relationship between stress and medical malpractice, and the potential value of stress manage- ment programs in reducing malpractice risk. First, hospital departments with malpractice suits reported higher levels of job stress than departments with fewer suits. Second, workplace stress levels predicted frequency of malpractice claims. Third, a longitudinal study showed a drop in medica- tion errors following the introduction of a stress management program. Fourth, hospitals implementing stress management initiatives had fewer malpractice claims than matched hospitals without the stress management training over a two-year period.

Costs of Fatigue and Sleepiness

Rogers (2008) cited costs of fatigue in the US as US$18 billion per year in lost productivity and accidents, with more than 1500 fatalities, 100 000 crashes and 76 000 injuries yearly attributed to fatigue-related driving on highways. She notes that two major nuclear power plant accidents (Chernobyl and Three Mile Island) and the running aground of the *Exxon Valdez* and subse- quent oil spill disaster occurred at night, with sleep deprivation seen as one of the causal factors. In addition, sleep deprivation was cited as a factor in poor decisions made the night before the launch of the Space Shuttle *Challenger*.

Fatigue and Sleepiness in Nursing

Rogers (2008) reviewed the effects of sleepiness, a tendency to fall asleep, fatigue, a strong sense of tiredness, low energy and high levels of exhaustion that reduce physical and/or cognitive functioning and nurse performance and patient safety. Both sleepiness and fatigue result from sleep deprivation. Her review of hundreds of studies indicated adverse health effects of sleep deprivation. Sleep deprivation was found to be associated with medical errors and risk of errors in nursing. It also puts nurse's health and safety at risk in terms of reduced driving ability and more sleep-related driving accidents. Increasing numbers of nurses are now working 12-hour shifts in order to better meet the needs of patients, and about 20 percent now indicate falling asleep while working.

Long Work Hours

Kivimaki et al (2015) undertook a meta-analysis investigating the association of long work hours and risk of coronary artery disease and stroke. They considered 25 longitudinal studies from 24 cohorts in Europe, the US and Australia. Controlling for age, sex, socioeconomic status and coronary heart disease and stroke at baseline, in a sample of 603 838 men and women, those individuals working 55 hours or more a week, compared with those working 35–40 hours a week, had a 13 percent greater chance of developing coronary heart disease. Also, in a sample of 528 908 women and men, they found those working 55 or more hours a week had a 33 percent increased risk of stroke.

SAFETY ISSUES IN THE WORKPLACE

Drug and Alcohol Use in the Workplace

A female school bus driver in Toronto has been charged with impaired driving. Her bus had 22 young students on board – fortunately none were injured when her bus struck a car (Rizza 2017).

Canada legalized the sale of marijuana in 2018, and several US states (for example, Colorado and California) have also legalized the purchase and consumption of marijuana. This raises issues of whether the use of marijuana before entering the workplace might have effects on employee health and safety behaviors (Phillips et al. 2015). Drivers in Ontario under the influence of cannabis will face large penalties (Benzie 2017).

Most drug users are employed, and it has been found that at least 14 percent of the workforce used an illicit drug in the preceding year (Frone 2011). Employees using drugs can influence events in their larger work environment. Drug use reduces cognitive abilities and reaction times, and has been associated with lower levels of attendance at work, poorer job performance and higher levels of workplace accidents and injuries

A larger number of employees have used alcohol in the past year (Frone 2011). The majority of employed women and men have had some alcohol either with a meal or to relax. However, some of these people are unable to control the amount of alcohol they consume. Alcoholism costs organizations billions of dollars each year in absenteeism, on-the-job injuries, reduced productivity, and health compensation. Also, some colleagues may get injured because of the actions of a drunk colleague.

Domestic Violence in Workplaces

Employees, typically women, experience domestic violence at home, and sometimes in their workplaces (54 percent), with a majority reporting that domestic violence reduced their workplace job performance (82 percent), and many noting missing work or coming late to work (38 percent), according to a recent Canadian study of 8000 workers (Wathen et al. 2014). Employing organizations in Canada lose almost C$80 million annually as a result (Zhang et al. 2012).

Guns in the Workplace

Many states in the US allow employees to bring guns with them to their workplaces. This probably increases the risk of an active-shooter scenario in workplaces (for example, the San Francisco UPS shootings and the Orlando shooter who killed five co-workers). Organizations need to develop an active-shooter emergency action plan and train employees on how to respond to an active-shooter scenario. In addition, planning on the best responses for all employees at the conclusion of an active-shooter incident must be undertaken.

Hazardous Materials

Mesothelioma is a rare type of cancer that develops in the lungs, abdomen or heart caused by coming into contact with asbestos. Mesothelioma has no known cure and therefore a poor prognosis. About 2400–2800 people in the US are diagnosed with mesothelioma each year, symptoms sometimes taking 20 to 30 years to appear. Life expectancy of mesothelioma victims is short. Mesothelioma patients and their families have received compensation through legal settlements from organizations that produced, made or distributed asbestos products. The average mesothelioma settlement was between US$1 million and US$1.4 million.

Grzywacs et al. (2011) document the effects of pesticide exposure on the health of Latino farm workers. Sparling et al. (2017) reported on the link between pesticide exposure and Parkinson's disease.

Transporting Hazardous Materials

On 5 July 2013, at 10.50 p.m., a train carrying 7.7 million liters of crude petroleum arrived in Antes, Quebec. The engineer parked the train on an incline. A small fire first broke out. Then the brakes were unable to hold the train and it rolled down toward Lac-Magantic, derailing almost the entire train at 1.15 a.m. About 6 million liters of crude were released from most of the 63 derailed tank cars. A larger fire and explosions broke out, leaving 47 people dead and about 2000 homeless as most of downtown Lac-Megantic was demolished. Interestingly, the crude in the tank cars was more volatile than indicated on shipping documents. Two years after this disaster, many residents still exhibited signs of post-traumatic stress disorder, levels of anxiety being very high. The plan is to again allow similar trains to follow this route. Survivors received financial compensation from a C\$435 million total compensation agreement. Six former railway employees faced criminal charges. The Transportation and Safety Board of Canada investigated this disaster and made several recommendations to prevent future accidents (for example, strengthen tank cars, the creation of emergency response assistance plans, and additional defenses to prevent runaway trains).

UNIQUE NEEDS OF SMALL BUSINESSES IN OCCUPATIONAL HEALTH AND SAFETY

About half of working Canadians in the private sector work in small businesses. Small-business owners are legally required to provide and maintain a safe and healthy workplace for their employees, customers or clients, and visitors. Unfortunately, small businesses have higher fatality rates than large businesses. Small businesses need support so they can understand occupational health and safety regulations, address them despite economic and information limitations, the lack of occupational health and safety systems, and information and accesses specifically targeted at small business. Evaluations of studies of intervention effectiveness in small business (unfortunately only five studies) had positive effects in safety-related outcomes such as employee beliefs and health outcomes (McCoy et al. 2014), and a larger study of 260 companies reported that increasing worker wellness improved worker safety (Newman et al. 2015).

The unique occupational and safety needs of small businesses typically receive little attention. Small businesses (for example, convenience stores, gas stations and 24-hour operations) are particularly vulnerable to robberies and burglaries. Small-business owners can take several steps to protect staff and prevent robberies and burglaries. These would involve the use of more adequately locked doors, installing an alarm system, improved lighting, installing a panic button to push in case of robberies, the removal of expensive products from windows at night, training the staff, having two staff members at all times, keeping cash held on the premises to a minimum, developing relationships with other small-business owners in the same area to foster information sharing and to share the costs of lighting and private security, in some cases, and to talk with police before an incident occurs.

OSHA *Small Business Handbook*

The US Occupational Safety and Health Administration (OSHA 2005) created a handbook specifically for small-business owners and managers of small businesses. The OSHA lays out a four-point program: management commitment and employee involvement, worksite analysis, hazard prevention and control, and training employees, supervisors and managers. They go into great detail in each of these areas, offering guidance on how to start, designating responsibility, organizing the workplace, fact gathering, creating a plan, the use of checklists, action plan worksheets and how to obtain further OSHA assistance.

DEALING WITH OCCUPATIONAL SAFETY AND HEALTH DISASTERS

Hurricane Katrina, the 11 September 2001 (9/11) attack on the World Trade Center, and the *Deepwater Horizon* explosion and oil spill in the Gulf of Mexico would qualify as occupation health and safety disasters. Other workplace disasters would include fires, explosions, oil spills, mine collapses, workplace shootings, and terrorist attacks on hotels, restaurants and airports.

Organizations are sometimes affected by natural disasters, such as Hurricane Katrina, as well as by disasters associated with employee behavior and workplace cultures such as serious workplace accidents, fires, explosions, crashes and injuries (for example, the BP *Deepwater Horizon* explosion). There are also disasters resulting from acts of terrorism, armed robberies and domestic violence incidents in the workplace.

Emergencies and disasters are unplanned, but employers can create safety and health management systems that shape their ability to handle such disasters. Here are some guidelines.

- Create a written emergency response plan.
- Communicate and discuss this plan with all employees. The plan should identify escape procedures and escape routes, responsibilities, who does what, the shutting down of plant operations (if relevant), a system of accounting for all employees, rescue and medical activities for staff who undertake them.
- Create emergency response teams who are fully trained in first aid, use of fire extinguishers, chemical spill control, search and emergency rescue procedures, and trauma counseling.
- Contacts with outside organizations such as fire departments, police, ambulance services and hospitals.

In all these disaster incidents, organizations also needed to deal with post-traumatic stress symptoms, to hold a post-incident debriefing for witnesses of the events and, in some cases, bereavement sessions in the case of deaths.

The BP *Deepwater Horizon* Disaster

The BP *Deepwater Horizon* disaster in April 2010 has received substantial analysis. This disaster killed 11 employees, created the worst oil spill in US history and cost BP billions of dollars in settlements. BP itself has come under criticism for cutting corners over several years, valuing speed over safety, and not having a safety plan to cope with a large oil spill. BP had no accidents for a few years before the 2010 disaster and became overconfident. A series of small decisions made over a long period of time led to the disaster, and once the disaster began, the crew failed to act quickly enough. Merji and DeWolf (2013), using a wide variety of secondary sources, concluded that the BP management of his disaster fell short in several critical ways. Studies of disasters usually use a three-stage model: pre-crisis – the crisis of management in terms of preparation and prevention, operational crisis – how the organization responds to the immediate crisis, and post-crisis – what it does to recover and legitimate its reputation and return to normal business routines. Merji and DeWolf concluded that BP did not perform well at any of these three crisis stages.

Other major disasters, such as the 1988 *Piper Alpha* North Sea oil rig explosion, the Montara explosion off Western Australia, the 1986 explosion of the Space Shuttle *Challenger*, followed by the 2003 explosion of *Columbia*, the 2011 Fukushima nuclear plant disaster, had a human dimension. All were human-made disasters that could have been prevented by acknowledging that accidents can happen in any organization, monitoring events and identifying and changing aspects of workplace culture that increased risks and hazards. It is also important that various levels of government and industry leaders work together to better prepare, train for, and address similar disasters.

Supporting First or Emergency Responders

There is considerable evidence that first responders often pay a heavy price for their efforts. Using data from 27 449 World Trade Center rescue and recovery workers, Wisnevesky et al. (2011), in a longitudinal study, found higher rates of physical and psychological health disorders. In another longitudinal study of the World Trade Center disaster, Jordan et al. (2011) reported higher than expected rates of mortality among first responders.

In the 9/11 terrorist attack on the World Trade Center in New York City, 450 emergency responders died, with hundreds more seriously injured. The Rand Corporation shortly thereafter convened a series of expert meetings on ways to better equip, train and support first responders. This work resulted in four volumes, the first published in 2002 (Jenkins et al. 2002), featuring ways of protecting emergency responders. These included, among others, personal protective equipment, information and training and site management strategies for increasing personal protection.

IRRESPONSIBLE AND EXPLOITATIVE WORKING CONDITIONS

Rana Plaza, an eight-story building collapsed in Savar, a district on the outskirts of Dhaka, capital of Bangladesh, collapsed in 2013 (Aulakh 2015). The death toll reached 1129, with 2515 people injured. Cracks in the building emerged early the previous day, yet workers were forced back to work the next day. The top floors were built without a permit, the building was not intended to be factories and substandard construction materials were used. Police in Bengladesh charged 41 people with murder.

In addition, there are some organizational sectors in which other types of irresponsible behavior that reduce health and safety are being undertaken which affect large numbers of individuals (Burke 2018). These include human trafficking, human slavery, sweatshops and child labor.

Violence in the Workplace

A disgruntled UPS employee in San Francisco killed two employees then himself. A man who was fired from a Florida awning factory returned and killed five people before taking his own life. The phrase 'going postal' emerged about 20 years ago to capture an increase in instances where disgruntled postal employees came to work with weapons and killed some of their colleagues. The postal service today does not have a monopoly on such events.

There often are warning signs of such behavior but these are usually ignored. Violence-prone behaviors include increased use of alcohol and drugs, increased lateness and absenteeism, depression and withdrawal, heightened levels of anger, making threats or verbally abusive comments to co-workers, making comments about suicidal thoughts and mood swings.

Workplace violence refers to any act in a workplace in which a person is abused, threatened, intimidated, assaulted or killed (Statistics Canada 2009). In the US from 2006 to 2010, an average of 551 workers were killed in work-related homicides annually (Bureau of Labor Statistics 2015). In 2010, there were 518 workplace homicides, accounting for 10 percent of all fatal workplace injuries. Shootings represented 78 percent of all workplace homicides in 2014, with most (83 percent) occurring in the private sector. Most workplace homicides involved men (72 percent). Types of assailants varied depending on whether the victim was man or a woman. More women were killed by relatives or personal acquaintances than men (39 percent women and 3 percent men). In 2004, workplace homicides were more likely to occur in the retail trade (17 percent), government (17 percent) and leisure and hospitality (15 percent).

Murder is the leading cause of workplace fatalities, most being robbery related. Incidents of workplace violence negatively affect organizations through of lost productivity, higher costs of insurance, costs of additional security, and image concerns.

Some types of work or working conditions increase the risk of workplace violence. These include having direct contact with clients, handling cash, valuables or prescription drugs, working alone or in small numbers, working with unstable or volatile people, working where alcohol is served, working in a community-based setting, working in high-crime areas, securing or protecting valuable goods and transporting people or goods. Every country in the developed world, and probably beyond, has created legislation addressing workplace violence (Ontario Ministry of Labour 2016). All jurisdictions have legislation against workplace violence typically related to their criminal code.

Suicide in the Workplace

Job stress emerges as a key factor in workplace suicide. Suicides account for about 31 percent of all deaths at work. The average age of an American suicide is 44 years, a midlife working age. Stalk and Bowman (2012) found that at least 11 percent of suicides were related to job problems. These job problems included poor performance reviews, tensions with co-workers, increased job pressures, and fear of layoffs and unemployment. An Australian study (Routley and Ozanne-Smith 2012) reported that 17 percent of suicides were work related, and causes included job demotions, conflict with co-workers, the result of being bullied and having a low level of control at work. Worker suicide, then, is the interaction between vulnerable individuals (for example, those with mental health problems), stress at work (for example, job insecurity, psychosocial job stressors, a boring work, high job demands and low job control) and living conditions that are stressful.

Milner et al. (2013), using meta-analysis of 34 studies, examined suicide rates in various occupations. They found that workers in the lowest skilled occupations (for example, laborers and cleaners) were at higher risk than workers in higher skilled occupations (for example, managers and clerical support workers). High-risk occupations include workers who are disadvantaged in their level of education, access to services, limited social resources, low quality of jobs, exposure to other job risks and limited access to community support.

Temporary Workers

Mojtehedzadeh and Kennedy (2017a) focus on the rights of temporary workers in Toronto, Canada. Mojtehedzadeh went undercover via a temporary employment agency to begin employment as a temporary worker at Fiera Foods, an industrial bakery. She and others placed circular pastry dough into soft plastic trays. She received 5 minutes' training. Temporary workers were told not to wear jewelry. She was paid in cash, and considered to be an employee of the placement agency rather than Fiera Foods. Supervisors shouted at them to work faster. Fiera employs about 1200 workers and says it values health and safety and helping immigrants to Canada. In general, temporary employees receive little training, have few rights, are given no promises of future commitments, no job security, can be fired with no notice and often work in the most risky jobs. Mojtehedzadeh was paid 10 cents below Ontario's minimum wage. She was not told where the fire extinguishers were, nor where the exits were. Temporary workers can refuse risky work but are reluctant to do so for fear of losing their jobs.

Though temporary workers get injured more often on their jobs than do permanent workers, temporary workers are afraid to admit their injuries, again for fear of losing their jobs.

Three Fiera employees have died at the workplace since 1999 (Mojtehedzadeh and Kennedy 2017b). One died while cleaning the inside of a vat and was crushed when someone started the mixer. A second employee was killed when struck by a car driven by another employee going home after work. A third employee died after only two weeks on the job when her hijab got caught in some machinery (Kennedy and Mojtehedzadeh 2017). After investigating this last incident, Fiera was charged with 39 health and safety violations. Fiera has been charged with 191 orders for health and safety violations over the past 20 years.

Fiera Foods was fined US$300 000 under the Occupational Health and Safety Act (Mojtehedzadeh and Kennedy 2017c). This fine was double that given to Fiera in 2002 for the death of 17-year-old temporary worker, Ivan Golyashov. Fiera was also required to hire an independent auditor of their health and safety and temporary agency processes. Grocers that sell Fiera products have asked to meet with the company to discuss Fiera's health and safety and employment standards that the grocers expect of their suppliers.

Precarious Work

Vives et al. (2013) examined the association of precarious employment and poor mental health in a sample of 5679 permanent and temporary workers in Spain. Their data were collected during 2004–05. Six dimensions of precariousness were considered: instability, low wages, employee rights, disempowerment, vulnerability and capacity to exercise rights (for example, access to vacation time). Both men and women in more precarious jobs reported poorer mental health (for example, nervousness, anxiety, depression and psychological distress).

CREATING SAFETY CULTURES

A safety culture is a product of positive workplace attitudes and behaviors of all employees at every level, acceptance of these attitudes and behaviors as vital, the presence of health and safety goals, policies and procedures supporting health and safety improvements, and employees accepting responsibility for levels of safety and safety improvements. A safety culture reflects how safety is managed in the organization. We know a lot about what an effective safety culture looks like, and research over the past decade in high-risk occupations (for example, off-shore rigs and mining) have supported the benefits of these efforts.

Workplace injuries, illnesses and deaths result in physical, financial and emotional costs to individuals and their families. There are also costs to employing organizations as well. The OSHA (2012) developed a White Paper on injury and illness prevention programs. These programs are proactive processes to help employers identify and fix workplace hazards before injuries or illness occurs. They indicate the benefits of developing such programs and their success in reducing injuries, illness and deaths. These programs typically include common features: management, leadership, employee participation, hazard identification and their addressing, training workers in how the program works, and periodic evaluation of the program efforts to identify potential improvements. They cite evidence of the success of these programs in reducing illness, injuries and deaths in various US states and in several industrial sectors in the US, including the Department of Defense. Small businesses have also benefitted from such programs.

Wachter and Yorio (2014), using manager and employee surveys, examined the relationship of safety management practices with objective safety statistics (for example, accident rates), and how these practices improved safety results through levels of employee engagement. The ten safety management practices were employee involvement and influence, pre-and post-task safety reviews, safe work procedures, hiring for safety, cooperation facilitation, safety training, communication and information sharing, accident investigation, detection and monitoring, and safe task assignments (task-employee matching). They observed: a negative relationship between the presence of the ten individual safety management practices, and their composite, with accident rates; a negative relationship with level of safety-focused worker emotional and cognitive engagement and accident rates; safety management systems that were positively associate with worker engagement ratings; and worker engagement moderated the relationship of safety management practices and safety outcomes (for example, accident rates).

Investigating Occupational Safety and Health Incidents

It is important that all occupational safety and health incidents are both reported and investigated, a requirement in many jurisdictions, and very useful in identifying causes and potential remedies. However, as Probst and Graso (2011) have reported, accidents tend to be under-reported and under-investigated. Several consulting organizations in the occupational safety and health area have created software to make investigations more efficient, standardized and effective.

The following steps need to be taken in a full investigation:

● Bring all the people needed to conduct the investigation together.
● Close off the area in which the incident occurred and seal it off.
● Identify people who might have seen the incident and bring them together.
● Interview the worker involved.
● Interview all witnesses.
● Photograph the incident scene for future reference.
● Write a report stating what caused the incident and what could prevent future incidents.
● Engage in a future follow-up of these suggestions.

INTERVENTIONS TO INCREASE OCCUPATIONAL HEALTH AND SAFETY

The good news is that occupational safety and health researchers, consultants and organizational managers, working together on occasion, have made strides in reducing workplace stress, injuries and deaths, and have improving well-being. A sample of these initiatives follows.

Increasing Employee Voice

In many workplaces, employees are reluctant to speak up about what they observe happening, even when it gets in the way of performance, for fear of disagreeing with higher levels of management or appearing stupid. Fear of speaking up has been found to be associated with higher levels of unsafe behavior and adverse outcomes in several sectors. Studies in the healthcare sector in particular have documented barriers in nursing staff in exercising their voice with doctors, sometimes resulting in unsafe work practices and accidents.

Belyansky et al. (2011), in a study of surgical resident or trainees and surgeons, asked them to recall an incident when a trainee spoke up and prevented an adverse event. Over 70 percent of trainees indicated that speaking up prevented such an event. Sayre et al. (2012a, 2012b) undertook two interventions to increase speaking up by nursing staff. Both involved training sessions to encourage nurses to speak up. The training in their 2012a study began with discussions of speaking up and why it was vital, showing in vignettes that speaking up increased patient safety and effectiveness, indicated support from senor hospital leadership for speaking up, and had nurses develop a personal action plan to increase their speaking up. A control group received no training. Follow-up data showed an increase in speaking-up levels by the trained nurses. Sayre et al.'s 2012b study replicated these findings.

Reducing Workplace Violence

Employers have a legal and moral obligation to create a workplace free of threats and violence.

The US Department of Justice (2002, p. 15) suggests the following roles:

- Adopt a zero-tolerance workplace violence policy and prevention program and communicate the policy and program to all employees.
- Identify areas within the workplace that have a higher potential for violence (for example reception and warehouse).
- Provide regular training in preventative measures for all new and current employees, supervisors and managers. Define what workplace violence is and identify violence-prone behaviors.
- Support victims of workplace or domestic violence.
- Adopt and practice fair and consistent disciplinary practices.
- Foster a climate of trust and respect among workers and between employees and management.
- Stabilize the workplace after an incident.
- Help to prevent future incidents.

Organizations can also undertake preventative measures. These include workplace design, administrative practices and work practices. Workplace design could involve creating physical barriers, reducing the number of entrances, using keys or access cards to control entrance, using better exterior lighting and placing employees closer to exits. Administrative practices might include keeping cash at a minimum, using locked safe deposit boxes. Work practices could include having a buddy system to provide support and protection, knowing where co-workers should be, and checking the credentials of clients and customers.

It is also important to address the aftermath of violence at work. The effects of violence remain after the violent act is over, and others besides the victim(s) are affected. Employee distress is heightened and must be dealt with quickly through the provision of information about what happened. Short-term and long-term psychological support is likely to be needed. Critical incident debriefing, trauma counseling and employee assistance programs can help victims as well.

After a violent incident organizations should undertake the following (Bishop et al. 2006):

- Bring all employees together immediately.
- Ensure that senior levels of management are present.
- Make counseling services available to those feeling a need.
- Have a social event a few weeks later and invite family members.

Dealing with an Active Shooter

An increasing number of active workplace shooters has occurred in the US, most recently in both San Francisco and Orlando. An active shooter is a person(s) actively involved in killing or trying to kill people in a workplace setting. Employers need to improve the protection of their employees from risks of an active shooter, and should be on the look-out for early warning signs that an employee might become an active shooter. Employers need to develop workplace violence plans that deal with potential active shooters. This involves training employees in how best to respond to active-shooter incidents. In addition, employers need to develop plans and programs for addressing the trauma to their employees.

Suicide Prevention Initiatives

There are a variety of organizations working in mental health and suicide prevention to help employers counsel employees and to help them prepare for and respond to suicide. Organizations of all sizes need to create and implement policies and programs that develop a mentally healthy workforce and prevent suicidal thoughts and behaviors. Most suicides are committed by people of working age and suicide is the leading cause of death for men aged 25–44 years and women aged 25–34 years.

Milner et al. (2014) reviewed published and unpublished writing on suicide prevention initiatives in the workplace. Very few of these efforts had been evaluated. Reducing suicide has individual, workplace and economic benefits. Milner et al. reviewed 13 interventions that included education and training, training the trainers, training and community awareness, how to assist co-workers after the suicide of a fellow employee, training to help a potential suicide victim stay safe, training and education of administrators and teaching individuals to identify a personal risk. Suicide prevention programs for the most part were short-lived standalone training programs in high-risk occupations and industries (for example, mining).

Boccio and Macari (2014) developed their safe haven model (SHM) for managers interested in initiating suicide prevention initiatives. The model has three emphases: creating a positive and supportive workplace that fosters psychological well-being; improving managers' and co-worker's understanding of risk factors for suicide, and recognizing them; and encouraging the use of mental health support services.

Drug- and Alcohol-Testing Programs

Very little research has examined drug testing in the workplace as a safety strategy. Most of the studies had methodological flaws, making it difficult to draw any conclusions. One rigorous study did show that drug-testing efforts reduced fatal accidents in the transportation industry (Pidd and Roche 2014). A study of employees in a large transportation company in Portugal (not included in the Pidd and Roche 2014 review) investigating the relationship of alcohol and drug testing and accidents over a five-year period, found that alcohol and drug testing reduced the risk of accidents (Marques et al. 2014).

Drug testing and educational programs can potentially reduce workplace accidents, decrease employee theft, increase productivity, and reduce turnover and absenteeism.

Toxic Employees

Many organizations have one or more toxic employees – employees that are always stirring up trouble with others. These employees are more likely to treat co-workers poorly and, in extreme cases, engage in violent behavior. Organizations need to take steps to identify their toxic employees before they can bully others, exhibit hostile behaviors and criticize others. Sometimes toxic employees can be identified in reference checks and not even hired. Toxic employees need to be talked with to identify why they behave this way and encouraged to change (that is, stop) this behavior. A meeting should be held with them a short time later to see if their behavior has changed. Some toxic employees can be coached as well. Supervisors have a major role to play in this effort.

Resilience Training

McCraty and Nila (2017) describe their ongoing research program on the effects of resilience training on police officer behavior and job performance. The overall goals of this training is to build and sustain officer resilience and reduce stress and psychological and physical health problems. The goals of this program are to:

- increase officers' ability to think clearly and identify appropriate responses to problems under pressure;
- increase their ability to be fully aware of their situation;
- reduce officers' stress symptoms;
- increase officers' resilience and stress tolerance;
- support intelligent use of their energy and resources and regain these more quickly; and
- increase officer's ability to minimize ineffective thoughts, feelings and behaviors.

They have found that their training programs reduce stress reactions, increase officer confidence in stressful situations, increase job performance and increase vitality and officer relationships with their families.

Mindfulness at Work

Mindfulness is a state of non-judgmental attentiveness to, and awareness of, our moment-to-moment experiences (Brown and Ryan 2003) – a psychological state in which individuals focus attention on events occurring in the here and now. Mindfulness treatments and training, including mindfulness meditations, have been found to be associated with improved health outcomes (Moore 2017). Mindfulness reduces stress levels and is now included in an increasing number of wellness programs.

Mindfulness is increasingly seen as relevant to both health and workplace performance, and is positively related to elements of work engagement and job satisfaction and negatively related to quit intentions (Dane and Brummel 2013). Reb et al. (2014) collected data from supervisors and their subordinates and reported that supervisors' scores on mindfulness were negatively correlated with subordinates' levels of exhaustion and positively correlated with subordinates' levels of work–life balance and ratings of their job performance.

Corporate Wellness Programs

Corporate wellness programs are long-term organizational initiatives aimed at supporting organizational practices and personal behavior associated with higher levels of employee physiological, emotional and social well-being (Wolfe and Parker 1994). Berry et al. (2010, p. 106) define a corporate wellness program as 'an organized, employer-sponsored program that is designed to support employees (and sometimes their families) as they adopt and sustain behaviors that reduce health risks, improve quality of life, enhance personal effectiveness, and benefit the organization's bottom line'. Healthy employees help organizations achieve healthy profits (Goetzel and Ozminkowski 2008).

Isaac and Ratzan (2013), from Johnson & Johnson, describe its program, which showed that for every dollar invested in wellness, Johnson & Johnson received nearly US$4 in lower healthcare costs, less absenteeism and more productivity. Four central beliefs made their program successful:

- a definition of health that includes well-being;
- prevention-focused educational offerings;
- rewards for healthy behaviors; and
- creating a workplace environment supporting employees in engaging in healthy behaviors.

Trauma Counseling

Employees in various occupations witness injuries and deaths. These occupations include police officers, first responders, firefighters, and transit employees working on subways that witness individuals committing suicide by jumping in front of a moving train. Organizations need to create programs that address the trauma that usually results from witnessing these events.

Post-traumatic stress disorder (PTSD) is a psychological reaction to a very stressful and physically challenging event or experience that commonly produces anxiety, depression, flashbacks, hypervigilance, suicidal thoughts and other psychological health issues. There are both physical and behavioral signs of PTSD that individuals and supervisors need to pay attention to. In addition, a pro-counseling culture is a necessity. Training courses should include content on PTSD, its signs and the potential value of treatment. Several forms of therapy have been found to be effective in addressing PTSD.

Following such a workplace event, the worksite should ideally be closed, with senior management indicating that action. Supervisors then should meet with every employee to assess their responses and whether additional professional help might be required. Witnesses should be given a break from work to recuperate, encouraged to talk about the incident and not rushed back to work. Engaging outside professionals may be required in cases of PTSD. Critical incident stress responses may last several weeks.

Engaging Patients in Patient Safety

In-patients and out-patients can experience 'adverse effects' or 'adverse events' defined as 'unintended harm to the patient by an act of commission or omission rather than the underlying disease or condition of the patient' (Institute of Medicine 2000, p. 317). Adverse events can include: adverse events per se, that is, injuries from medical care rather than the normal history of the illness; near misses, that is, errors with the potential for injury but no harm resulted as the error was quickly identified; and medical errors with minimal risk of harm (Weingart et al. 2005).

Weingart et al. (2011), in a survey of 2015 patients, reported that patients having greater participation in their care indicated greater satisfaction with the quality of their care and fewer adverse effects. Weingart et al. (2005), in a sample of 228 hospital in-patients, found they were able to identify adverse events or near misses affecting their care (for example, errors or injuries). Some of these were seen as potentially life-threatening. Interestingly, none of these incidents were present on hospital incident reporting sheets. This requires changes in the attitudes and behaviors of patients and their families, and of healthcare providers. Patients need to take more active roles in decisions about their health, and providers need to relinquish control and listen to patients more carefully while exchanging ideas with each other. Patients and families of patients should be encouraged to speak about concerns and suggestions. Healthcare providers need to support the creation of learning cultures. A Canadian guide for improving patient safety through engagement, offers guidance on how to move forward on these issues (Canadian Patient Safety Institute 2017).

Increasing Workplace Civility

Workplace incivility including abusive supervision, toxic leadership, sexual harassment, bullying and rudeness seems to be on the increase in workplaces (Perrewe et al. 2015). Pearson and Porath (2005) reported that 89 percent of professionals in the US had been insulted or bullied at work and 78 percent said their organizational commitment declined as a result. Leiter et al. (2011), in a longitudinal study of 1173 healthcare workers working in 41 nursing units, had respondents complete measures of their social relationships at work, burnout, work engagement and intent to quit before and after a six-month intervention to improve civility, respect and work engagement undertaken in some units but not in other comparison units. Measures improved significantly for the intervention units, compared with the comparison units, in workplace civility, respect, job satisfaction, management trust and co-worker incivility, and absences declined.

A NECESSARY FUTURE

This collection offers a partial review of the countless and daunting occupational safety and health challenges currently facing workplaces. While some workplaces are well positioned to tackle these challenges, having developed effective occupational safety and health initiatives, most workplaces are still falling short. Research and practice have identified critical initiatives and how best to introduce them. We, as a community, need to make these a higher priority – a need this collection begins to address.

NOTE

* Preparation of this chapter was supported in part by York University.

REFERENCES

Aulakh, P. (2015), 'Lessons of Rana Plaza go unheeded', *Toronto Star*, 22 April, A11.

Belyansky, I., Martin, T.R., Prabhu, A.S., Tsirline, V.B., Howley, L.D., Phillis, R. et al. (2011), 'Poor resident-attending intraoperative communication may compromise patient safety', *Journal of Surgical Research*, **171** (2), 386–94.

Benzie, R. (2017), 'Driving high will net stiff penalties', *Toronto Star*, 19 September, A3.

Berry, L.L., Mirabito, A.M. and Baun, W.B. (2010), 'What's the hard return on employee wellness programs?', *Harvard Business Review*, **88** (12), 104–12.

Biddle, E.A. (2004), 'The economic cost of fatal occupational injuries in the United States, 1980–97', *Contemporary Economic Policy*, **22** (3), 370–81.

Bishop, S., McCullough, B., Thompson, C. and Vasi, N. (2006), 'Resiliency in the aftermath of repetitive violence in the workplace', *Journal of Workplace Behavioral Health*, **21** (3–4), 101–18.

Boccio, D.E., and Macari, A.M.(2014), 'Workplace as safe haven: how managers can mitigate risk for employee suicide', *Journal of Workplace Behavioral Health*, **29** (1), 32–54.

Bogner, M.S. (2010), *Human Errors in Medicine*, 2nd edn, Boca Raton, FL: CRC Press.

Brown, K.W., and Ryan, R.M. (2003), 'The benefits of being present: mindfulness and its role in psychological a well-being', *Journal of Personality and Social Psychology*, **84** (4), 822–48.

Bureau of Labor Statistics (2915), *Census of Fatal Occupational Injuries 2014*, Washington, DC: US Department of Labor.

Burke, R.J. (2018), 'Violence and abuse in the workplace: an increasing challenge', in R.J. Burke and C.L. Cooper (eds), *Violence and Abuse in and Around Organizations*, New York: Routledge, pp. 1–38.

Canadian Patient Safety Institute (2017), *Engaging Patients in Patient Safety: A Canadian Guide*, Ottawa: Canadian Patient Safety Institute.

Committee on Quality of Health Care in America (2000), *To Err is Human: Building a Safer Health Care System*, Washington, DC: National Academies Press.

Dane, E. and Brummel, B.J. (2013), 'Examining workplace mindfulness and its relations to job performance and turnover intentions', *Human Relations*, **67** (1), 105–28.

Frone, M.R. (2011), 'Alcohol and illicit drug use in the workforce and workplace', in J.C. Quick and L. Tetrick (eds), *Handbook of Occupational Health Psychology*, Washington, DC: American Psychological Association, pp. 277–96.

Goetzel, R.Z and Ozminkowski, R.J. (2008), 'The health and cost benefits of worksite health promotion programs', *Annual Review of Public Health*, **29**, 303–23.

Grzyacz, J.G., Quandt, S.A. and Arcury, T.A. (2011), 'Job stress and pesticide exposure among immigrant Latino farmworkers', in R.J. Burke, S. Clarke and C.L. Cooper (eds), *Occupational Health and Safety*, Aldershot: Gower, pp. 277–94.

Institute of Medicine (2000), *To Err is Human: Building a Safer Health Care System*, Washington, DC: Institute of Medicine.

Isaac, F.W. and Ratzan, S.C. (2013), 'Corporate wellness programs: why investing in employee health and well-being is an investment in the health of the company', in R.J. Burke and C.L. Cooper (eds), *The Fulfilling Workplace: The Organization's Role in Achieving Individual and Organizational Health*, Aldershot: Gower, pp. 301–14.

Jenkins, B.A., Peterson, D.J., Bartis, J.T, LaTourrette, T., Brahmakulam, I., Houser, A, et al. (2002), *Protecting Emergency Responders: Lessons Learned from Terrorist Attacks*, Santa Monica, CA: Rand Corporation.

Jones, J.W., Barge, B.N., Steffy, B.D., Fay, L.M, Kunz, L.K. and Wuebker, L.J. (1988), 'Stress and medical malpractice: organizational risk assessment and intervention', *Journal of Applied Psychology*, **73** (4), 727–35.

Jordan, H.T., Brackbill, R.M., Cone, J.E., Debchoudhury, I., Farfel, M.R., Greene, C.M. et al. (2011), 'Mortality among survivors of the Sept. 11, 2001, World Trade Center disaster: results from the World Trade Center Health Registry cohort', *The Lancet*, **378** (9794), 879–87.

Kennedy, B. and Mojtehedzadeh, S. (2017), 'Where is Amina? Why can't anyone tell me what happened? A year after temp agency worker Amina Diaby died at Fiera Foods, the family still searchers for answers', *Toronto Star*, 8 September.

Kivimaki, M., Jokela, M., Nyuberg, S.T., Singh-Manoux, A., Fransson, E.I., Alfredsson, L. et al. (2015), 'Long working hours and risk of coronary heart disease and stroke: a systematic review and meta-analysis of published and unpublished data for 603838 individuals', *The Lancet*, **386** (10005), 1739–46.

Leiter, M.P., Laschinger, H.K.S., Day, A. and Oore, D.G. (2011), 'The impact of civility interventions on employee social behavior, distress, and attitudes', *Journal of Applied Psychology*, **96** (6), 1258–74.

Marques, P.H., Jesus, V. Olea, S.A., Vairinhos, V. and Jacinto, C. (2014), 'The effect of alcohol and drug testing at the workplace on individual's occupational accident risk', *Safety Science*, **68** (October), 108–20.

McCoy, K., Stinson, K, Scott, K., Tenney, L. and Newman, L.S. (2014), 'Health promotion in small business: a systematic review of factors influencing adoption and effectiveness of worksite wellness programs', *Journal of Occupational and Environmental Medicine*, **56** (6), 579–87.

McCraty, R. and Nila, M. (2017), 'The impact of resilience training on officers' well-being and performance', in R.J. Burke (ed.), *Stress in Policing: Sources, Consequences and Interventions*, New York: Routledge, pp. 257–76.

Mejri, M. and DeWolf, D. (2013), 'Crisis management: lessons learnt from the BP Deepwater Horizon oil spill', *Business Management and Strategy*, **4** (2), 67–90.

Milner, A., Page, K., Spencer-Thomas, S. and Lamontage, A.D. (2014), 'Workplace suicide prevention: a systematic review of published and unpublished activities', *Health Promotion International*, **30**, (1), 29–37.

Milner, A., Spittal, M.J., Pirkis, J. and La Montagne, A.D. (2013), 'Suicide by occupation: systematic review and meta-analysis', *British Journal of Psychiatry*, **203** (6), 409–16.

Mojtehedzadeh, S. and Kennedy, B. (2017a), 'Little training. Few rights. No promises. This is the new reality for a growing number of temporary employees. I went undercover as a temp agency worker. This is what I found', *Toronto Star*, 9 September, A1, A20.

Mojtehedzadeh, S. and Kennedy, B. (2017b), 'Two other workers have died at Fiera Foods', *Toronto Star*, 10 September, A18.

Mojtehedzadeh, S. and Kennedy, B. (2017c), 'What happened to Amina Diaby was a tragedy: we have to do better and we will', *Toronto Star*, 15 September, A1, A16.

Moore, K.A. (2017), 'Mindfulness at work', in R.J. Burke and K.M. Page (eds), *Research Handbook on Work and Well-Being*, Cheltenham, UK and Northampton, MA, USA: Edward Elgar, pp. 453–67.

Murphy, L R., DuBois, D. and Hurrell, J.J. (1986), 'Accident reduction through stress management', *Journal of Business and Psychology*, **1** (1), 5–18.

Newman, L.S., Stinson, K.E,., Metcalf, D., Fang, H., Brockbank C.S., Jinnett, K. et al. (2015), 'Implementation of a worksite wellness program targeting small business', *Journal of Occupational and Environmental Medicine*, **57** (1), 14–21.

Occupational Safety and Health Administration (OSHA) (2005), *Small Business Handbook*, Washington, DC: Occupational Safety and Health Administration.

Occupational Safety and Health Administration (OSHA) (2012), 'Injury and illness prevention programs', White Paper, Washington, DC, Occupational safety and Health Administration.

Ontario Ministry of Labour (2016), *Health and Safety Guidelines. Workplace Violence and Harassment: Understanding the Law*, Toronto: Ontario Ministry of Labour.

Pearson, C.M. and Porath, C.L. (2005), 'On the nature, consequences, and remedies of workplace incivility: no time for "nice"? Think again', *Academy of Management Executive*, **19** (1), 7–18.

Perrewe, P.L., Halbesleben, J.R.B. and Rosen, C.C. (2015), *Mistreatment in Organizations*, Bingley: Emerald.

Peters, G.A. and Peters, B.J. (2007), *Medical Error and Patient Safety: Human Factors in Medicine*, Boca Raton, FL: CRC Press.

Phillips, J.A., Holland, M.G., Baldwin, D.D., Meuleveld, L.G., Mueller, K.L., Perkison, B. et al. (2015), 'Marijuana in the workplace: guidance for occupational health professionals and employers', *Journal of Occupational and Environmental Medicine*, **57** (4), 459–75.

Pidd, K. and Roche, A.M. (2014), 'How effective is drug testing as a workplace safety strategy? A systematic review of the evidence', *Accident Analysis and Prevention*, **71** (October), 154–65.

Probst, T.M. and Graso, M. (2011), 'Reporting and investigating accidents: recognizing the tip of the iceberg', in R.J. Burke, S. Clarke and C.L. Cooper (eds), *Occupational Health and Safety*, Aldershot: Gower, pp. 71–93.

Reb, J.M., Narayaman, J. and Chaturvedi, S. (2014), 'Leading mindfully: two studies of the influence of supervisor trait mindfulness on employee well-being and performance', *Mindfulness*, **5** (1), 36–45.

Rizza, A. (2017), 'School bus driver charged with impaired driving', *Toronto Star*, 31 August, GT2.

Rogers, A.E. (2008), *The Effects of Fatigue and Sleepiness on Nurse Performance and Patient Safety*, Rockville, MD: Agency for Healthcare Research and Quality.

Routley, V.H. and Ozanne-Smith, J.E. (2012), 'Work related suicide in Victoria Australia: a broad perspective', *International Journal of Injury Control and Safety Promotion*, **19** (2), 131–4.

Sayre, M.M., McNeese-Smith, D., Leach, L.S. and Phillips, L.R. (2012b), 'An educational intervention designed to increase "speaking up" behaviors in nurses and improve patient safety', *Journal of Nursing Care*, **27** (2), 154–60.

Sayre, M.M., McNeese-Smith, D., Phillips, L.R. and Leach, L.S. (2012a), 'A strategy to improve nurses speaking up and collaborating for patient safety', *Journal of Nursing Administration*, **42** (10), 458–60.

Shanafelt, T.D., Bradley, K.A., Wipf, J.E. and Back, A.L. (2010), 'Burnout and self-reported patient care in an internal medicine residency program', *Annals of Internal Medicine*, **136** (5), 358–67.

Sneddon, A., Mearns, K. and Flin, R. (2013), 'Stress, fatigue, situation awareness and safety in offshore drilling crews', *Safety Science*, **56** (July), 80–83.

Sparling, A.S., Marin, D.W. and Posey, L.B. (2017), 'An evaluation of the proposed worker protection standard with respect to pesticide exposure and Parkinson's disease', *Internal Journal of Environmental Research and Public Health*, **14** (6), 640–53.

Stalk, S. and Bowman, B. (2012), *Suicide Movie: Social Patterns, 1900–2009*, Cambridge, MA: Hogrefe.

Statistics Canada (2009), *Violence in the Workplace*, Ottawa: Statistics Canada.

Steffy, B.D., Jones, J.W., Murphy, L.R. and Kunz, L. (1986), 'A demonstration of the impact of stress abatement programs on reducing employee accidents and their costs', *American Journal of Health Promotion*, **1** (5), 25–32.

US Department of Justice (2002), *Workplace Violence: Issues in Response*, Washington, DC: US Department of Justice.

Vives, A., Amable, M., Ferrer, M., Moncada, S., Llorens, C., Muntaner, C. et al. (2013), 'Employment precariousness and poor mental health: evidence from Spain on a new social determinant of health', *Journal of Environmental and Public Health*, art. 978656, 1–10, accessed 23 April 2019 at https://www.hindawi.com/journals/jeph/2013/978656/.

Wachter, J.K. and Yorio, P.L. (2014), 'A system of safety management practices and worker engagement for reducing and preventing accidents: an empirical and theoretical investigation', *Accident Analysis and Prevention*, **68** (July), 117–30.

Wathen, C.N., MacGregor, J.C.D. and MacQuarrie, B.J., with the Canadian Labour Congress (2014), *Can Work be Safe, When Home Isn't? Initial Findings of a Pan-Canadian Survey on Domestic Violence and the Workplace*, London, ON: Centre for Research and Education on Violence against Women and Children.

Weingart, S.N., Pagovich, O, Sands, D.Z., Li, J.M., Aronson, M.D., Davis, R.B. et al. (2005), 'What can hospitalized patients tell us about adverse events? Learning from patient-reported incidents', *Journal of General Internal Medicine*, **20** (9), 830–36.

Weingart, S.N., Zhu, J., Chiappetta, L., Stuver, S.O., Schneider, E.C., Epstein, A.M. et al. (2011), 'Hospitalized patients' participation and its impact on quality of care and patient safety', *International Journal of Quality in Health Care*, **23** (3), 269–77.

Wisnivesky, J.P., Teitelbaum, S.L., Todd, A.C., Boffetta, P., Crane, M., Crowley, L.

et al. (2011), 'Persistence of multiple illnesses in World Trade Center rescue and recovery workers: a cohort study', *The Lancet*, **378** (9794), 888–97.

Wolfe, R.A. and Parker, D.E. (1994), 'Employee health management: challenges and opportunities', *Academy of Management Executive*, **8** (2), 22–31.

Zhang, T., Hoddenbagh, J., McDonald, S. and Scrim, K. (2012), *An Estimation of the Economic Impact of Spousal Violence in Canada, 2009*, Ottawa, ON: Department of Justice Canada, Research and Statistics Division.

2. Accident under-reporting in the workplace

Tahira M. Probst, Erica L. Bettac and Christopher T. Austin

In his speech at the twentieth World Congress on Safety and Health at Work, ILO Director-General Guy Ryder stated that 'work claims more victims every year than war' (World Congress on Safety and Health at Work 2014; p. 1). One worker dies from a work-related accident or disease and 160 workers experience a work-related accident every 15 seconds. The statistics for work-related injury and illness are staggering, and on the rise. In 2017, three years after Ryder's address, estimates of fatal occupational incidents and diseases worldwide increased from 2.3 million to 2.78 million workers (ILO 2017). Also, there are nearly 374 million non-fatal work-related injuries and illnesses annually worldwide, with many of these prompting extended work absences. With the increase in such workplace incidents comes greater financial burden: the cost of illnesses, injuries and deaths combined reached 3.94 percent of global gross domestic product, or $2.99 trillion (ILO 2017). In the US alone, approximately 2.8 million non-fatal workplace injuries and illnesses were reported (Bureau of Labor Statistics 2018), representing a price tag of over $1 billion per week for US employers (Liberty Mutual Workplace Safety Index 2018).

Despite these sobering statistics, research also increasingly indicates that these statistics may greatly under-represent the frequency of non-fatal occupational injuries (for example, Lowery et al. 1998; Leigh et al. 2004; Hämäläinen et al. 2006; Rosenman et al. 2006; Probst et al. 2008; Probst and Estrada 2010; Probst and Graso 2013; Probst, 2015; Petitta et al. 2017). While there are probably numerous contributing factors to this underestimation, perhaps the most prominent include individual-level under-reporting (that is, employees failing to report work-related illnesses and injuries to their employer) and organizational-level under-reporting (that is, organizations neglecting to accurately report employee illnesses and injuries to regulatory surveillance authorities). Figure 2.1 illustrates

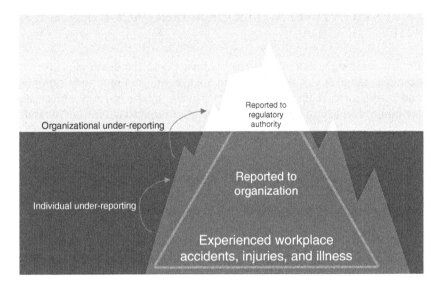

Figure 2.1   The accident reporting iceberg

how these contributing factors can distort the statistics provided by national and international surveillance systems, thus reflecting only the tip of the proverbial iceberg. Such injury surveillance methods can be implicated at the employee and/or organizational level (Weddle 1996; Probst and Graso 2011; Petitta et al. 2017).

The purpose of this chapter is to examine the organizational and psychosocial factors linked to the under-reporting of workplace accidents. For this purpose, a workplace accident is defined as any unplanned and uncontrolled event that led to injury to persons, damage to property, plant or equipment, or some other loss to the company. Thus, we expand beyond a sole focus on injury under-reporting to also include other significant events that organizations would typically expect their employees to report (that is, damage to property or equipment). We first provide a selected review of regularly utilized processes that organizations and countries have implemented for reporting, investigating and tracking employee injuries and illnesses. We then define the concept of accident under-reporting and review research on the frequency of and organizational and individual correlates of under-reporting. Finally, we propose several potential interventions organizations can use to improve the accuracy of accident reporting as well as propose directions for future research in this area.

INVESTIGATING AND REPORTING ACCIDENTS IN THE WORKPLACE

From a seemingly innocent near-miss incident to the death of a worker, accidents range greatly in severity, but are generally required to be reported and investigated if they meet certain criteria. In the hope of preventing (or minimizing) future incidents, investigations aim to determine the individual, organizational and/or job-related factors relevant to the accident's cause. This section begins by describing the standard reporting process organizations use and subsequently examines the procedure of reporting work-related accidents to the proper governmental regulatory authority.

Investigating and Reporting Accidents to Employers

Although accident investigations are conducted in response to an incident, organizations must be proactive in establishing guidelines and procedures well before they are required. Moreover, organizations should have an established protocol where an accident investigation team can properly examine the accident from varying perspectives. This team is typically comprised of any witnesses, the injured worker's immediate supervisor, the company designated safety officer and an employee representative elected by the workers to represent them (Washington State Department of Labor and Industries 2015). Thus, the accident investigation team should include an intentionally diverse team that amplifies the objectivity and diligence of the investigation procedure.

The initial step following an injury is attending to the affected employee(s). Following medical attention, the preliminary phase of an investigation can commence. Analogous to a fact-finding mission (Geller 2016), this stage should include a thorough examination where the accident scene is secured, photographed, and witnesses interviewed to answer the following types of questions (Washington State Department of Labor and Industries 2015):

- Where and when did the accident occur?
- Who was present?
- What was the employee doing prior to and at the time of the accident?
- What occurred during the accident?

After the accident scene is secured and data collected, the accident investigation needs to establish the chain of events leading to the accident and identify the root causes of the accident. Causal factors could involve individual (for example, lack of safety skills, low safety motivation and/or distorted risk perceptions), job (for example, ergonomic factors, workload and other environmental characteristics), and/or organizational variables (for example, poor communication, few resources, a poor safety climate and/or lack of safety leadership; Christian et al. 2009). It is also necessary to acknowledge there are both proximal and distal causes of an accident (Christian et al. 2009). Examples leading to failure in this may include a lack of employee safety knowledge and/or safety motivation (proximal cause), or this lack of knowledge and/or motivation could in turn be traced to, say, poor organizational safety climate, a lack of training provided and/or the employees' general propensity to make risky behaviors (distal causes). Note that an accident is often not attributable to a single cause but to a merging of factors (Christian et al. 2009; Geller 2016).

Following the identification of the accident's proximal and distal causes, the third and final stage of the accident investigation is the development and implementation of recommendations to prevent or minimize such events in the future. Results of the investigation may reveal changes needed on the part of the employee (for example, attendance at safety training), the job task environment (for example, installation of automatic shut-off switches) and/or the organization itself (for example, long-term efforts to improve the safety climate). Although not technically part of an accident investigation, a formal attempt should be made to collect follow-up information concerning the effectiveness of the implemented recommendations at improving the individual, job and/or organizational risk factors identified during the investigation.

As with accidents, a similar process should be used in investigating and reporting near-miss incidents. The rationale behind near-miss reporting systems is centered on the injury triangle model. Initially proposed by Heinrich (1931), the model has two assumptions. The first is that for every one serious accident or death, there are roughly 30 accidents with lost days, and 300 near misses. Some debate surrounds the precise ratio of near-miss to incurred accident. For instance, Bird and Germain (1996) estimated for every one death or catastrophic property loss, there are 30 minor injuries and/or property damage incidents, and 600 near misses. Regardless of the exact ratios, the underlying principles and assumptions are similar (Nielsen et al. 2006; Bellamy 2015). Together, these models can be understood as iceberg models; that is, severe safety incidents are often the most evident, but numerous minor and potential incidents lurk beneath the surface.

The second assumption is that near misses and severe injuries share equivalent underlying causal processes. Near misses and accidents are argued to merely differ by the slightly varying circumstances surrounding each event. It is therefore critical to attend as diligently to the reporting/investigation of near misses as it is the accidents themselves. Moreover, considering the significantly larger number of near misses relative to the number of accidents, near misses provide many low-cost, injury-free opportunities to learn from the conditions in which they occurred. Necessary precautions implemented before these circumstances can prevent a full-blown lost-time injury or worse (Barach and Small 2000; Nielsen et al. 2006; Bellamy 2015). Despite the practicality of near-miss reporting and investigation, individual and organizational responses to near misses typically lack the same sense of urgency as a real injury. Thus, employees often bypass near-miss reporting processes and organizations seldom conduct thorough investigations of such incidents.

Organizational Reporting of Accidents to Regulatory Authorities

Accurate surveillance estimates require not only that employees report incidents to their companies, but also that organizations accurately report workplace injuries and illnesses to the appropriate governmental regulatory authorities. Reporting these incidents plays an important role, as the reports are used for many purposes, such as performance measurement, inspection targeting, resource allocation, safety standards development, identification of high-risk and low-hazard industry sectors and compilation by national surveillance programs (OSHA 2018).

In the US, the Occupational Safety and Health Administration (OSHA) maintains the reporting and recordkeeping system to monitor job-related injuries and illnesses. Established by the Occupational Safety and Health Act 1970, the OSHA facilitates a nationwide surveillance system through annual logs of workplace injuries and illnesses required from businesses with ten or more employees. Records of the log data serve three primary functions: summaries are posted for employees' viewing, they are maintained for five years in case inspection is requested by the OSHA and/or any state regulators, and they are used to compute injury rates by employer size, industry and other classifications.

Outside the US, a similar regulatory body is used in the European Union (EU). Known as the European Agency for Safety and Health at Work, it is the primary bureaucratic body in developing guidelines for occupational injury surveillance. Although each EU member country has its own distinctive reporting requirements, there is some standardization directed by the Framework Directive on Health and Safety in the Workplace, where all accidents leading to an absence of more than four calendar days must be integrated in the European Statistics on Accidents at Work (ESAW) database (Eurostat 2016). Many developing countries lack any regulation in enforced occupational injury monitoring (Hämäläinen et al. 2006). The dearth of monetary resources is the most commonly cited reason for this, and with the absence and/or unreliability of data, it is difficult to obtain accurate global estimates of occupational accident rates (Hämäläinen et al. 2006).

DEFINITION AND PREVALENCE OF ACCIDENT UNDER-REPORTING

Even when extensive reporting requirements exist at both the organizational and national levels, research increasingly suggests these numbers represent significant undercounts of the true extent of work-related injuries and illnesses. In this section we define the concept of under-reporting and examine its prevalence and potential causes.

Organizational-Level Accident Reporting

Definitional terms
While specific reporting requirements differ from country to country, in the US a reportable event (that is, an event that must be recorded in the log of workplace illnesses and injuries and reported to the OSHA) is any work-related injury or illness that results in death, loss of consciousness, days away from work, restricted job duty or transfer, or medical treatment beyond first aid (Bureau of Labor Statistics 2016). Logged events are referred to as a recordable events. A standardized incidence rate, known as the organization's recordable rate, can then be calculated using the following equation: $(N / EH) \times 200000$, where N = the number of recordable events, EH = total hours worked by all employees during the calendar year and 200000 = base number of hours for 100 equivalent full-time workers (that is, 40 hours per week for 50 weeks a year). The resulting standardized incidence rate (that is, the number of annual injuries per 100 workers) can then be used to compare injury rates across companies and industries regardless of number of employees or hours worked (Bureau of Labor Statistics 2008).

Research indicates that not all events meeting the definition of a reportable event are logged. These events are called unreported events. Organizational under-reporting occurs when there is a discrepancy between the number of reportable events and the number of recordable events appearing in the OSHA log. Thus, organizational under-reporting is defined as a function of (1) the total number of experienced reportable events and (2) the number reported to the regulatory authority (that is, recordables). As the discrepancy between reportable and recordable events increases, organizational-level under-reporting can be said to increase.

Prevalence of organizational under-reporting

Precise estimates of organizational under-reporting vary; however, several studies have documented its prevalence (see Glazner et al. 1998; Pransky et al. 1999; Leigh et al. 2004). For example, in a study conducted during the construction of Denver International Airport, injury rates were found to be more than double the published national rates for that industry (Glazner et al. 1998). Similar rates of under-reporting have been suggested by comparing figures between work-related injuries and illnesses (for example, worker's compensation claims or medical records) and national data from the Bureau of Labor Statistics' Survey of Occupational Injuries and Illnesses (Pransky et al. 1999; Leigh et al. 2004).

A more comprehensive study by Rosenman et al. (2006) compared figures from all individuals and companies in Michigan sampled during three years of the annual Bureau of Labor Statistics' survey against the injury and illness data reported to other Michigan databases (for example, workers' compensation) by the same individuals and companies. This comparison revealed that the national surveillance system failed to reflect accurately up to 68 percent of work-related injuries and illnesses. In a later study, Probst et al. (2008), using a sample of 38 construction contractors, revealed nearly 78 percent of all incidents meeting the definition of a recordable event went unreported in the contractors' OSHA logs.

Given these findings, even the most conservative estimates appear to indicate that half of all reportable events are not accurately communicated to the appropriate regulatory authority.

Individual-Level Under-Reporting

Definition and operationalization

Individual-level under-reporting is defined similarly to organizational under-reporting in that it involves a comparison of the number of workplace injuries and illnesses that are reported to the company with the number of injuries and illnesses that are experienced. Thus, as with organizational under-reporting, as the discrepancy between the number of reported and experienced accidents increases, individual-level under-reporting increases.

Despite their similar definitions, precise measurement of under-reporting at the individual level is more challenging than at the organizational level. At the organizational level, the number of events reported to the regulatory authorities (that is, recordable events) can be found in the company's OSHA logs. These data can then be compared with workers' compensation data and/or to medical records data providing the number of reportable events. Thus, relatively objective figures of reportable versus recordable events can be found.

However, it is much more difficult to accurately capture the number of events that employees should have reported to their organization. First, the criteria for what constitutes a reportable event varies from company to company. In some companies, there is a 'report everything' policy which includes close calls, near misses and unsafe conditions and/or behaviors. Others require only that actual injuries be reported. Thus, there is no standardized measure of what constitutes a reportable event as there is at the national level.

More problematic, even if we define a reportable event using the same criteria as the OSHA recordable events, there is no objective measure of what gets reported to the employer (that is, reported events). As noted previously, at least half of all reportable events do not make it into the official log of workplace illnesses and injuries. Thus, relying on these logs would severely underestimate the number of events that employees had reported to their organization.

Therefore, researchers typically rely upon self-report data from employees to estimate the number of reported and experienced events. For example, Probst and Estrada (2010) used a free-recall measure derived from Smecko and Hayes (1999), which asked employees to indicate how many safety incidents they experienced and reported and how many they experienced but did not report to their supervisor over the past 12 months. A safety incident could include any unplanned event that led to injury, property damage and/or other loss to the company. Together, these two items enable a comparison of total reportable events (reported and unreported) with the number eventually reported. A newer recognition-based measure (Probst and Graso 2013) calculates experienced and reported events based on the US Bureau of Labor Statistics' Occupational Injury and Illness Classification System (OIICS; Bureau of Labor Statistics 2014) by asking employees to indicate whether they experienced and subsequently reported exposure to 17 unique events (for example, slip, trip, fall, hit by object and electrical shock). By comparing the levels of experienced and reported events, the extent of under-reporting can be determined.

Reliance on self-report safety data can also raise methodological concerns. First, self-report data can be inaccurate simply owing to an inability to correctly recall safety incidents. For example, the literature suggests that many minor accidents might be forgotten owing to extended recall periods. Self-report measures may also be misleading owing to impression management goals of the employee.

Prevalence of individual-level under-reporting

Given the measurement challenges noted above, precise estimates of individual-level under-reporting are difficult. Nevertheless, research (for example, Probst and Estrada 2010; Probst and Graso 2013; Probst et al. 2013a) indicates it may be quite pervasive, with estimates from those studies suggesting 57–80 percent of all accidents experienced by employees go unreported to their company. More recently, data from a large public transit agency (Byrd et al. 2018) revealed 77 percent of experienced accidents were not properly reported.

INDIVIDUAL AND ORGANIZATIONAL CORRELATES OF REPORTING BEHAVIOR

Despite the challenges in determining the exact pervasiveness of under-reporting, there is a general consensus that both organizational- and individual-level under-reporting exists. Thus, it is important to understand the causes and consequences of such under-reporting.

At the organizational-level, organizations with high levels of accident under-reporting do not have fewer accidents; they just have fewer reported accidents. While superficially having fewer apparent accidents may have short-term benefits (for example, a lower workers' compensation loss rate), organizations incur substantial costs associated with the long-term health and safety of their employees, as well as costs related to this failure to address the source of work-related injuries or accidents (Pransky et al. 1999). These costs may be compounded with government-imposed penalties and fines stemming from this organizational under-reporting.

Similarly, the costs associated with accident under-reporting extend to the employee. Failure to utilize worker's compensation presents an enduring consequence of under-reporting and places the financial burden of medical claims solely on the employee. Subsequently, this financial responsibility may result in injuries going untreated, leading to greater problems for the employee and potentially their co-workers, for example, a decline in productivity owing to attending work while ill or injured (Gallagher and Myers 1996; Loeppke et al. 2003).

Considering the prevalence of under-reporting and the serious nature of its consequences, it is vital to identify factors that predict and contribute to individual- and organizational-level under-reporting. Table 2.1 summarizes a variety of psychosocial factors that have been theorized and/or empirically demonstrated to be predictive of under-reporting.

Predictors of Individual-Level Under-Reporting

The majority of under-reporting research has focused on root causes for employee under-reporting rather than on organizational-level

Table 2.1   Antecedents of individual and organizational under-reporting

Predictors of individual-level under-reporting	Predictors of organizational-level under-reporting
Individual variables: Psychological safety climate Job insecurity Improper diagnosis or causal attributions Fear of reprisals Consideration of future safety consequences Safety-related moral disengagement Perceived safety-production conflict Experienced psychological and physical aggression Employment status (contingent versus permanent) *Organizational and work group variables:* Organizational safety climate Supervisor safety enforcement Lack of training Misguided safety incentive programs Work group norms and peer pressure Production pressure Onerous and/or lack of reporting systems Organizational justice perceptions Lack of trust in management Punitive versus non-punitive reporting consequences	*Organizational variables:* Organizational safety climate Collective bargaining agreements Organizational size Industry sector *Regulatory and other external variables:* Government penalties and fines Insurance premiums and Experience Modification Rates Fear of litigation Fear of negative publicity exposure

under-reporting. Furthermore, many of the studies that assess correlates of individual under-reporting do not measure under-reporting as discussed in this chapter. Instead, many researchers simply gather descriptive data reflecting employee justifications for not reporting an accident to their employer. However, this approach results in little empirical analysis of the extent to which these variables are predictive of discrepancies between reported and experienced events.

Despite the aforementioned barriers, researchers have documented numerous predictors of individual-level under-reporting: demographic characteristics such as age, organizational tenure and employment status (Palali and van Ours 2017; Probst et al. 2018); personality characteristics, particularly, consideration of future safety consequences (Probst et al. 2013b); level of trust in the employee–employer relationship (Carmeli and Gittell 2009); perceptions of an organization's workplace and gender climate (Tei-Tominaga and Nakanishi 2018; Austin et al. forthcoming); safety-related moral disengagement (Petitta et al. 2017); perceived lack of management responsiveness (Clarke 1998); fear of reprisals or loss of workplace perks and pay incentives (Webb et al. 1989; Pransky et al. 1999; Sinclair and Tetrick 2004); and an acceptance that injuries are common-place in certain lines of work (Pransky et al. 1999). In addition, employees may have issues classifying whether or not an incident meets the definition of an accident, which may stem from a lack of adequate safety training and/ or improper attributions regarding the event's cause (Pransky et al. 1999). Furthermore, laborious (Glendon 1991) or punitive reporting systems can dissuade employees from reporting an incident. If employees anticipate that they will not receive just treatment from their employer, the likelihood of them reporting an incident decreases substantially (Weiner et al. 2008).

Misguided safety incentive systems also may help explain the prevalence of individual-level under-reporting. Although well intended, many serve to reward employees for being accident-free instead of safety compli-ant. These systems inadvertently encourage employees to under-report experienced accidents in order to preserve their ability to receive company-administered safety rewards (Pransky et al. 1999). These apparent social pressures, combined with work group norms, serve as deterrents for employee to report experienced accidents to their supervisors (Sinclair and Tetrick 2004). Similarly, research (Jiang et al. 2018) also indicates that employees who are victims of physical and psychological aggression at work tend to not only experience more injuries at work, but also under-report those experiences, potentially out of fear of retaliation.

Individual perceptions of an organization's safety climate (also referred to as psychological safety climate; Clarke 2009) and the degree to which supervisors enforce safety policies represent additional predictors of individual under-reporting. Specifically, under-reporting diminished when employees perceived a positive organizational safety climate (Probst and Estrada 2010). Furthermore, when supervisors continually enforced safety policies, employees experienced fewer accidents and more fully reported those that were experienced. In contrast, a poor safety climate and/or lax enforcement was associated with a higher ratio of unreported to reported accidents (greater than 3:1).

Production pressure represents an additional factor that may have a det-rimental impact on employee health and safety (for example, Landsbergis et al. 1999; McLain and Jarrell 2007). In support of this, Probst and Graso (2013) examined the extent to which production pressure was related to employee attitudes and behaviors regarding accident reporting. Results suggest that employees who perceived higher levels of production pressure also reported increased negative attitudes toward the reporting of accidents. In addition, as perceptions of production pressure increased, rates of accident under-reporting concurrently increased.

An apparent motivator for accident reporting may stem from consideration of potential negative consequences of reporting such events. Indeed, Probst and Estrada (2010) found that 64 percent of all respondents indicated they had experienced at least one negative consequence as a result of reporting an accident. These consequences ranged from poor interpersonal treatment (for example, being blamed for the incident) to adverse job performance outcomes (for example poor performance review). These concerns also increase fears regarding the sustainability of valued job incentives or someone's position within an organization (that is, job security; Probst et al. 2013a; Milch and Laumann 2016; Palali and van Ours 2017). For example, Probst et al. (2013a) found that job insecurity (that is, fear of job loss) was not only a significant predictor of the prevalence of accidents and injuries, but also of accident under-reporting decisions. Similarly, Probst and Graso (2013) found that prior experience with negative consequences of reporting accidents resulted in increased reluctance to fully report experienced accidents.

Predictors of Organizational-Level Under-Reporting

Though research mainly focuses on individual-level under-reporting and often include variables operating at the organizational level (for example, production pressure and safety climate), there are unique factors that are strictly associated with organizational-level under-reporting. For example, complex or bureaucratic safety management systems, pervasive fear of litigation and negative publicity, as well as economic pressures, government penalties for safety violations, higher insurance premiums and collective bargaining agreements, represent disincentives for accurate organizational reporting of experienced accidents and injuries (Conway and Svenson 1998; Barach and Small 2000; Zacharatos and Barling 2004; Jeffcott et al. 2006; Milch and Laumann 2016; Palali and van Ours 2017).

Influenced by such disincentives, Zahlis and Hansen (2005) argue that there is a continual disconnect between organizational measures (OSHA recordables) and valued outcomes (workers' compensation costs and expenses). Furthermore, these discrepancies highlight an over-reliance on poor (often lagging) indicators of organizational safety and the high stakes attached to an organization's OSHA injury rate.

Probst et al. (2008) utilized medical records data provided by an Owner-Controlled Insurance Program, finding significant differences between the number of reported injuries in OSHA logs and the number of injuries experienced. Specifically, for every three injuries that were properly recorded, an additional 11 went unrecorded in the log. Furthermore, this rate of under-reporting was significantly predicted by the organizational safety climate, such that organizations with a poor safety climate had a significantly higher rate of unreported injuries than organizations with a positive safety climate (17 versus 3 per 100 workers). Interestingly, there was no significant difference in the injury rates that were reported to the OSHA (3.98 versus 3.64), indicating that organizations with poor and positive safety climates reported comparable injury rates. Yet, those reports were far more distorted (that is, inaccurate) for the companies with poor safety climates. These results further highlight the unreliability of using OSHA logs as the primary metric of safety behaviors within an organization.

Addressing their findings, Probst et al. (2008) acknowledged that the discrepancies could possibly be caused by intentional manipulation of the numbers. However, it is equally plausible that these discrepancies could be attributable organizations with a poor safety climate perhaps not devoting adequate resources to accurately tracking recordables, not providing proper training for safety personnel regarding proper identification of recordable injuries, and perhaps having implicit incentive systems that reward managers and administrators for coding injuries as something other than a recordable. These and other findings (for example, Probst 2015; Zadow et al. 2017), emphasize that safety climate represents a significant predictor of organizational-level under-reporting.

IMPLICATIONS AND FUTURE RESEARCH NEEDS

A growing body of research has documented and identified numerous individual and organizational factors associated with accident under-reporting. Although the vast majority of this research has considered these factors in isolation from each other, collectively the findings from the extant literature indicate the need to develop comprehensive multilevel models of accident reporting behavior including variables operating at the individual, work group, organizational and national levels that contribute to the problem of under-reporting.

Workplace interventions (and evaluations of the extent to which these are successful) are also needed to improve the accuracy of accident reporting. While recommendations for improving safety incentive systems and reporting systems have been suggested, the effectiveness of such interventions are rarely empirically tested. For example, Reason (1997) argued that there are five preconditions necessary for organizations to report accurately: (1) indemnity against disciplinary action, (2) confidentiality of reporting, (3) separating the agency that collects and analyzes the data from the regulatory authority, (4) rapid and useful availability of feedback and (5) ease in using the reporting system. Based on the literature review summarized in Table 2.1, many of these same recommendations would apply at the individual level: (1) no-fault reporting where accident investigations are seen as fact finding as opposed to fault finding; (2) confidential and/or anonymous reporting; (3) a visible and positive organizational response to reports of hazardous situations and/or injuries; and (4) a simple straightforward reporting system.

It is important to also understand what types of injuries and illnesses go unreported, in order to better target interventions. The inclusion of actual loss data in future studies (that is, workers' compensation claims rates and costs) would allow for a more in-depth analysis of the types of injuries that are being unreported in workplace injury and illness logs. Similarly, in self-report surveys, researchers should determine if employees fail to report some types of accidents more often than others. For example, in their study of copper miners, Probst and Graso (2013) used open-ended questions to ask employees how many total accidents they had experienced and reported, and how many they had experienced but not reported. They then asked workers to indicate how many of the accidents within each category had resulted in lost time, first aid, no injury and property/equipment damage. Using this metric, they found that nearly 52 percent of all accidents went unreported. However, the highest rates of under-reporting occurred with events requiring only first aid (89 percent unreported) or that resulted in a self-assessment of 'no injury' (61 percent unreported). However, only 33.3 percent of all lost-time injuries and 24 percent of all damage to property or equipment went unreported. This would seem to indicate that employees are more willing to report more severe injuries (particularly those requiring time off work) and those resulting in property damage, perhaps owing to the higher visibility of such events. Smaller, less noticeable events that apparently cause little or no harm tend not to get reported. This comports with a study conducted by Nielsen et al. (2006) which asked metalworkers to indicate their willingness to report several types of hypothetical events. In that study, employees indicated they would be most willing to report lost-time incidents followed by minor injuries and then near misses.

CONCLUSION

Frequently cited national and international injury, illness and fatality statistics can appear staggering; yet, a large and growing body of research suggests these data may only represent the tip of the iceberg. This chapter presented evidence of major discrepancies between accidents experienced by employees, what gets reported to organizations and, ultimately, what organizations report to the regulatory authorities. Using conservative estimates, it appears that for every injury captured in the national surveillance system data, at least two incidents were reported to organizations, and at least four were experienced by employees. Extrapolating from the 2.9 million officially reported work-related illnesses and injuries in the US in 2017, this would imply that there were an additional 13 million events hidden below the tip of that iceberg. Given the prevalence of this under-reporting, we reviewed research on empirically established individual and organizational correlates of under-reporting and discussed the need for more comprehensive multi-level research examining the interplay among these correlates, as well as more research on practical interventions that organizations can utilize to improve the accuracy of accident reporting.

REFERENCES

Austin, C.T., Probst, T.M., Petitta, L. and Barbaranelli, C. (forthcoming), 'Gender climate and accident underreporting: the role of gender congruence', manuscript in preparation.

Barach, P. and Small, S.D. (2000), 'Reporting and preventing medical mishaps: lessons from non-medical near miss reporting systems', *British Medical Journal*, **320** (7237), 759–63.

Bellamy, L.J. (2015), 'Exploring the relationship between major hazard, fatal and non-fatal accidents through outcomes and causes', *Safety Science*, **71** (January), 93–103.

Bird, F.E. Jr and Germain, G.L. (1996), *Practical Loss Control Leadership*, Loganville, GA: International Loss Control Institute.

Bureau of Labor Statistics (2008), 'Workplace injury and illness summary', United States Department of Labor, Washington, DC.

Bureau of Labor Statistics (2014), 'Occupational injury and illness classification manual', accessed 1 May 2019 at http://www.bls.gov/iif/oshoiics.htm.

Bureau of Labor Statistics (2016), 'Occupational safety and health definitions', accessed 1 May 2019 at http://www.bls.gov.iif.oshdef.htm.

Bureau of Labor Statistics (2018), 'Employer-reported workplace injuries and ill-nesses, 2017', accessed 1 May 2019 at https://www.bls.gov/news.release/archives/osh_11092017.pdf .

Byrd, J., Gailey, N.J., Probst, T.M, and Jiang, L. (2018), 'Explaining the job insecurity-safety link in the public transportation industry: the mediating role of safety-production conflict', *Safety Science*, **106** (July), 255–62.

Carmeli, A. and Gittell, J.H. (2009), 'High-quality relationships, psychological safety, and learning from failures in work organizations', *Journal of Organizational Behavior*, **30** (6), 709–29.

Christian, M.S., Bradley, J.C., Wallace, J.C. and Burke, M.J. (2009), 'Workplace safety: a meta-analysis of the roles of person and situation factors', *Journal of Applied Psychology*, **94** (5), 1103–27.

Clarke, S. (1998), 'Organizational factors affecting the incident reporting of train drivers', *Work & Stress*, **12** (1), 6–16.

Clarke, S. (2009), 'Accidents and safety in the workplace', in S. Cartwright and C.L. Cooper (eds), *The Oxford Handbook of Organizational Well-Being*, Oxford: Oxford University Press, pp. 31–54.

Conway, H. and Svenson, J. (1998), 'Occupational injury and illness rates, 1992–1996: why they fell', *Monthly Labor Review*, **121** (11), 34–58.

Eurostat (2016), *Accidents at Work Statistics*, Luxembourg: Office for Official Publications of the European Communities.

Gallagher, R.M. and Myers, P. (1996), 'Referral delay in back pain patients on worker's compensation: cost and policy implications', *Psychosomatics: Journal of Consultation Liaison Psychiatry*, **37** (3), 270–84.

Geller, E.S. (2016), *The Psychology of Safety Handbook*, Boca Raton, FL: CRC Press.

Glazner J.E., Borgerding J., Lowery J.T., Bondy J., Mueller K.L. and Kreiss K. (1998), 'Construction injury rates may exceed national estimates: evidence from the construction of Denver International Airport', *American Journal of Industrial Medicine*, **34** (2), 105–12.

Glendon, A.I. (1991), 'Accident data analysis', *Journal of Health and Safety*, **7**, 5–24.

Hämäläinen, P., Takala, J. and Saarela, K.L. (2006), ‚Global estimates of occupational accidents', *Safety Science*, **44** (2), 137–56.

Heinrich, H.W. (1931), *Industrial Accident Prevention*, New York: McGraw-Hill.

International Labour Organization (ILO) (2017), 'Safety and health at work', accessed 1 May 2019 at http://www.ilo.org/global/topics/safety-and-health-at-work/lang--en/index.htm.

Jeffcott, S., Pidgeon, N., Weyman, A. and Walls, J. (2006), 'Risk, trust and safety culture in U.K. train operating companies', *Risk Analysis*, **26** (5), 1105–21.

Jiang, L., Probst, T.M., Benson, W. and Byrd, J. (2018), 'Voices carry: effects of verbal and physical aggression on injuries and accident reporting', *Accident Analysis and Prevention*, **108** (September), 190–99.

Landsbergis, P.A., Cahill, J. and Schnall, P. (1999), 'The impact of lean production and related new systems of work organization on worker health', *Journal of Occupational Health Psychology*, **4** (2), 108–30.

Leigh, J.P., Marcin, J.P. and Miller, T.R. (2004), 'An estimate of the U.S. government's undercount of nonfatal occupational injuries', *Journal of Occupational and Environmental Medicine*, **46** (1), 10–18.

Liberty Mutual Insurance Safety Index (2018), 'Workplace injuries cost U.S. companies over $1 billion per week', *PRN Newswire*, 8 May, accessed 1 May 2019 at https://www.prnewswire.com/news-releases/workplace-injuries-cost-us-companies-over-1-billion-per-week-300644539.html.

Loeppke, R., Hymel., P.A., Loafland, J.H., Pizzi, L.T., Konicki, D.L., Anstadt, G.W. et al. (2003), 'Health-related workplace productivity measurement: general and migraine-specific recommendations from the ACOEM expert panel', *Journal of Occupational and Environmental Medicine*, **45** (4), 349–59.

Lowery, J.T., Borgerding, J.A., Zehn, B., Glazner, J.E., Bondy, J. and Kreiss, K. (1998), 'Risk factors for injury among construction worker at Denver International Airport', *American Journal of Industrial Medicine*, **34** (2), 105–12.

McLain, D. L., and Jarrell, K. A. (2007), 'The perceived compatibility of safety and production expectations in hazardous occupations', *Journal of Safety Research*, **38**, 299–309.

Milch, V. and Laumann, K. (2016), 'Interorganizational complexity and organizational accident risk: A literature review', *Safety Science*, **82** (February), 9–17.

Nielsen, K.J., Carstensen, O. and Rasmussen, K. (2006), 'The prevention of occupational injuries in two industrial plants using an incident reporting scheme', *Journal of Safety Research*, **37** (5), 479–86.

Occupational Safety and Health Administration (OSHA) (2018), 'Employer rights and responsibilities following a federal OSHA inspection', OSHA 3000-04R 2018, accessed 1 May 2019 at https://www.osha.gov/Publications/osha3000.pdf.

Palali, A. and van Ours, J.C. (2017), 'Workplace accidents and workplace safety: on under-reporting and temporary jobs', *Labour*, **31** (1), 1–14.

Petitta, L., Probst, T.M. and Barbaranelli, C. (2017), 'Safety culture, moral disengagement, and accident underreporting', *Journal of Business Ethics*, **141** (3), 489–504.

Pransky, G., Snyder, T., Dembe, A. and Himmelstein, J. (1999), ,Under-reporting of work-related disorders in the workplace: a case study and review of the literature', *Ergonomics*, **42** (1), 171–82.

Probst, T.M. (2015), 'Organizational safety climate and supervisor safety enforcement: multilevel explorations of the causes of accident underreporting', *Journal of Applied Psychology*, **100** (6), 1899–907.

Probst, T.M. and Estrada, A.X. (2010), 'Accident under-reporting among employees: testing the moderating influence of psychological safety climate and supervisor enforcement of safety practices', *Accident Analysis & Prevention*, **42** (5), 1438–44.

Probst, T.M. and Graso, M. (2011), 'Reporting and investigating accidents: recognizing the tip of the iceberg', in S. Clarke, C. Cooper and R. Burke (eds), *Occupational Health and Safety: Psychological and Behavioral Challenges*, Farnham: Gower, pp. 71–94.

Probst, T.M. and Graso, M. (2013), 'Pressure to produce = pressure to reduce accident reporting?', *Accident Analysis & Prevention*, **59** (August), 580–87.

Probst, T.M., Barbaranelli, C. and Petitta, L. (2013a), 'The relationship between job insecurity and accident under-reporting: a test in two countries', *Work & Stress*, **27** (4), 383–402.

Probst, T.M., Brubaker, T.L. and Barsotti, A. (2008), 'Organizational underreporting of injury rates: an examination of the moderating effect of organizational safety climate', *Journal of Applied Psychology*, **93** (5), 1147–54.

Probst, T.M., Graso, M., Estrada, A.X. and Greer, S. (2013b), 'Consideration of future safety consequences: a new predictor of employee safety', *Accident Analysis & Prevention*, **55** (June), 124–34.

Probst, T.M., Petitta, L., Barbaranelli, C. and Lavaysse, L.M. (2018), 'Moderating effects of contingent work on the relationship between job insecurity and employee safety', *Safety Science*, **106** (July), 285–93.

Reason, J.T. (1997), *Managing the Risks of Organizational Accidents*, Aldershot: Ashgate.

Rosenman, K.D., Kalush, A., Reilly, M.J., Gardiner, J.C., Reeves. M. and Luo, Z. (2006), 'How much work-related injury and illness is missed by the current national surveillance system', *Journal of Occupational and Environmental Medicine*, **48** (4), 357–65.

Sinclair, R.R. and Tetrick, L.E. (2004), 'Pay and benefits: the role of compensation systems in workplace safety', in J. Barling, and M. Frone (eds), *Psychology of Workplace Safety*, Washington, DC: American Psychological Association, pp. 181–201.

Smecko, T. and Hayes, B. (1999), 'Measuring compliance with safety behaviors at work', paper presented at the fourteenth annual conference of the Society for Industrial and Organizational Psychology, Atlanta, GA, April.

Tei-Tominaga, M. and Nakanishi, M. (2018), 'The influence of supportive and ethical work environments on work-related accidents, injuries, and serious psychological distress among hospital nurses', *International Journal of Environmental Research and Public Health*, **15** (2), 240.

Washington State Department of Labor and Industries (2015), 'Washington Department of Labor and Industries claims management performance audit', report, June, accessed 1 May 2019 at http://leg.wa.gov/jlarc/reports/2015/Workers CompPhase2/pf2/doc/WCS_FullReport.pdf.

Webb, G.R., Redman, S., Wilkinson, C. and Sanson-Fisher, R.W. (1989), 'Filtering effects in reporting work injuries', *Accident Analysis and Prevention*, **21** (2), 115–23.

Weddle, M.G. (1996), 'Reporting occupational injuries: the first step', *Journal of Safety Research*, **27** (4), 217–23.

Weiner, B.J., Hobgood, C. and Lewis, M.A. (2008), 'The meaning of justice in safety incident reporting', *Social Science and Medicine*, **66** (2), 403–13.

World Congress on Safety and Health at Work (2014), 'ILO Director-General: "Work claims more victims than war"', 26 August, accessed 1 May 2019 at http://www.ilo.org/global/about-the-ilo/newsroom/news/WCMS_302517/lang--en/index.htm.

Zacharatos, A. and Barling, J. (2004), 'High performance work systems and occupational safety', in J. Barling, and M. Frone (eds), *Psychology of Workplace Safety*, Washington, DC: American Psychological Association, pp. 203–22.

Zadow, A.J., Dollard, M.F., Mclinton, S.S., Lawrence, P. and Tuckey, M.R. (2017), 'Psychosocial safety climate, emotional exhaustion, and work injuries in healthcare workplaces', *Stress and Health*, **33** (5), 558–69.

Zahlis, D.F. and Hansen, L.L. (2005), 'Beware the disconnect', *Professional Safety*, **50** (11), 18–24.

3. Stress, human errors and accidents

Astrid M. Richardsen, Monica Martinussen and Sabine Kaiser

Work is an important part of most people's lives, and can be a source of engagement and challenge. However, work may also be a source of stress. Over the years, research has uncovered the characteristics of job stress, antecedents as well as consequences in terms of reduced psychological well-being and physical health. The effects of occupational stressors on workplace accidents and injuries are less well known (Clarke 2012). It is generally agreed that the way in which work stressors influence employee's behavior is by inducing stress symptoms or strain, leading to reductions in productivity as well as health and well-being (Karasek and Theorell 1990; Lazarus 1991). The emotional, cognitive, and behavioral responses to stress may increase the risk of errors and accidents and the potential for injury (Halbesleben 2010; Souza et al. 2014). According to Halbesleben (2010), the accumulation of work stressors may lead to exhaustion, representing a state of depleted resources, which may directly affect vulnerability to accidents. Responses to stress may include cognitive problems, such as lowered concentration and increased distractibility, which may affect safety behavior and increase the likelihood of mistakes and accidents (Halbesleben 2010; Clarke 2012; Mathisen and Bergh 2016).

The reasons why the effects of work stressors on accidents and injuries are difficult to establish have partly to do with the theory and definition of stress. Some theories describe stress as the factors or elements that have a negative impact on the individual, for example, distracting noise or pressure at work (stimulus-based theories), while other theories are concerned with the consequences of stress, such as various emotional and physical reactions (response-based theories) (Cooper et al. 2001). The latter tradition is exemplified by Selye (1984, p. 55), who defined stress as 'the nonspecific response of the body to any demand', a general stress response that is valid for everyone and consists of three phases: the alarm phase, the resistance phase and the exhaustion phase. Selye (1984) was the first to point out the detrimental effects of acute and prolonged stress, both physiologically and cognitively. He was able to show that a number of diseases of adaptation, such as cardiovascular diseases, digestive disorders and mental derangements, were not caused by direct damage, by germs or poisons, but were largely due to errors in our adaptive response to stressors and failures in the stress-fighting mechanism. Evidence suggests that the experience of workplace stress is directly or indirectly related to seven of the ten leading causes of death in developed nations (Quick et al. 2003): heart disease, cancer, stroke, injuries, suicide/homicide, chronic liver disease and emphysema or chronic bronchitis.

A more modern understanding requires stress to be regarded as the interaction or transaction between demands and the resources available to the individual (Lazarus 1991). Stress, accompanied by both psychological and physiological reactions, develops as a result of a dynamic process in which demands in the environment (for example, intense workloads, time constraints, shift work, problems or conflicts) outweigh the individual's coping resources (including experience, physical and mental health, personal abilities and, potentially, external support). An important point is that the person must evaluate the demands and consider whether these demands exceed his or her resources. Owing to this cognitive evaluation, what one individual considers a stressor is not necessarily considered a stressor by someone else (Lazarus 1991).

This chapter focuses on the consequences of experiencing stress at work, both acute situations where a danger or immediate threat is clearly present, and more long-term stressors such as a prolonged high workload. A large amount of research has focused on potential work stressors, including factors intrinsic to the job itself (such as noise, workload, long work hours, risks, hazards and emergencies), roles in the organization, relationships at work, career development issues, atypical employment, organizational structure and climate, and the home–work interface (Cartwright and Cooper 1997). The consequences of stress have mainly been studied in relation to employee health and well-being, such as burnout and engagement, but less research has focused on the consequences of stress in terms of safety, work performance, accidents and errors. A great deal of the research on accidents and errors has focused on the practice in operative professions such as medicine and aviation where the consequences of human error are dramatic, very expensive and may involve the loss of lives. We therefore focus on research conducted among healthcare professionals and pilots.

CONSEQUENCES OF STRESS

Stress has both short-term and long-term consequences that may affect an individual's job performance and safety at work. A critical stress situation may arise when something unusual takes place, such as a patient becoming critically ill or a mechanical error occurring while the plane is descending. It is difficult to measure how an acute stress situation affects various cognitive functions in a real-life setting for a number of reasons. For example, a study may interfere with normal operations, and it may be difficult to predict when a critical situation will occur. Another option is to use laboratory experiments, for example, a flight simulator, where it is possible to introduce different types of emergencies and study how this affects pilot performance. Studies of incidents that have already taken place, reconstructing the decisions made and observing how stress contributed to the event, have the drawback of being based on human recollection and perception of the event and what happened (Martinussen and Hunter 2017). Situations may appear very different in hindsight compared with how the situation actually occurred and was experienced by the people involved. Some have attempted quasi-experimental designs to try to understand the relationship between an individual's level of cognitive failure, psychological stress and workplace accident occurrence (Day et al. 2012). Retrospective analysis of accident-case individuals versus control-matched individuals on measures of general health and cognitive failures, showed that individuals in the accident-case group had higher scores when compared with matched-controls, suggesting that individuals who are stressed are more likely to have an accident in the workplace because of a propensity for cognitive failures (Day et al. 2012). While this is generally accepted, some variations in the severity of incidents and accidents have been found in different industry settings (Defraia 2015).

There are some consistent findings on the immediate effect of stress on cognitive functions and decision-making capacity. According to a review provided by Orasanu (1997), stress may have the following effects: people make more errors; attention is reduced, causing tunnel vision or selective hearing; the capacity to take in and monitor information diminishes, affecting communication between those involved; scanning (vision) becomes more chaotic; short-term memory is reduced; and there is often a simplification of strategies, such as preferring speed over accuracy. In general, cognitive processes that involve retrieval of information from long-term memory are resilient to stress, while processes that require the use of short-term memory are more vulnerable. Experienced employees will make fewer mistakes under pressure than inexperienced employees because they have a greater number of experiences stored in their long-term memory, and are more likely to use an established rule-based approach rather than having to consider several options or even improvise new solutions (Orasanu 1997).

Dismukes et al. (2018) have suggested that stress may influence working memory and attention in three different ways. First, attention is more easily captured by stimuli that are salient but less relevant to the task at hand. Secondly, the ability to shift attention efficiently between separate tasks is reduced. Also, the ability to update and monitor information in working memory is impaired (Dismukes et al. 2018). Thus, cognitive functions are subject to a number of stress-related consequences affecting how we perceive our surroundings, process information, and make decisions in critical situations. A simulation study of trainee pilots (Morris and Leung 2006) indicated that as the mental workload and auditory input increased, participants experienced considerable difficulty in carrying out a prioritization task. Also, when under a high workload, their ability to comprehend more than two chunks of auditory data deteriorated rapidly (Morris and Leung 2006).

Long-term workplace stress affects workplace health and safety as well as employee well-being. Work stress is widespread and can influence individual feelings, perceptions, thought processes, performance, as well as physiological and biochemical responses (Murphy et al. 1986). For example, a national survey of Finnish employees found higher rates of occupational injuries in employees working in haste or reporting to be stressed at work (Salminen et al. 2017).

An increasing number of researchers in workplace safety, agree that there is a direct association between increased levels of worker stress and increases in workplace accidents (Murphy et al. 1986; Mora et al. 2002; Welker-Hood 2006; Halbesleben 2010; Clarke 2012; Nielsen et al. 2013; Rasmussen et al. 2014; Tomei et al. 2015; Kim et al. 2016; Rodrigues et al. 2017). This relationship can be attributed to cognitive impairment, such as preoccupations and distractions resulting from elevated stress levels and higher levels of fatigue. Work stress may also lead to reduced safety behavior, as indicated in a study of firefighters, where stress mediated by burnout was associated with lower levels of communicating safety concerns (Smith et al. 2018). As a result, firefighters were less likely to use personal protective equipment properly and follow standard operating procedures, which in turn could result in injuries (Smith et al. 2018).

One of the long-term consequences of work stress is burnout, often defined as a syndrome of exhaustion, cynicism and reduced professional efficacy (Leiter and Maslach 2009). Again, the health consequences of burnout are well documented, and there are also some studies that have looked at burnout and risk of injury and accidents. In one study, administrative employees of a large US organization completed two annual surveys of burnout, workload and supervision (Leiter and Maslach 2009). These surveys were linked to workplace injury rates for the work units, recorded by the organization's human resources office during the subsequent year, for each of the two years of the study. The longitudinal analysis found that workload and exhaustion predicted the incidence of injuries during the subsequent year. Multiple regression analyses established that exhaustion mediated the relationship of workload with injury rate. Thus job stress can impair workers' job performance by reducing their capacity for complex physical skills and impairing cognitive functioning, and may place them at greater risk of an accident on the job (Leiter and Maslach 2009).

A meta-analysis by Nahrgang et al. (2011) found that burnout was negatively related to working safely, whereas work engagement motivated employees to follow safety regulations. Depending on the industry, burnout was also significantly correlated with higher incidents of accidents and injuries and adverse events (near misses and errors) (Nahrgang et al. 2011). Another meta-analysis examined the difference between hindrance and challenge stressors on occupational accidents (Clarke 2012). Hindrance stressors such as work conflict or role conflict were assumed to reduce motivation and influence worker attitudes negatively, whereas challenge stressors present workers with a possibility to expand their skills and knowledge. The findings indicated a link between hindrance stressors and accidents, but not for challenge stressors. The association between hindrance stressors and accidents was fully mediated by safety behaviors such as compliance with safety rules and participation in safety-related activities (Clarke 2012).

STRESS AND MEDICAL ERRORS

It is estimated that up to 98 000 people die annually from medical errors in hospitals, more than deaths from motor vehicle accidents, breast cancer and AIDS combined (Committee on Quality of Health Care in America 2000). More people die from medical errors than from workplace accidents. Medical errors by care providers lead to adverse health outcomes for patients, in addition to other factors such as improper use of medical equipment, poor or fatigued decision making, stress and cognitive overload (Peters and Peters 2007; Bogner 2010). One study investigated the occupational health hazards that may lead to medical errors and injury among healthcare workers (doctors, nurses and ward orderlies) in an obstetrics unit (Orji et al. 2002). The common occupational health hazards were work-related stress, needle-stick injuries, bloodstains on skin, sleep disturbance, skin reactions, assault from patients and hepatitis. A greater percentage of doctors, compared with nurses and ward orderlies, used safety precautions such as gloves, facemasks and aprons. While all the staff employed regular handwashing after various procedures, proper disposal of needles and sharps into separate puncture-resistant containers was inadequate among all categories of staff.

A number of studies have found relationships between long-term stress and cognitive impairment, and workplace accidents or injuries. Elfering et al. (2017) did a study among surgery nurses, and found that both task-related and emotional demands were related to cognitive stress symptoms, including problems in concentrating, deciding, memorizing and reflecting. Such cognitive stress symptoms are assumed to constitute risks to patient safety. In a study of hospital nurses in Korea, Park and Kim (2013) found that factors affecting patient safety incidents were cognitive failure and job stress in the form of low job autonomy and job instability. Similarly, Renouard et al. (2017) note the impact of cognitive and behavioral factors in complications in dental procedures. Lawton and Parker (1998) reviewed research on the relationship between accident liability and individual differences, particularly on workplace accidents. Personality, cognitive and social factors were considered. Accident involvement included both errors and/or violations. Errors were primarily associated with cognitive factors, while safety violations were associated with social psychological factors. Stress reactions were seen as having effects of increasing risk-taking behaviors and reducing information-processing capacity of others (Lawton and Parker 1998).

A number of studies have investigated the role of burnout in patient safety in various medical settings. Physician burnout has been shown to increase medical errors (Shanafelt et al. 2010a). In several systematic reviews of physician burnout, researchers found that there was a significant relationship between physician and resident burnout and safety-related quality of care (Hall et al. 2016; Dewa et al. 2017a, 2017b). Hall et al. (2016) reviewed 46 studies investigating the relationship between healthcare staff well-being, burnout and patient safety, and found significant associations between poor well-being, moderate to high burnout and poor patient safety, such as medical errors. Similar results were found in a study of internal medical residents in the US (Shanafelt et al. 2010b). A study by Williams et al. (2007) found that stressed, burned out and dissatisfied physicians reported greater likelihood of making errors and more frequent instances of suboptimal patient care. Another study of surgeons (Shanafelt et al. 2010a) found that all three aspects of burnout were significant predictors of self-reported recent medical errors. In a longitudinal and multicenter study of burnout and error in junior doctors, it was found that doctors who were burned out reported making more errors than doctors who were not burned out (O'Connor et al. 2017). Other studies have found that depression, rather than burnout, was related to risk of medical errors among intensive care unit workers (Garrouste-Orgeas et al. 2015).

However, some of the methodological issues in these studies, such as the use of self-perceived errors and self-reported accident propensity, make the results and implications uncertain. A study of hospital employees linked at the team level to the hospital workplace injury register showed that the concordance between survey-reported and registered injury rates was low, indicating that many injuries go unreported (Zadow et al. 2017). Emotional exhaustion was the strongest predictor of survey-reported total injuries and under-reporting. Psychosocial safety climate was the strongest predictor of registered injury rates over time. The authors conclude that these results underscore the need to consider both an individual and psychosocial safety explanation of injury events, and a psychosocial explanation of injury under-reporting (Zadow et al. 2017).

Medical errors among nurses is also a critical issue in patient safety, as they need to have good concentration, sound judgement, and quick reaction times in emergency situations. Nursing shortages often lead to increased nurse workloads, working consecutive shifts and taking on overtime work. In a study of nurses in 90 hospitals in Thailand, it was found that extended work hours were predictive of nurse outcomes of emotional exhaustion and depersonalization, and was also associated with detrimental patient outcomes such as patient identification errors, communication errors and patient complaints (Kunaviktikul et al. 2015). Another study found that time pressure and burnout had an interactive effect on patient safety (Teng et al. 2010).

In an integrative review of the literature on stress and burnout of hospital nursing professionals and the interface with patient safety, the results showed that stress and burnout were associated with vulnerability to provide unsafe care to patients (Rodrigues et al. 2017). The working environment with precarious working conditions and the excessive workload of nurses were the main contributing factors to the experiences of stress and burnout in the study populations. In another review of studies investigating nurse burnout and patient safety, Ross (2016) concluded that healthcare provider burnout was significantly associated with higher incidence of errors and near misses. The rates of burnout in the nursing populations were high, and burnout scores were also related to poor well-being, for example, depression, anxiety and poor quality of life. Ross (2016) argues that organizational efforts to reduce nurse burnout would not only increase patient safety, but reduce healthcare costs for hospitals and improve nurses' quality of life. Similar arguments for the need to develop and implement effective organizational remedies have been advanced in studies of physician burnout (West et al. 2018). Some researchers have advocated stress reduction as an intervention to reduce accident levels (Murphy et al. 1986; Sneddon et al. 2013). They specifically highlight fatigue and sleep deprivation as reducing levels of alertness and attention, which increases accident frequencies.

SHIFT WORK

Shift work is commonplace for most people in both healthcare and aviation, and has several consequences in relation to health issues, sleep, work achievements, the risk of accidents, increased work-to-home conflicts, and participation in social activities taking place on weekday evenings and weekends. Pilots who travel across time zones also experience problems owing to jet lag. Working shifts may also influence an individual's relationship with the workplace, for example in the form of reduced work satisfaction (Demerouti et al. 2004). Sleep difficulties are one of the most common problems associated with shift work in general, and with night shifts in particular (Pallesen 2006). Working night shifts influences the internal clock (or body clock) and its adjustment, and prevents the important restorative function that sleep has on the body (especially the brain) and the production of certain types of hormones (Pallesen 2006). While there are some individual differences in tolerance to night and shift work, several studies have found adverse health implications in shift workers, including increased exposure to cardiovascular diseases, problems with digestion and cancer (Costa 1996; Tüchsen et al. 2006). Mitler et al. (1988) did a report for the Association of Professional Sleep Societies and concluded that night-time fatigue and sleepiness among operators contributed to two major nuclear power plant accidents (Chernobyl and Three Mile Island). In addition, the crash of the *Exxon Valdez* occurred at night, with sleep deprivation seen as one of the causal factors. Sleep deprivation was also cited as a factor in poor decisions made the night before the launch of the Space Shuttle *Challenger*.

Working shifts, and working night shifts in particular, is associated with an elevated risk of accidents (Folkard and Tucker 2003; Rogers 2008). Studies show that people make more mistakes if they are sleep deprived and that they need longer time to perform basic tasks; and that drowsiness is associated with reduced attention and lack of concentration, which are major factors in accidents and accident proneness (Folkard and Tucker 2003; Åkerstedt 2007).

A number of studies have examined how sleep deprivation affects performance and how it relates to accidents (Lusa et al. 2002; Åkerstedt 2007; Sugden et al. 2010). In an overview by Åkerstedt (2007), studies revealed that tasks requiring constant attention are more affected by sleep deprivation, while more advanced tasks, such as reasoning, were less affected. A study of firefighters found that sleep disturbance was reported to occur as soon as the total working time exceeded 50 hours per week, and working more than 70 hours per week increased the risk of occupational accidents almost fourfold (Lusa et al. 2002). The consequences of sleep deprivation generally increase during an extended period without sleep, however, even after a prolonged period of being awake, performance will improve somewhat during the time period in which a person is normally awake, and performance improves during the day relative to the night (Åkerstedt 2007).

Sneddon et al. (2013) examined the effects of both work stress and fatigue on work situation awareness among drillers working off shore on the UK continental shelf. Drillers reporting high levels of stress, sleep disruption and fatigue indicated lower levels of work situation awareness, with stress being the strongest predictor. In an earlier study of oil rig drillers and situation awareness, Sneddon et al. (2006) found that being isolated and away from home, stress and fatigue were associated with lower levels of situation awareness.

Rogers (2008) reviewed the literature on the effects of fatigue and sleepiness on nurse performance and patient safety. She defined sleepiness as 'the tendency to fall asleep', whereas fatigue refers to 'an overwhelming sense of tiredness, lack of energy, and a feeling of exhaustion associated with impaired physical and/or cognitive functioning' (Rogers 2008, p. 1). The review showed that individuals working nights and rotating shifts rarely obtain optimal amounts of sleep, and the detrimental effects of sleep deprivation are cumulative. Studies over the past 15 years unanimously revealed that insufficient sleep is associated with cognitive problems, mood alterations, reduced job performance, reduced motivation, increased safety risks, and physiological changes (Rogers 2008). Many studies showed that inadequate sleep is an important contributor to medical errors, as well as nurses' own health and safety (for example, reduced immune function and more traffic accidents). Extended work hours and long shifts are associated with risk of errors for interns and resident physicians, but recent studies have shown that this is also the case for nurses (Rogers 2008). Rogers (2008) cites a number of research studies using a variety of research designs, all documenting the adverse effects of insufficient sleep on medical errors and injuries.

Arimura et al. (2010) conducted a study of medical errors among hospital nurses in Japan. In a survey of almost 500 nurses in two general hospitals, they found that shift work and poor mental health were significantly related to medical errors (Arimura et al. 2010). Other studies have shown that nurses who worked rotating schedules were twice as likely to report committing medication errors than nurses who worked day shifts (Gold et al. 1992). In a study of nurses completing daily logbooks for one month, it was found that stress and struggling to stay awake during work hours were the primary predictors of errors, including medication errors, charting information, procedural deviations and physical injuries (Dorrian et al. 2008).

One study assessed the real-time influence of emotional stress, workload and sleep deprivation on self-reported medication errors among physicians in academic hospitals (Dollarhide et al. 2014). A longitudinal design was used, and results indicated that medication errors were associated with higher perceived workload and higher emotional stress scores, but were not significantly associated with lack of sleep. A survey among emergency medical services workers indicated that almost half of the workers were classified as fatigued, mainly as a result of sleep deprivation (Patterson et al. 2012). The results indicated 1.9 greater odds of injury, 2.2 greater odds of medical error and 3.6 greater odds of safety-compromising behavior among fatigued versus non-fatigued workers.

In aviation, fatigue has been identified as an important contributing factor in many accidents (Fanjoy et al. 2010). A study of Portuguese pilots indicated relatively high levels of reported sleep problems and fatigue among the respondents (Reis et al. 2016), and a large proportion (91 percent) reported they have made mistakes in the cockpit as a direct consequence of fatigue (Reis et al. 2013). Similar findings were documented in an interview study of European commercial pilots, where especially tiredness and fatigue were related to self-reported incidents (Loewenthal et al. 2000). In a study by Sexton et al. (2000) the objective was to survey operating theatre and intensive care unit staff about attitudes concerning error, stress and teamwork, and to compare these attitudes with those of airline cockpit crews. Participants were 1033 doctors, nurses, fellows and residents working in operating theatres and intensive care units in various countries, and over 30 000 cockpit crew members (captains, first officers and second officers) of major airlines around the world. Pilots were least likely to deny dire effects of fatigue on performance compared with healthcare personnel, which may be a result of the strong focus in aviation on crew resource management (CRM) training. Crew resource management training includes the non-technical skills needed to perform safely, for example, getting along with crew members, knowing when and how to take charge effectively in critical situations and maintaining situational awareness (Martinussen and Hunter 2017).

STRESS, AVIATION ACCIDENTS AND INCIDENTS

The aviation industry is a business characterized by a strong focus on safety, tough competition between airlines and a setting where human error may have serious consequences. In the past the pilot profession was often portrayed as glamorous, but it is now characterized by long working hours and time away from home, time pressure and, often, temporary contracts (atypical employment) (Bennett 2011). Several surveys have described the working conditions as stressful, and CareerCast (2018) concluded in their evaluation of 11 core stress factors in 200 jobs (for example, risk, deadlines, physical demands and travel), that being a pilot was ranked as the third most stressful job after firefighters and enlisted military personnel. This rating was based on an objective evaluation of stressors and working conditions, and not on workers' ratings of their experienced levels of stress. While stress results from an appraisal process in which the same situation may affect individuals differently, there are some stressors that may be perceived as problematic by many professionals working in aviation. A comprehensive review by Albuquerque and Fonseca (2017) lists a number of stressors that may affect pilots as well as other types of personnel in aviation, including physical/physiological stressors, and job-related stressors such as difficult working conditions.

A study of Norwegian pilots (Mjøs 2001) examined the relationship between communication and operational failures during a formal proficiency test in a full flight simulator. Cockpit communications of 13 crews were recorded in situations with high workload and stress. When tasks became more demanding, anxiety and stress led to inefficient communication and rigidity in problem solving. Under these conditions, the crews building a shared mental model were most successful in reducing operational risk. The communication method used during the normal operation of the aircraft, mainly passing on information related to the task, did not work under high psychological stress (Mjøs 2001).

A review by Dismukes et al. (2015) examined the effects of stress on pilot performance, and found that only a few studies had investigated experts performing highly practiced tasks, whereas the majority of studies were simulation studies conducted in a laboratory setting. Despite these limitations, the literature provided a general picture of how acute stress impairs basic cognitive processes such as attention and working memory (Dismukes et al. 2018). This impairment may, in turn, impact more complex tasks such as calculations, decision making and team performance (see Dismukes et al. 2018 for an overview). Well-practiced tasks may be less susceptible to the effects of stress than new tasks that require information processing.

In order to understand how acute stress may affect experienced pilots, Dismukes et al. (2018) analyzed 12 major airline accidents to determine the type of errors that had occurred. Owing to the serious situation that the pilots and crew members were involved in, it is reasonable to assume that they experienced high levels of stress. A categorization of 212 different types of errors resulted in seven main error categories. The three most frequent were inadequate assessment of the situation, inadvertent omission of required actions and poor management of competing task demands. The least frequent types of errors were failure to acquire information, and inadequate physical execution of action (Dismukes et al. 2018). Some of these types of errors may be the result of reduced working memory, which makes prioritizing and rapid shifts between tasks more demanding. Also, crews under stress tend to provide and seek less information, and verbalization tends to be truncated and becomes less specific (Dismukes et al. 2018). The authors suggest that emphasis should be paid to developing tools including training, procedures and cockpit systems that help crews overcome these types of errors when in a critical and stressful situation (Dismukes et al. 2018).

A survey of commercial airline pilots ($N = 1145$) sampled from a European pilots' professional association indicated relatively high levels of burnout among the respondents (40.2 percent) (Demerouti et al. 2018). The job demands–resources (JD-R) model was tested using simulator performance and happiness as outcome variables. Burnout was predicted by job demands such as job insecurity and work–life conflict in line with the JD-R model. A significant negative relationship between burnout and happiness was detected, whereas the effect of burnout on simulator performance was mediated by job crafting (Demerouti et al. 2018). Resources such as social support and possibilities for future development was positively related to happiness and job crafting. The response rate was relatively low in this study (13 percent), which may have influenced the findings, especially the prevalence estimates of burnout. An earlier, smaller study on burnout showed high levels of burnout among approximately one-third of the respondents in a sample of US regional airline pilots (Fanjoy et al. 2010). That study also indicated a relationship between burnout and work-related factors such as perceived management pressure to continue flight with shortened rest periods, critical equipment problems and severe weather conditions (Fanjoy et al. 2010).

INTERVENTIONS TO MINIMIZE THE NEGATIVE CONSEQUENCES OF STRESS

Several national governments (for example, Australia, the US and Canada) have advocated stress-reduction materials to organizations in efforts to improve employee well-being and safety performance. Preventing medical errors, accidents and work injuries requires a clear understanding of how they occur, how they are recorded and the accuracy of injury surveillance.

Many suggestions for helping employees in operative occupations that deal with work-related stress are non-specific. A report by Members of

Professional Wellbeing Committee et al. (2013) made several recommenda-tions for reducing the impact of occupational fatigue and exhaustion among anesthesiologists on the safety of surgical patients. The recommen-dations included an assessment of fatigue-related risks such as shift hours and staffing levels, as well as staff input on work schedules. They also recommended that the process in which patients were transferred between caregivers should be carefully examined. In addition to the organizational recommendations, it was suggested that the staff should receive education about the importance of sleep and consequences of fatigue, and that a fatigue management plan should be developed and implemented among the staff. Rogers (2008) reviewed efforts by hospitals to reduce the negative effects of insufficient sleep on medical errors and injuries. Interventions have included education on sleep, altering the start time of shifts, rest breaks, napping rooms, exercise, bright lights and encouraging coffee consumption (Rogers 2008).

For military pilots, mindfulness training has been examined as a way of reducing stress and increasing performance (Meland et al. 2015a). The intervention took place over a period of 12 months, and included basic exercises such as yoga, body scan and meditation. The effectiveness of the program was examined in a small-scale pre-post study of 21 fighter pilots and mission support personnel. The results indicated positive findings on some of the outcome variables, and high user satisfaction with the inter-vention. Another study using a similar intervention in a military helicopter unit, indicated favorable outcomes on cortisol levels and perceived mental demands compared with the control group (Meland et al. 2015b). Both studies had some shortcomings related to sample size and design, but sug-gest that mindfulness training may be used in high-performance groups, such as military pilots, for reducing stress.

Different coping strategies for managing stress among pilots have also been outlined by Bor et al. (2017), and include individual strategies for reducing stress such as living healthily, sleep hygiene, exercise, relaxation techniques and having interests outside work. In addition, they describe work-related strategies that may be used to manage workload and establish boundaries between paid work and private life (Bor et al. 2017). Such coping strategies will also have relevance for other occupations, for exam-ple, in the medical profession.

Peer-support programs have also been introduced in aviation as a way for pilots to receive help to deal with various problems and stressors. The overall goal is to improve mental health and resilience, and to ensure sup-port for pilots in need of more help. Many organizations and airlines offer this type of support program (see, for example, Stiftung Mayday (english. stiftung-mayday.de) or the European Pilot Peer Support Initiative (eppsi.

eu). Earlier efforts within peer-support focused on helping people deal with the effects of accidents or incidents. One of these approaches is critical incident stress management (CISM) and may be used after experiencing a critical or highly stressful incident or accident. According to Mitchell and Leonhardt (2010), the main objectives of a CISM program are to mitigate the impact of a traumatic event, facilitate normal recovery processes, restore individuals, groups and organizations to adaptive function, and identify people who would benefit from additional help and support. Usually this will include information, training and support. Studies examining the effectiveness of such interventions seem to be disappointing in terms of reducing long-term reactions to trauma, such as post-traumatic stress disorder (PTSD) symptoms, based on two systematic reviews (Rose et al. 2002; Roberts et al. 2009). However, the content of the interventions reviewed seemed to vary, and the number of studies reviewed was limited. Treatment aimed specifically at people with PTSD symptoms seems to be more effective, as indicated by a meta-analysis of studies which found that up to 67 percent of those who complete the treatment no longer satisfy the criteria to be re-diagnosed with PTSD (Bradley et al. 2005).

However, the majority of the summarized interventions have focused on the individual, but it is important to keep in mind is that an accident or incident is caused by a number of factors, including factors related to both the organizational and system levels, as outlined in the Swiss-cheese model by Reason (1997). This model suggests that governments, organizations and people create barriers to the occurrence of accidents, but that these barriers may have holes which, if they line up, can result in an accident. This way of understanding why accidents occur is frequently used for analyzing aviation accidents, but has also been advocated in medicine. Welker-Hood (2006) claims that the critical implication of nurses' job-related stress is the risk of accidents and errors, including improper use of equipment, not following safety procedures and medication errors. She claims that instead of looking at an individual's responsibility, there is a need to use systems-based approaches to reducing workplace errors based on understanding which aspects of the work environment contribute to nurses' stress and affect their ability to perform their jobs.

CONCLUSION

No professional is immune to the effects of external demands such as a heavy workload, time pressure, shift work, unfavorable working conditions or emergencies, even if they have been carefully selected and trained and have a high professional standard. Stress occurs when the demands exceed the individual's available resources, both personal and in the work environment. Acute stress which may occur in a critical or traumatic event, is associated with cognitive failures and reduced situation awareness and erroneous decision making. Long-term cumulative stress in the work environment reduces both mental and physical energy, which in turn may affect cognitive functions and increase the risk of errors and accidents. The mechanisms for how stress is related to accidents and errors are not fully understood, but there are probably both direct effects of stress as well as effects mediated by variables such as burnout, fatigue and safety behavior.

The consequences of stress are many, for the individual experiencing stress but also for the organization and third parties such as patients and passengers. Several interventions focusing on the individual, have been suggested, but a more comprehensive systems approach is needed that also addresses workplace culture, safety attitudes and behaviors, and the organization as a whole. Preventive measures may include a favorable work environment that promotes work crafting and autonomy, make relevant resources available that buffer the negative consequences of stressors, and ensure that the work schedule is designed so that there is enough time for rest and sleep.

REFERENCES

Åkerstedt, T. (2007), 'Altered sleep/wake patterns and mental performance', *Physiology and Behavior*, **90** (2–3), 209–18.

Albuquerque, C. and Fonseca, M. (2017), 'Psychosocial stressors associated with being a pilot', in R. Bor, C. Eriksen, M. Oakes and P. Scragg (eds), *Pilot Mental Health Assessment And Support. A Practitioner's Guide*, New York: Routledge, pp. 287–308.

Arimura, M., Imai, M., Okawa, M., Fujimura, T. and Yamada, N. (2010), 'Sleep, mental health status, and medical errors among hospital nurses in Japan', *Industrial Health*, **48** (6), 811–17.

Bennett, S. (2011), 'The pilot life-style: a sociological study of the commercial pilot's work and home life', Institute of Lifelong Learning, University of Leicester.

Bogner, M.S. (2010), *Human Errors in Medicine*, 2nd edn, Boca Raton, FL: CRC Press.

Bor, R., Eriksen, C., Oakes, M. and Scragg, P. (2017), *Pilot Mental Health Assessment and Support: A Practitioner's Guide*, New York: Routledge.

Bradley, R., Greene, J., Russ, E., Dutra, L. and Westen, D. (2005), 'A multidimensional meta-analysis of psychotherapy for PTSD', *American Journal of Psychiatry*, **162** (2), 214–27.

CareerCast (2018), 'The most stressful jobs of 2018', accessed 28 December 2018 at https://www.careercast.com/jobs-rated/2018-most-stressful-jobs?page=0.

Cartwright, S. and Cooper, C.L. (1997), *Managing Workplace Stress*, Thousand Oaks, CA: Sage.

Clarke, S. (2012), 'The effect of challenge and hindrance stressors on safety behavior and safety outcomes: a meta-analysis', *Journal of Occupational Health Psychology*, **17** (4), 387–97.

Committee on Quality of Health Care in America (2000), *To Err Is Human: Building a Safer Health Care System*, L.T. Kohn, J.M. Corrigan and M.S. Donaldson (eds), Washington, DC: National Academies Press.

Cooper, C.L., Dewe, P.J. and O'Driscoll, M.P. (2001), *Organizational Stress: A Review and Critique of Theory, Research, and Applications*, Thousand Oaks, CA: Sage.

Costa, G. (1996), 'The impact of shift and night work on health', *Applied Ergonomics*, **27** (1), 9–16.

Day, A.J., Brasher, K. and Bridger, R.S. (2012), 'Accident proneness revisited: the role of psychological stress and cognitive failure', *Accident Analysis & Prevention*, **49** (November), 532–5.

Defraia, G.S. (2015), 'Psychological trauma in the workplace: variation of incident severity among industry settings and between recurring vs isolated incidents', *International Journal of Occupational and Environmental Medicine*, **6** (3), 155–68.

Demerouti, E., Geurts, S.A., Bakker, A. and Euwema, M. (2004), 'The impact on shift work on work–home conflict, job attitudes and health', *Ergonomics*, **47** (9), 1125–37.

Demerouti, E., Veldhuis, W., Coombes, C. and Hunter, R. (2018), 'Burnout among pilots: psychosocial factors related to happiness and performance at simulator training', *Ergonomics*, online, 18 June, 1–13, doi:org/10.1080/00140139.2018.1464667.

Dewa, C.S., Loong, D., Bonato, S. and Trojanowski, L. (2017a), 'The relationship between physician burnout and quality of healthcare in terms of safety and acceptability: a systematic review', *BMJ Open*, **7** (6), e015141, accessed 28 December 2018 at http://dx.doi.org/10.1136/bmjopen-2016-015141.

Dewa, C.S., Loong, D., Bonato, S., Trojanowski, L. and Rea, M. (2017b), 'The relationship between resident burnout and safety-related and acceptability-related quality of healthcare: a systematic literature review', *BMC Medical Education*, **17** (1), 195.

Dismukes, K., Goldsmith, T.E. and Kochan, J. (2015), 'Effects of acute stress on aircrew performance: Literature review and analysis of operational aspects', NASA Technical Memorandum TM2015-218930, NASA Ames Research Center, Moffett Field, CA, accessed 28 December 2018 at https://www.researchgate.net/profile/Key_Dismukes2/publication/299578388_Effects_of_Acute_Stress_on_Aircrew_Performance_Literature_Review_and_Analysis_of_Operational_Aspects/links/56fff4a708aea6b77469b3b5/Effects-of-Acute-Stress-on-Aircrew-Performance-Literature-Review-and-Analysis-of-Operational-Aspects.pdf.

Dismukes, K., Kochan, J.A. and Goldsmith, T.E. (2018), 'Flight crew errors in challenging and stressful situations', *Aviation Psychology and Applied Human Factors*, **8** (1), 35–46.

Dollarhide, A.W., Rutledge, T., Weinger, M.B., Fisher, E.S., Jain, S., Wolfson, T. et al. (2014), 'A real-time assessment of factors influencing medication events', *Journal for Healthcare Quality*, **36** (5), 5–12.

Dorrian, J., Tolley, C., Lamond, N., van den Heuvel, C., Pincombe, J., Rogers, A.E. et al. (2008), 'Sleep and errors in a group of Australian hospital nurses at work and during the commute', *Applied Ergonomics*, **39** (5), 605–13.

Elfering, A., Grebner, S., Leitner, M., Hirschmuller, A., Kubosch, E.J. and Baur, H. (2017), 'Quantitative work demands, emotional demands, and cognitive stress symptoms in surgery nurses', *Psychology Health & Medicine*, **22** (5), 604–10.

Fanjoy, R.O., Harriman, S.L. and DeMik, R.J. (2010), 'Individual and environmental predictors of burnout among regional airline pilots', *International Journal of Applied Aviation Studies*, **10** (1), 15–30.

Folkard, F. and Tucker, P. (2003), 'Shift work, safety, and productivity', *Occupational Medicine*, **53** (2), 95–101.

Garrouste-Orgeas, M., Perrin, M., Soufir, L., Vesin, A., Blot, F., Maxime, V. et al. (2015), 'The Iatroref study: medical errors are associated with symptoms of depression in ICU staff but not burnout or safety culture', *Intensive Care Medicine*, **41** (2), 273–84.

Gold, D.R., Rogacz, S., Bock, N., Tosteson, T.D., Baum, T.M., Speizer, F.E. and Czeisler, C.A. (1992), 'Rotating shift work, sleep, and accidents related to sleepiness in hospital nurses', *American Journal of Public Health*, **82** (7), 1011–14.

Halbesleben, J.R. (2010), 'The role of exhaustion and workarounds in predicting occupational injuries: a cross-lagged panel study of health care professionals', *Journal of Occupational Health Psychology*, **15** (1), 1–16.

Hall, L.H., Johnson, J., Watt, I., Tsipa, A. and O'Connor, D.B. (2016), 'Healthcare staff wellbeing, burnout, and patient safety: a systematic review', *PLoS ONE*, **11** (7), e0159015, accessed 28 December 2018 at https://doi.org/10.1371/journal.pone.0159015.

Karasek, R. and Theorell, T. (1990), *Healthy Work: Stress, Productivity, and the Reconstruction of Working Life*, New York: Basic Books.

Kim, Y.K., Ahn, Y.S., Kim, K., Yoon, J.H. and Roh, J. (2016), 'Association between job stress and occupational injuries among Korean firefighters: a nationwide cross-sectional study', *BMJ Open*, **6** (11), e012002, accessed 28 December 2018 at http://dx.doi.org/10.1136/bmjopen-2016-012002.

Kunaviktikul, W., Wichaikhum, O., Nantsupawat, A., Nantsupawat, R., Chontawan, R., Klunklin, A. et al. (2015), 'Nurses' extended work hours: patient, nurse and organizational outcomes', *International Nursing Review*, **62** (3), 386–93.

Lawton, R. and Parker, D. (1998), 'Individual differences in accident liability: a review and integrative approach', *Human Factors*, **40** (4), 655–71.

Lazarus, R.S. (1991), 'Psychological stress in the workplace', in P.L. Perrewe (ed.), *Handbook on Job Stress*, Corte Madeira, CA: Select Press, pp. 1–13.

Leiter, M.P. and Maslach, C. (2009), 'Burnout and workplace injuries: a longitudinal analysis', in A.M. Rossi, J.C. Quick and P.L. Perrewé (eds), *Stress and Quality of Working Life: The Positive and the Negative*, Charlotte, NC: Information Age, pp. 3–18.

Loewenthal, K.M., Eysenck, M., Harris, D., Lubitsh, G., Gorton, T. and Bicknell, H. (2000), 'Stress, distress and air traffic incidents: job dysfunction and distress in airline pilots in relation to contextually assessed stress', *Stress Medicine*, **16** (3), 179–83.

Lusa, S., Hakkanen, M., Luukkonen, R. and Viikari-Juntura, E. (2002), 'Perceived physical work capacity, stress, sleep disturbance and occupational accidents among firefighters working during a strike', *Work & Stress*, **16** (3), 264–74.

Martinussen, M. and Hunter, D.R. (2017), *Aviation Psychology and Human Factors*, 2nd edn, Boca Raton, FL: CRC Press.

Mathisen, G.E. and Bergh, L.I.V. (2016), 'Action errors and rule violations at offshore oil rigs: the role of engagement, emotional exhaustion and health complaints', *Safety Science*, **85** (June), 130–38.

Meland, A., Fonne, V., Wagstaff, A. and Pensgaard, A.M. (2015a), 'Mindfulness-based mental training in a high-performance combat aviation population: a one-year intervention study and two-year follow-up', *International Journal of Aviation Psychology*, **25** (1), 48–61.

Meland, A., Ishimatsu, K., Pensgaard, A.M., Wagstaff, A., Fonne, V., Garde, A.H. et al. (2015b), 'Impact of mindfulness training on physiological measures of stress and objective measures of attention control in a military helicopter unit', *International Journal of Aviation Psychology*, **25** (3–4), 191–208.

Members of Professional Wellbeing Committee, World Federation of Societies of Anesthesiologists, Moore, R., Gupta, P. and Duval Neto, G.F. (2013), 'Occupational fatigue: impact on anesthesiologist's health and the safety of surgical patients. As anesthesiologists we are frequently working in a stressful environment. Do you disagree with this?', *Brazilian Journal of Anesthesiology*, **63** (2), 167–9.

Mitchell, J.T. and Leonhardt, J. (2010), 'Critical Incident stress management (CISM): an effective peer support program for aviation industries', *International Journal of Applied Aviation Studies*, **10** (1), 99–116 accessed 22 May 2019 at https://www.academy.jccbi.gov/ama-800/Summer_2010.pdf.

Mitler, M.M., Carskadon, M.A., Czeisler, C.A., Dement, W.C., Dinges, D.F. and Graeber, R.C. (1988), 'Catastrophes, sleep, and public policy: consensus report', *Sleep*, **11** (1), 100–109.

Mjøs, K. (2001), 'Communication and operational failures in the cockpit', *Human Factors and Aerospace Safety*, **1** (4), 323–40.

Mora, P., Segovia, A. and Lopez, G. (2002), 'The influence of stress and job satisfaction on propensity to leave organization, absenteeism, and accident rate', *Ansiedad y Estres*, **8** (2–3), 275–84.

Morris, C.H. and Leung, Y.K. (2006), 'Pilot mental workload: how well do pilots really perform?', *Ergonomics*, 49 (**15**), 1581–96.

Murphy, L.R., DuBois, D. and Hurreu, J.J. (1986), 'Accident reduction through stress management', *Journal of Business and Psychology*, **1** (1), 5–18.

Nahrgang, J.D., Morgeson, F.P. and Hofmann, D.A. (2011), 'Safety at work: a meta-analytic investigation of the link between job demands, job resources, burnout, engagement, and safety outcomes', *Journal of Applied Psychology*, **96** (1), 71–94.

Nielsen, K.J., Pedersen, A.H., Rasmussen, K., Pape, L. and Mikkelsen, K.L. (2013), 'Work-related stressors and occurrence of adverse events in an ED', *American Journal of Emergency Medicine*, **31** (3), 504–8.

O'Connor, P., Lydon, S., O'Dea, A., Hehir, L., Offiah, G., Vellinga, A. et al. (2017), 'A longitudinal and multicentre study of burnout and error in Irish junior doctors', *Postgraduate Medical Journal*, **93** (1105), 660–64.

Orasanu, J. (1997), 'Stress and naturalistic decision making: strengthening the weak links', in R. Flin, E. Salas, M. Strub and L. Martin (eds), *Decision Making under Stress. Emerging Themes and Applications*, Aldershot: Ashgate, pp. 43–66.

Orji, E.O., Fasubaa, O.B., Onwudiegwu, U., Dare, F.O. and Ogunniyi, S.O. (2002), 'Occupational health hazards among health care workers in an obstetrics and gynaecology unit of a Nigerian teaching hospital', *Journal of Obstetrics & Gynaecology*, **22** (1), 75–8.

Pallesen, S. (2006), 'Søvn' ('Sleep'), in J. Eid and B.H. Johnsen (eds), *Operativ psykologi (Operational Psychology)*, Bergen: Fagbokforlaget, pp. 196–215.

Park, Y.M. and Kim, S.Y. (2013), 'Impacts of job stress and cognitive failure on patient safety incidents among hospital nurses', *Safety and Health at Work*, **4** (4), 210–15.

Patterson, P.D., Weaver, M.D., Frank, R.C., Warner, C.W., Martin-Gill, C., Guyette,

F.X. et al. (2012), 'Association between poor sleep, fatigue, and safety outcomes in emergency medical services providers', *Prehospital Emergency Care*, **16** (1), 86–97.

Peters, G.A. and Peters, B.J. (2007), *Medical Error and Patient Safety: Human Factors in Medicine*, Boca Raton, FL: CRC Press.

Quick, J.C., Cooper, C.L., Nelson, D.L., Quick, J.D. and Gavin, J.H. (2003), 'Stress, health, and well-being at work', in J. Greenberg (ed.), *Organizational Behavior: The State of the Science*, Mahwah, NJ: Lawrence Erlbaum Associates, pp. 53–89.

Rasmussen, K., Pedersen, A.H., Pape, L., Mikkelsen, K.L., Madsen, M.D. and Nielsen, K.J. (2014), 'Work environment influences adverse events in an emergency department', *Danish Medical Journal*, **61** (5), A4812.

Reason, J. (1997), *Managing the Risks of Organizational Accidents*, Aldershot: Ashgate.

Reis, C., Mestre, C. and Canhão, H. (2013), 'Prevalence of fatigue in a group of airline pilots', *Aviation Space and Environmental Medicine*, **84** (8), 828–33.

Reis, C., Mestre, C., Canhão, H., Gradwell, D. and Paiva, T. (2016), 'Sleep complaints and fatigue of airline pilots', *Sleep Science*, **9** (2), 73–7.

Renouard, F., Amalberti, R. and Renouard, E. (2017), 'Are "human factors" the primary cause of complications in the field of implant dentistry?', *International Journal of Oral & Maxillofacial Implants*, **32** (2), e55–e61.

Roberts, N.P., Kitchiner, N.J., Kenardy, J. and Bisson, J. (2009), 'Multiple session early psychological interventions for the prevention of post-traumatic stress disorder', *Cochrane Database of Systematic Reviews*, 8 July, (3), CD006869.

Rodrigues, C., Santos, V.E.P. and Sousa, P. (2017), 'Patient safety and nursing: interface with stress and burnout syndrome', *Revista Brasileira de Enfermagem*, **70** (5), 1083–8.

Rogers, A.E. (2008), 'The effects of fatigue and sleepiness on nurse performance and patient safety', in R.G. Hughes (ed.), *Patient Safety and Quality: An Evidence-Based Handbook for Nurses*, Rockville, MD: Agency for Healthcare Research and Quality, pp. 2-511–2-545, accessed at https://www.ncbi.nlm.nih.gov/books/NBK2645/.

Rose, S., Bisson, J., Churchill, R. and Wessely, S. (2002), 'Psychological debriefing for preventing post traumatic stress disorder (PTSD)', *Cochrane Database of Systematic Reviews*, 22 April (2), CD000560.

Ross, J. (2016), 'The connection between burnout and patient safety', *Journal of PeriAnesthesia Nursing*, **31** (6), 539–41.

Salminen, S., Perttula, P., Hirvonen, M., Perkio-Makela, M. and Vartia, M. (2017), 'Link between haste and occupational injury', *Work: Journal of Prevention, Assessment & Rehabilitation*, **56** (1), 119–24.

Selye, H. (1984), *The Stress of Life*, revd edn, New York: McGraw-Hill.

Sexton, J.B., Thomas, E.J. and Helmreich, R.L. (2000), 'Error, stress, and teamwork in medicine and aviation: cross sectional surveys', *British Medical Journal*, **320** (7237), 745–9.

Shanafelt, T.D., Balch, C.M., Bechamps, G., Russell, T., Dyrbye, L., Satele, D. et al. (2010a), 'Burnout and medical errors among American surgeons', *Annals of Surgery*, **251** (6), 995–1000.

Shanafelt, T.D., Bradley, K.A., Wipf, J.E. and Back, A.L. (2010b), 'Burnout and self-reported patient care in an internal medicine residency program', *Annals of Internal Medicine*, **136** (5), 358–67.

Smith, T.D., Hughes, K., DeJoy, D.M. and Dyal, M.A. (2018), 'Assessment of relationships between work stress, work–family conflict, burnout and firefighter safety behavior outcomes', *Safety Science*, **103** (March), 287–92.

Sneddon, A., Mearns, K. and Flin, R. (2006), 'Situation awareness and safety in offshore drill crews', *Cognition, Technology & Work*, **8** (4), 255–67.

Sneddon, A., Mearns, K. and Flin, R. (2013), 'Stress, fatigue, situation awareness and safety in offshore drilling crews', *Safety Science*, **56** (July), 80–88.

Souza, K., Cantley, L.F., Slade, M.D., Eisen, E.A., Christiani, D. and Cullen, M.R. (2014), 'Individual-level and plant-level predictors of acute, traumatic occupational injuries in a manufacturing cohort', *Occupational and Environmental Medicine*, **71** (7), 477–83.

Sugden, C., Aggarwal, R. and Darzi, A. (2010), 'Re: Sleep deprivation, fatigue, medical error and patient safety', *American Journal of Surgery*, **199** (3), 433–4.

Teng, C.-I., Shyu, Y.-I.L., Chiou, W.-K., Fan, H.-C. and Lam, S.M. (2010), 'Interactive effects of nurse-experienced time pressure and burnout on patient safety: a cross-sectional survey', *International Journal of Nursing Studies*, **47** (11), 1442–50.

Tomei, G., Capozzella, A., Rosati, M.V., Tomei, F., Rinaldi, G., Chighine, A. et al. (2015), 'Stress e infortuni sul lavoro' ('Stress and work-related injuries'), *Clinica Terapeutica*, **166** (1), 7–22.

Tüchsen, F., Hannerz, H. and Burr, H. (2006), 'A 12 year prospective study of circulatory disease among Danish shift workers', *Occupational & Environmental Medicine*, **63** (7), 451–5.

Welker-Hood, K. (2006), 'Does workplace stress lead to accident or error? Many nurses feel the pressure', *American Journal of Nursing*, **106** (9), 104.

West, C.P., Dyrbye, L.N. and Shanafelt, T.D. (2018), 'Physician burnout: contributors, consequences and solutions', *Journal of Internal Medicine*, **283** (6), 516–29.

Williams, E.S., Manwell, L.B., Konrad, T.R. and Linzer, M. (2007), 'The relationship of organizational culture, stress, satisfaction, and burnout with physician-reported error and suboptimal patient care: results from the MEMO study', *Health Care Management Review*, **32** (3), 203–12.

Zadow, A.J., Dollard, M.F., McLinton, S.S., Lawrence, P. and Tuckey, M.R. (2017), 'Psychosocial safety climate, emotional exhaustion, and work injuries in healthcare workplaces', *Stress and Health*, **33** (5), 558–69.

PART II

Workplace Health and Safety Factors

4. Drug use and workplace safety: issues and good practice responses

Ken Pidd, Ann Roche and Vinita Duraisingam

INTRODUCTION

There is increasing concern over the potential implications of employee alcohol and other drug use for workplace safety and well-being. This concern is largely driven by growing awareness of two issues. First, there is a large body of evidence indicating that alcohol and other drug use (including prescribed drugs) can result in human performance decrements, with negative implications for workplace safety. Second, data collected over the past few decades provide clear evidence that, in most developed countries, the overwhelming majority of drinkers and drug users are employed. Furthermore, particular workforce groups have disproportionately higher prevalence of use.

Managing the risk of alcohol and other drug-related harm in the workplace is not always straightforward. While employee drinking and drug use during work hours has obvious implications for workplace safety, so can patterns of consumption that occur away from the workplace and out of working hours. In addition, there is a body of evidence indicating that employee consumption patterns are influenced by workplace factors, which in turn pose a risk to workplace safety. The aim of this chapter is to review the evidence concerning these issues, identify effective evidence-based interventions and describe good practice strategies for minimising the risk of alcohol and other drug-related harm in the workplace.

THE EFFECTS OF ALCOHOL AND OTHER DRUG USE ON HUMAN PERFORMANCE AND COGNITION

Most psychoactive drugs[1] are classified as either depressants or stimulants.[2] Depressants (for example, alcohol, opioids and benzodiazepines) depress human central nervous system function, while stimulants (for example, methamphetamine, amphetamine and cocaine) do the opposite by stimulating central nervous system function.

Alcohol

Globally, the most commonly used drug is alcohol (World Health Organization 2018). Alcohol is a depressant that slows down the body's motor and sensory systems, resulting in impaired balance, coordination, perception and decision making, and slowed reflex and cue response times. While the effects of a given dose will vary according to individual differences such as body size, gender, health status and drug tolerance level, the effects of alcohol are dose dependent with increases in the amount consumed resulting in increased impairment (Martin et al. 2013; Fell and Voas 2014). There is a large body of evidence indicating that the acute effects of alcohol have a substantial negative impact on performing complex divided-attention tasks, such as driving a motor vehicle (Martin et al. 2013; Irwin et al. 2017).

Post-intoxication (hangover) effects can also negatively impact performance and cognition. While some studies have produced conflicting findings (for example, Stephens et al. 2008), a recent systematic review and meta-analysis of relevant research identified that short- and long-term memory, sustained attention, and psychomotor speed are impaired the day after heavy drinking (Gunn et al. 2018). In addition to post-intoxication effects, alcohol use can also have substantial negative health outcomes which may indirectly impact on workplace safety. Alcohol use is a substantial contributor to the global burden of morbidity and mortality, contributing to more than 200 chronic diseases and injury-related health conditions (World Health Organization 2018).

Illicit Drug Use

As in the case of alcohol, illicit drug use is also a significant contributor to the global burden of disease (Degenhardt et al. 2013). The negative health effects of illicit drug use include: the acute toxic effects of the drug (for example, overdose) and intoxication (for example, injury); drug dependence; and disease or illness associated with regular chronic use (Degenhardt and Hall 2012).

Globally, the most commonly used illicit drug is cannabis (marijuana) (United Nations Office on Drugs and Crime 2016). While cannabis can be classified as a depressant, its pharmacological effects can be complex (Ashton 1999). The acute effects of cannabis include euphoria (although naive users may initially experience anxiety and paranoia), drowsiness, distorted perception, impaired psychomotor and gross motor function, impaired cognition memory, and impaired performance when undertaking complex skilled activities (Ashton 1999; Ramaekers et al. 2004, 2006) such as driving a motor vehicle (Hartman and Huestis 2013; Bondallaz et al. 2016). The health-related effects of chronic cannabis use can include cardiovascular problems, respiratory illness (if cannabis is smoked) and potential mental health issues (Sachs et al., 2015).

The second most common type of illicit drug used globally is amphetamine-type stimulants such as amphetamine, methamphetamine and ecstasy (United Nations Office on Drugs and Crime 2016). The acute effects of methamphetamine (a more potent form of amphetamine) include increased arousal, alertness and confidence, and a sense of euphoria. There is also a dose-dependent negative effect on cognition, reasoning and psychomotor performance, together with a depressant-type effect during withdrawal (Logan 2002). While the use of amphetamine-type substances has been linked to increased risk of a motor vehicle accident (Sheridan et al. 2006; Hayley et al. 2016), the evidence for this link is generally less robust than evidence for alcohol and cannabis use (Kelly et al. 2004). However, as is the case for alcohol and cannabis, the health effects of chronic stimulant use may also indirectly impact workplace safety via employee health. Chronic use can result in psychotic episodes, anxiety and depression, mood swings, insomnia, malnutrition, and elevated blood pressure and heart rate (Murray 1998; Sheridan et al. 2006).

Recent years have also seen an increasing focus on the use and misuse of prescribed drugs, such as benzodiazepines and opioids. Both benzodiazepines and opioids are depressants that have legitimate medical uses. Benzodiazepines are a group of drugs classified as minor tranquillisers and are usually prescribed for sleep or anxiety disorders. While heroin is an opioid, most types of opioids (for example, morphine, methadone, fentanyl and oxycodone) are pharmaceutical drugs prescribed to alleviate acute pain. Despite the medicinal purpose of these drugs, they are also commonly used recreationally or for non-prescribed self-medication purposes. In 2016, 4.3 per cent of the US population over the age of 12 years misused prescription painkillers or opioids and 2.2 per cent misused prescription tranquillisers (Centers for Disease Control and Prevention 2018). In the same year, 3.6 per cent of the Australian population (aged 14 years or older) misused prescription painkillers or opioids and 1.6 per cent misused prescription tranquillisers (Australian Institute of Health and Welfare 2017). Both of these drugs can produce dose-dependent effects that are similar to other central nervous system depressants. The effects of benzodiazepine use can include delayed reaction times, reduced hand-eye coordination and substantial cognitive impairment (Drummer 2002). There is also clear evidence that benzodiazepine use impairs driving performance (Daurat et al. 2013) and increases the risk of a motor vehicle accident (Rapoport et al. 2009). Similarly, opioids can impair psychomotor functioning, which is important for the performance of complex, divided attention tasks (Stout and Farrell 2003). However, there is less evidence of a relationship between either the therapeutic or illicit use of opioids on driving performance (Lenne et al. 2000; Ferreira et al. 2018).

Evidence concerning the negative effects of alcohol, cannabis, amphetamine-type stimulants, benzodiazepines and opioids on both human performance and health indicates their potential to adversely affect workplace safety. Moreover, use may affect safety directly, through intoxication and/or hangover-related impairment, or indirectly, through the negative effects of chronic use on employee health. However, evidence concerning the nature and extent of alcohol and other drug-related workplace accidents and injuries is less clear.

Alcohol and Other Drug-Related Workplace Accidents and Injury

Over the past few decades there have been several reviews of research concerning alcohol and other drug-related workplace accidents and injuries (Stallones and Kraus 1993; Frone 2004; Ramchand et al. 2009). The earliest of these reviews concluded that there was insufficient evidence to establish a causal relationship between alcohol use and workplace injuries (Stallones and Kraus 1993). A decade later, Frone (2004) reviewed studies that examined both alcohol and other drugs and found little credible data on the extent and nature of alcohol and other drug-related workplace accidents and injuries. Moreover, much of the evidence available at that time was either dated and/or suffered from methodological weaknesses that limited conclusions regarding causality (Frone 2004).

A more recent review (Ramchand et al. 2009) examined 33 studies published between 1994 and 2008. Mixed findings were reported, with some studies demonstrating a relationship between alcohol and other drug use and workplace accidents or injuries, and others finding no relationship. Ramchand et al. (2009) concluded that while there was an association between use and occupational injury, the proportion of injuries attributed to use was relatively small. In addition, Ramchand et al. (2009) identified that the association between use and occupational injury varied according to gender, age, and industry. They also indicated that other potentially confounding factors, such as individual differences in risk-taking disposition, were omitted from the reviewed studies. Therefore, the potential for unmeasured variables to partially account for the weak empirical associations between use and occupational injury could not be overruled.

Since Ramchand et al.'s (2009) review, several additional studies have been published. McNeilly et al. (2010) reviewed coronial records of 355 work-related fatalities in Victoria, Australia, between 2001 and 2006. Of 43 employee deaths that tested positive for alcohol or other drugs (12 per cent of all deaths), alcohol was present in 26 and cannabis or amphetamines in 20. A similar study that examined occupational fatalities in the US state of Iowa from 2005 to 2009 (Ramirez et al. 2013) found that of the 280 cases where toxicology reports were available, 22 per cent (61) were positive for alcohol and/or other drugs. Alcohol and cannabis were the drugs most commonly detected, and the positive toxicology rate varied widely between industry groups (Ramirez et al. 2013).

In a study of positive post-accident tests for illicit drug use in the US aviation industry from 1995 to 2005, Li et al. (2011) found that cannabis was the drug most commonly detected and that those who tested positive (for any illicit drug use) were 2.9 times more likely to be involved in an accident. However, as less than 1 per cent of those tested were positive, the attributable risk of drug use to an accident was small (1.2 per cent), indicating that drug use played only a very small role in aviation accidents (Li et al. 2011). A similar case controlled study that examined post-accident testing across a range of US industries found no association between cannabis use and work-related accidents (Price 2014). A more recent self-report survey of young US workers found the frequency of alcohol intoxication was significantly related to injury rates; however, age, gender and hours worked were also significant predicates of injury (Parish et al. 2016).

The results of these more recent studies, and the earlier reviews of research concerning alcohol and other drug use and workplace accidents and injuries, have several important implications. First, owing to the methodological limitations of most of these studies, few conclusions can be drawn regarding causality. Nearly all of the studies conducted to date are of a cross-sectional, self-report design, or retrospective studies using administrative and/or hospital data. While measured potential confounders can be statistically controlled in such study designs, definitive conclusions regarding causality cannot be drawn.

Second, the available research has produced mixed results, and where an association between drug use and injuries has been identified the proportion of alcohol or other drug-related injuries was relatively low. However, it is important to note that studies examining the toxicology results of work-related fatalities have tended to identify higher proportions of alcohol and/or other drug-related causes compared with studies of non-fatal injury. With the caveat that accidents resulting in fatalities are likely to be subjected to a higher rate of drug testing, this nonetheless suggests that drug or alcohol-related accidents may result in more serious outcomes. Another important finding of these studies was that the most commonly detected drug in the workplace was alcohol.

Third, the proportion of alcohol or other drug-related injuries identified across some studies varied according to age, gender, and industry. It is likely that the relationship between alcohol or other drug use and workplace injury is complex, confounded by a range of factors including the working environment, level of safety risk in the workplace and individual differences in risk-taking propensity. No studies reported analyses that accounted for such potential confounders. The complexity of the relationship between drug use and workplace accidents is highlighted in a recent US study that examined the relationship between the legalisation of medicinal cannabis and workplace fatalities (Anderson et al. 2018). Contrary to expectations, the recent introduction of legislation that legalised medicinal use of cannabis across 29 US states was associated with a 19.5 per cent reduction in expected workplace fatalities among workers aged 25–44 years. The precise reason for this reduction was not clear, however Anderson et al. (2018) speculated that legalisation of medicinal cannabis may have led to a reduction in the use of alcohol and other substances that may have a greater impact on cognitive and motor skill function.

THE PREVALENCE OF EMPLOYEE ALCOHOL AND OTHER DRUG USE

Any examination of the potential risk of alcohol and other drugs to workplace safety also needs to consider the prevalence of employee use. In doing so, it is important to recognise the distinction between workforce use and workplace use, as this distinction has implications for understanding the relationship between use and workplace safety. Workforce use refers to the overall consumption patterns of employees regardless of when and where use occurs. Workforce use mainly occurs away from the workplace outside working hours. Thus, the relationship between workforce use and workplace safety can be direct, via attending work with a post-intoxication hangover, or indirect, via alcohol and other drug-related health or behaviour problems. By contrast, workplace use refers to use that occurs at work and/or during work hours (including lunch and other breaks). The relationship between workplace use and workplace safety is direct, that is, via performance and cognitive impairments that result from use.

PREVALENCE OF WORKFORCE USE

Across most developed countries, a substantial proportion of the working population engage in illicit drug and/or harmful alcohol use. For example, across European countries, the proportion of employees who reported using an illicit drug in the past 12 months ranged from 0.6 per cent to 13.4 per cent (Corral et al. 2012). National estimates of harmful alcohol use indicated 5 per cent to 20 per cent of European employees are either addicted to, or at risk of becoming addicted to, alcohol (Corral et al. 2012). In the UK, 13 per cent of the workforce had used an illicit drug in the past year and 35 per cent drank more than the recommended limits of 14 units[3] of alcohol per week (Smith et al. 2004). Among US employees, 19.2 per cent of full-time employees had used an illicit drug in the past 12 months, while 32.7 per cent of full-time employees had engaged in binge drinking[4] in the past month (Center for Behavioral Health Statistics and Quality 2017). In Australia, 18 per cent of the workforce drink at short-term risk levels[5] at least weekly (NCETA 2015) and 17.5 per cent have used an illicit drug at least once in the past 12 months (Pidd et al. 2008).

PREVALENCE OF WORKPLACE USE

While data concerning the prevalence of workplace alcohol and other drug use is more limited than data on workforce use, available data is of concern. Studies across European countries indicate that between 4 per cent and 11 per cent of the workforce drink alcohol during working hours, and workplace drug tests conducted in Italy and France detected illicit drugs in 1.8 per cent and 8.5 per cent (respectively) of the workforce tested (Corral et al. 2012). In 2016, 5.5 per cent of the US workforce subjected to a random workplace urine test returned a positive result for illicit drugs, as did 6.3 per cent of those subjected to a random workplace oral fluid test (Quest Diagnostics 2017). Australian research has identified that 8.7 per cent of the workforce report that they usually drink alcohol at work and 5.6 per cent reported attending work under the influence at least once in the past 12 months (Pidd et al. 2011). In addition, 0.9 per cent of the Australian workforce have reported usually using other drugs at work and 2 per cent reported attending work under the influence of other drugs in the past 12 months (Pidd et al. 2011).

INDUSTRY AND OCCUPATIONAL DIFFERENCES IN PREVALENCE

Prevalence data also indicate that employee alcohol and other drug consumption patterns vary widely across industry and occupational groups. In the US, 8.7 per cent of the total workforce had used alcohol heavily in the past month compared with 17.5 per cent of mining and 16.5 per cent of construction industry employees (Bush and Lipari 2015). While 8.6 per cent of the total US workforce had used an illicit drug in the past month, 16.9 per cent of those employed in hospitality (accommodation and food/ beverage services) had done so (Bush and Lipari 2015). These industry differences remained significant even after controlling for gender and age differences (Bush and Lipari 2015). Corral et al. (2012) noted occupational and industry variations across European countries, with higher prevalence among construction, transport, agriculture, hospitality (hotels and restaurants) and male blue-collar workers compared with other industries and occupations.

Similar differences in prevalence data among occupational and industry groups have been reported in the Australian workforce. Controlling for socio-demographic factors, the proportion of Australian employees who frequently (at least weekly) drank at levels associated with short-term risk of harm was lowest in the education sector (3.8 per cent), but significantly higher in the agriculture (14 per cent), hospitality (13.6 per cent), manufacturing (10.5 per cent) and construction (9.5 per cent) industries, and among blue-collar workers (12.6 per cent) (Berry et al. 2007). Variations between occupational and industry groups are also evident for workplace use. Australian hospitality industry workers have been identified as being 3.5 times more likely than other workers to drink alcohol and two to three times more likely to use drugs at work or attend work under the influence of alcohol or other drugs (Pidd et al. 2011). Other Australian workforce groups identified as high risk for workplace use included construction and financial services industry workers, tradespersons and unskilled workers, with significant differences in prevalence remaining after controlling for age, gender and other demographic variables (Pidd et al. 2011).

Prevalence data indicate that the alcohol and other drug use patterns of substantial proportions of the workforce may present a workplace safety risk, regardless of whether use occurs at work during work hours, or away from work and outside work hours. These prevalence data also show that consumption patterns, and therefore alcohol and other drug-related workplace safety risk, vary significantly across different occupations and industries. Moreover, these variations remain even when age, gender and other demographic variables known to be associated with alcohol and other drug use are controlled for, indicating other factors play a role (Berry et al. 2007; Pidd et al. 2011; Bush and Lipari 2015). Understanding why prevalence varies across workforce groups has important implications for the design and implementation of strategies to reduce risk.

EXPLANATIONS OF VARYING PREVALENCE

Individual differences such as age and gender, together with attitudes and beliefs concerning alcohol and other drug use play an important role in determining a person's alcohol and other drug consumption patterns. The attitudes, beliefs and behaviours of an individual's family and close social networks are also influential, as are the social norms and expectations of the wider community. In addition to these influences, reviews of research highlight the important role that workplace factors play in determining employee alcohol and other drug consumption patterns (Bennett and Lehman 2003; Pidd and Roche 2008).

A key workplace factor is work stress, which is known to affect employee alcohol and other drug consumption patterns (Frone 2008). Numerous factors relating to working conditions, such as hazardous or dangerous work, shift work, long and/or irregular hours, poor industrial relations, low pay, boredom, job insecurity, low job satisfaction and workplace events such as serious accidents, industrial disputation, and downsizing are all associated with increased levels of employee physical and/or psychological distress. Elevated levels of work-related stress are then alleviated by the use of alcohol or other drugs during and/or outside work hours.

Another important influence is the availability of alcohol and other drugs (Ames and Grube 1999). Low levels of supervision, and the lack of formal workplace policies can influence employees' beliefs about the availability and/or acceptability of alcohol or other drugs during working hours. Working in close proximity to bars, hotels and other sources of alcohol and other drugs, or working away from normal workplace and/or social controls, can influence alcohol and other drug availability and consumption during and after working hours. Normative support for use, in the form of co-worker behaviour and expectations, can also affect employees' consumption patterns both at and away from the workplace. For example, while drinking at a work-related social function might receive normative support from co-workers, there may be less normative support for drinking during working hours.

Workplace conditions that contribute to high levels of work stress, workplace alcohol and other drug availability, and workplace normative support for use can combine to create workplace social networks. These networks, or subcultures, develop their own norms for use that may differ from an employee's norms for use away from the workplace (Pidd et al. 2014). For example, employees may be pressured to join co-workers in regular, end of the working week drinking rituals, despite not normally drinking on a regular basis. Similarly, employees working long and/or irregular hours may be encouraged by co-workers to use stimulants to combat the effects of fatigue.

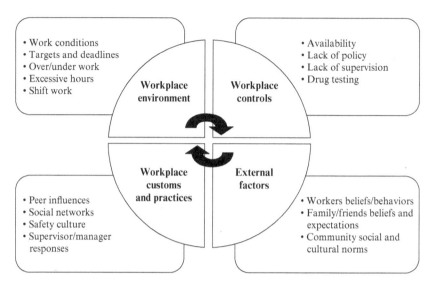

Source: Pidd and Roche (2008).

Figure 4.1   A cultural model of employee alcohol and other drug use

Building on the early work of Ames and Janes (1992), Pidd and Roche (2008) outlined an integrated model, proposing that the stressors, controls and subcultures contained within the workplace interact to result in an overall workplace culture that influences alcohol and other drug use (Figure 4.1). They argued that this workplace culture can shape the consumption patterns of individuals and social groups not only within the workplace, but also in settings external to the workplace. Elements of this integrated model include:

1. workplace customs and practices (for example, workplace subcultures and social networks, co-worker behavioural norms at work and in work-related settings, the workplace industrial relations climate and administrative/management culture);
2. workplace conditions (that is, environmental factors that impact on consumption patterns via workplace stressors);
3. workplace controls (that is, factors that contribute to the availability of alcohol and other drugs in the workplace); and
4. external factors (for example, employees' pre-existing attitudes, beliefs and behaviours regarding drinking and drug use, the values, behaviours and expectations of family members, and the social and cultural norms of the wider community).

A cultural perspective has two important implications for strategies designed to address alcohol and other drug-related risk to workplace safety. First, it highlights the complexity of the relationship between work and employee drug use. A range of factors both internal and external to the workplace can individually, or in combination, contribute to employee consumption patterns. Thus, an assessment of these factors needs to be conducted before designing and implementing specific interventions. Such an assessment not only helps identify the extent and precise nature of any potential risk to safety and well-being, but also allows for strategies to be tailored to deal with specific identified risks.

The second implication is that workplace strategies need to acknowledge the pivotal role of workplace culture. Workplace culture not only has a direct influence on employee consumption patterns but can also mediate the influence of workplace conditions, workplace controls and external factors. Central to this concept of workplace culture is the alcohol and other drug-related norms of both management and employees and the way in which the workplace deals with alcohol and other drug-related issues. Good practice responses go beyond a focus on individual 'problem' employees to include whole-of-workplace strategies that include all management, employees, visitors and guests at workplace functions.

GOOD PRACTICE RESPONSES

Traditional responses to alcohol and other drug-related harm in the workplace have been to focus on the identification and treatment of employees with a perceived 'problem'. While providing help to such employees is important, there is also a need to consider the much larger number of employees who experience only occasional alcohol and other drug-related problems, but collectively account for a much greater proportion of related harm in the workplace. For example, heavy drinkers who frequently drink at levels associated with high risk of harm take more days off owing to their drinking compared with employees who drink less (Pidd et al. 2006; Roche et al. 2008). However, the much larger number of employees who only occasionally drink at high-risk levels account for the majority of alcohol-related absenteeism (Pidd et al. 2006; Roche et al. 2008). While this example refers to a productivity issue, a similar scenario is likely to apply to workplace safety.

The past two decades have seen a shift away from the traditional approach to managing alcohol and other drug-related harm in the workplace to a broader primary prevention approach (for example, Bennett and Lehman 2003; ILO 2003; Pidd and Roche 2008). There are three basic components of such an approach that are central to effective responses. A recent Australian study has demonstrated that a cultural approach to the development and implementation of these three components of a good practice response can be effective as a workplace alcohol and other drug risk management strategy (Pidd et al. 2018). Furthermore, if co-developed with employees, this approach can overcome barriers to implementation (Cameron et al. 2018).

1. The Development and Implementation of a Formal Workplace Policy

A formal written policy forms the basis of any response to risk of alcohol and other drug-related harm in the workplace. Policies based on good practice are those that are informed by a workplace risk and needs assessment that identifies the extent and nature of the risk and the infrastructure that supports the policy and associated procedures. The policy and procedures most likely to be effective are those tailored to suit the specific needs and resources of individual workplaces.

The policy should be a comprehensive document that states objectives, the methods of achieving the objectives, and the roles and responsibilities of those who will implement the policy. It should outline the work organisation's position on alcohol and other drug use (including the legitimate use of pharmaceutical drugs) and guidelines and strategies for dealing with all aspects of alcohol and other drug-related issues in the workplace. This includes detailing under what conditions alcohol is made available at workplace functions and the incorporation of strategies to minimise any subsequent alcohol-related harm.

The policy should also outline procedures for approaching and dealing with an affected employee, provide information on treatment or counselling services and detail any disciplinary action that may be taken if the policy is breached. To be effective, policies also need to be widely disseminated, implemented throughout the entire workplace and be universal in application.

A key component of an effective policy is consultation with stakeholders such as management, unions, employees, supervisors and occupational health and safety staff. This consultation process allows for the development of mutually acceptable goals and procedures. Successful consultation with stakeholders during the policy development stage is often crucial for policy credibility, acceptance, and awareness. There is evidence that employee awareness of a workplace policy is associated with lower levels of alcohol and other drug use (Bush and Lipari 2014; Pidd et al. 2016).

2. The Provision of Education and Training

Informing employees about the policy content is necessary to ensure employees understand how to comply. Good practice awareness and education programmes contribute to the health and well-being of employees by providing information about: (1) alcohol and other drug-related harm in the workplace; (2) workplace factors that may contribute to increased risk of harm; and (3) general alcohol and other drug-related health information including access to rehabilitation and treatment. In most cases, the success of any workplace response is dependent on changing existing attitudes and behaviours relating to use. The provision of regular, ongoing employee education plays a crucial role in this regard.

The credibility, acceptance and success of any workplace policy is dependent on the attitudes and actions of supervisors, managers, safety personnel, employee representatives, and other key staff who may be responsible for the policy's implementation. Providing training for employees who implement the policy and manage workplace alcohol and other drug-related issues improves their confidence in carrying out their roles and builds their capacity to identify and respond to workplace alcohol and other drug-related harm. Training can also enhance supervision and management capability, as it develops skills on how to communicate with employees who perform poorly owing to alcohol or other drug-related issues. In order to contribute to a workplace culture that inhibits alcohol and other drug-related harm, training programmes need to be regular, ongoing and adaptable to changing circumstances.

3. Access to Counselling and Treatment

An important component of any workplace response is access to counselling and treatment services. Some employers provide these services via an employee assistance programme (EAP) or pay for private services; others use community-based non-profit services. Regardless of which type of service is utilised, the employer must ensure the service provider has appropriate and relevant counselling and treatment skills and knowledge. While access to counselling and treatment may be compulsory when employees breach conditions of the policy, employees should also be given the opportunity to access these services voluntarily. In either case, employees should be assisted to locate and access these services and be provided with paid or unpaid leave to attend. It is important that confidentially is assured. The provision of access to counselling and treatment helps avoid the financial costs and loss of morale among co-workers that is associated with dismissal. It is also less punitive than instant dismissal for policy breaches and therefore more likely to be accepted and endorsed by employees.

OTHER STRATEGIES

In addition to these three basic components of a good practice response to alcohol and other drug-related risk to workplace safety, a number of other strategies have been used in the workplace, with varying degrees of success.

Drug Testing

One strategy that is becoming increasingly commonplace is workplace drug testing (Pierce 2007; Walsh 2008). The aim of workplace testing is to improve workplace safety and productivity by deterring employee use and/or reducing alcohol and other drug-related workplace injuries. However, reviews of relevant research consistently identify that there is insufficient evidence to definitively conclude that workplace drug testing achieves these aims (Cashman et al. 2009; Macdonald et al. 2010; Frone 2013; Pidd and Roche 2014). While two of these reviews found that alcohol testing appeared to be an effective strategy for reducing fatalities in the transport industry (Cashman et al. 2009; Pidd and Roche 2014), all have concluded that the evidence for the effectiveness of testing overall is mixed, with most of the reviewed research of poor methodological quality.

More recent research has also produced mixed results. An examination of drug testing in the US construction industry found companies with drug-testing programmes had slightly, but not significantly, lower injury rates compared to non-testing companies (Schofield et al. 2013). By contrast, a study that involved a wider range of US industries found that testing was associated with a significant reduction (15 per cent) in minor injury rates involving no lost work, but no significant association was found for lost time injury rates (Waehrer et al. 2016). However, conclusions that can be drawn from these studies are restricted owing to the cross-sectional design utilised. A more rigorous study, utilising a longitudinal retrospective case control design, found that employees subjected to random testing had significantly lower rates of accidents when compared with non-tested employees (Marques et al. 2014).

While evidence concerning the effectiveness of workplace drug testing is limited, this does not imply that testing has no role to play in managing alcohol and other drug-related risk to workplace safety. Modern breath analysis devices are a relatively reliable indicator of blood alcohol concentration (Lindberg et al. 2007) and therefore represent a useful tool for detecting alcohol-related impairment that may be a risk to workplace safety. While urinalysis and oral fluid or saliva tests cannot detect impairment, they can be useful for providing relatively reliable and objective indicators of past and potentially recent use of other drugs (Moore 2011; Phan et al. 2012). However, poorly implemented testing programmes may not only fail to reduce alcohol and other drug-related risk to workplace safety but can result in unintended and counter-productive outcomes (Pidd and Roche 2014).

For example, rather than reducing or ceasing use, employees exposed to a drug-testing programme may change their behaviour to avoid detection. This could include changing patterns of use, using drugs that have a shorter window of detection (for example, methamphetamine), are less detectable (for example, synthetic drugs) or more easily explained (for example, pharmaceuticals), or using masking agents that are readily available on the Internet. There is also some limited evidence that post-accident testing may lead to the under-reporting of minor accidents and/or near misses (Morantz and Mas 2008). This may explain the results obtained by Waehrer et al. (2016), who found that testing resulted in a reduction of minor injuries associated with no lost time, but not more serious lost time injuries. Workplace testing programmes associated with punitive outcomes (for example, instant dismissal) may be more likely to result in unexpected counter-productive outcomes than testing programmes that result in counselling and/or treatment referral in the first instance.

Workplace drug-testing programmes may also be less effective if they are introduced as a single, stand-alone strategy. When this occurs, often the focus is placed on illicit drug use with alcohol use overlooked or downplayed. This is a serious issue for a workplace safety strategy given that, based on available evidence, alcohol is of greater concern than illicit drugs for most workplaces.

In addition, drug testing attempts to strengthen workplace controls by focusing on individual employees. On its own, such an approach is narrow in focus and inconsistent with a whole-of-workplace approach that also targets the much wider range of workplace factors associated with alcohol and other drug-related harm. Testing may be a useful tool for managing alcohol and other drug-related risk to workplace safety, but to maximise its effectiveness, it needs to be embedded within a much broader whole-of-workplace response, targeting workplace culture as described by Pidd and Roche (2008). At a minimum, workplace testing programmes need to include education and training programmes that not only raise awareness of the testing policy and process, but also raise employees', supervisors' and managers' awareness of alcohol and other drug-related risk to workplace safety and worker well-being, building their capacity to respond to this risk.

Good Practice Drug Testing

To maximise the potential effectiveness of workplace testing and mini-mise the likelihood of unexpected counter-productive outcomes, work-place testing programmes need to be based on good practice and be readily accepted by employees as a legitimate safety strategy. Good practice workplace drug testing is a two-step process. The first step involves an initial on-site screen, using a point of collection test device. Most point of collection test devices use immunoassay techniques that are less reliable and accurate than laboratory analysis. For this reason, the second step involves laboratory analysis to confirm the accuracy of any initial on-site screen that detects the presence of a drug. Laboratory analysis typically involves more reliable and accurate mass-spectrometry techniques.

Good practice also dictates that the testing process must comply with any relevant guidelines or standards. While these guidelines and standards may vary between countries they should at least cover:

1. specimen collection, storage, handling and dispatch/transport to a laboratory and chain of custody procedures;
2. qualifications and training required for on-site specimen collectors and laboratory personnel; and
3. cut point levels for a positive on-site screen and laboratory analysis.

To be accepted by employees, testing programmes should be justified as being needed to address an identified safety risk, and the associated policy and procedures need to be adequately disseminated and applied in a procedurally fair manner. The programme also needs to:

1. result in counselling, treatment, and rehabilitation rather than punitive outcomes;
2. target safety sensitive and integrity reliant work roles;
3. allow for employee input into the development and implementation of the programme;
4. allow for a right of appeal; and
5. incorporate appropriate education and training.

Health Promotion, Brief Interventions, Peer Interventions and Psychosocial Skills Training

Health promotion, brief intervention, peer intervention and psychosocial skills training have been identified as additional strategies that have potential for reducing alcohol and other drug-related risk to workplace safety and worker well-being (Webb et al. 2009). A comprehensive review of these strategies is beyond the scope of this chapter, however, a brief description is provided here.

Health promotion

Workplace health promotion programmes have a long history and in general, have been effective for improving employee well-being and productivity in the workplace (Kuoppala et al. 2008; Goetzel et al. 2014). The basic premise of health promotion programmes is that healthy lifestyles are incompatible with heavy alcohol consumption and other drug use. Incorporating alcohol and drug issues within the context of wider health concerns can motivate behavioural change, and has been shown to reduce levels of risky alcohol consumption (Heirich and Sieck 2000).

Brief interventions

Brief interventions aim to identify potential problems with an individual's alcohol and other drug use and motivate those identified as being at risk to change their consumption patterns (Babor and Higgins-Biddle 2000). Typically, brief interventions provide tailored feedback in the form of information and advice concerning risk of harm owing to the individual's alcohol and other drug use. Brief interventions have been shown to be an effective strategy for workplaces regardless if they are web based (Billings et al. 2008; Doumas and Hannah 2008) or delivered face to face (Araki et al. 2006). Workplace brief interventions appear particularly effective when incorporated into broader primary prevention interventions such as health promotion programmes (Heirich and Sieck 2000; Hermansson et al. 2010).

Peer interventions

Peer interventions involve the use of peers as agents of change and have demonstrated effectiveness for addressing a wide range of social and health-related behaviours (Rivera and Nangle 2008). Applied to the workplace, peer interventions are based on the premise that co-workers are in the best position to recognise and respond to employees with alcohol or other drug problems. Peer interventions involve the use of trained co-workers to recognise alcohol or other drug problems among their peers and intervene appropriately. Evaluations of these programmes indicate that they have been effective in identifying and addressing problem behaviours and have contributed to reducing use and related harm (Miller et al. 2007; Spicer and Miller 2005).

Psychosocial skills training

Psychosocial interventions use a range of strategies including motivational interviewing, cognitive behaviour therapy, problem solving, goal setting, social skills training, contingency management and improving coping strategies. The aim is to provide vulnerable, at-risk employees with the skills to deal with life, social and emotional problems that can lead to alcohol and other drug use. Evaluations of workplace psychosocial skills training indicates that it can reduce alcohol and other drug use and personal problems associated with use (Broome and Bennett 2011; Pidd et al. 2015).

CONCLUSIONS

While evidence of the causal role of employee alcohol and other drug use in workplace accidents and injuries is limited, the prevalence of use among the workforce and the resulting negative impact on performance indicates that employee alcohol and other drug use can be a potential workplace safety risk. In addition, prevalence of use and safety risk vary substantially between workforce groups. Moreover, working conditions and workplace environments that influence consumption patterns and related harm vary across different workplaces. Thus, good practice responses to alcohol and other drug-related harm in the workplace are likely to be most effective when they are tailored to suit the specific conditions, needs and resources of individual workplaces. The adoption of a multifaceted, whole-of-workplace approach that aims to create a workplace safety and well-being culture is key to minimising the risk of alcohol and other drug-related harm.

NOTES

1. Psychoactive drugs are chemical substances that act on the central nervous system to alter mood, perception, behaviour and/or consciousness.
2. The exception is hallucinogens, such as lysergic acid diethylamide (LSD) and mescaline, which alter perceptions of reality, with use resulting in hallucinations or subjective changes in thought, emotion and consciousness.
3. In the UK one single unit equals 8 grams of alcohol.
4. Binge alcohol use is defined in the US as drinking five or more drinks (containing 14 grams of alcohol per drink) (for males) or four or more drinks (for females) on the same occasion (that is, at the same time or within a couple of hours of each other) on at least one day in the past 30 days.
5. Current Australian alcohol guidelines state that drinking five or more standard drinks (10 grams of alcohol per standard drink) on any single occasion significantly increases short-term risk of alcohol-related injury.

REFERENCES

Ames, G.M. and Grube, J.W. (1999), 'Alcohol availability and workplace drinking: mixed method analyses', *Journal of Studies on Alcohol*, **60** (3), 383–93.

Ames, G.M. and Janes, C. (1992), 'A cultural approach to conceptualizing alcohol and the workplace', *Alcohol and Health Research World*, **16** (2), 112–19.

Anderson, D.M., Rees, D.I. and Tekin, E. (2018), 'Medical marijuana laws and workplace fatalities in the United States', *International Journal of Drug Policy*, **60** (October), 33–9.

Araki, I., Hashimoto, H., Kono, K., Matsuki, H. and Yano, E. (2006), 'Controlled trial of worksite health education through face-to-face counseling vs. e-mail on drinking behavior modification', *Journal of Occupational Health*, **48** (4), 239–45.

Ashton, C.H. (1999), 'Adverse effects of cannabis and cannabinoids', *British Journal of Anaesthesia*, **83** (4), 637–49.

Australian Institute of Health and Welfare (2017), 'National Drug Strategy Household Survey 2016: detailed findings. Misuse of pharmaceuticals infographic', Australian Institute of Health and Welfare, Canberra, accessed 31 August 2018 at https://www.aihw.gov.au/getmedia/2243e2cb-8847-4ee5-9afc-0e5169129035/aihw-infographic-ndshs-2016-misuse-pharmaceuticals.pdf.aspx.

Babor, T. and Higgins-Biddle, J. (2000), 'Alcohol screening and brief intervention: dissemination strategies for medical practice and public health', *Addiction*, **95** (5), 677–86.

Bennett, J. and Lehman, W. (2003), 'Understanding employee alcohol anbd other drug use: toward a multilevel approach', in J. Bennett and W. Lehman (eds), *Preventing Workplace Substance Use: Beyond Drug Testing to Wellness*, Washington DC: American Psychological Association, pp. 29–56.

Berry, J., Pidd, K., Roche, A. and Harrison, J. (2007), 'Prevalence and patterns of alcohol use in the Australian workforce: findings from the 2001 National Drug Strategy Household Survey', *Addiction*, **102** (9), 1399–410.

Billings, D.W., Cook, R.F., Hendrickson, A. and Dove, D.C. (2008), 'A web-based approach to managing stress and mood disorders in the workforce', *Journal of Occupational and Environmental Medicine*, **50** (8), 960–68.

Bondallaz, P., Favrat, B., Chtioui, H., Fornari, E., Maeder, P. and Giroud, C. (2016), 'Cannabis and its effects on driving skills', *Forensic Science International*, **268** (November), 92–102.

Broome, K.M. and Bennett, J.B. (2011), 'Reducing heavy alcohol consumption in young restaurant workers', *Journal of Studies on Alcohol and Drugs*, **72** (1), 117–24.

Bush, D.M. and Lipari, R.N. (2014), 'Workplace policies and programs concerning alcohol and drug use', The CBHSQ Report, 7 August, Center for Behavioral Health Statistics and Quality, Substance Abuse and Mental Health Services Administration, Rockville, MD, accessed 29 April 2019 at https://www.ncbi.nlm.nih.gov/books/NBK384657/.

Bush, D.M. and Lipari, R.N. (2015), 'Substance use and substance use disorder, by industry', The CBHSQ Report, 16 April, Center for Behavioral Health Statistics and Quality, Substance Abuse and Mental Health Services Administration, Rockville, MD, accessed 29 April 2019 at https://www.ncbi.nlm.nih.gov/books/NBK343542/#SR-159_RB-1959.ssuggcit.

Cameron, J., Pidd, K., Roche, A., Lee, N. and Jenner, L. (2018), 'A co-produced cultural approach to workplace alcohol interventions: barriers and facilitators', *Drugs: Education, Prevention and Policy*, doi:10.1080/09687637.2018.1468871.

Cashman, C.M., Ruotsalainen, J.H., Greiner, B.A., Beirne, P.V. and Verbeek, J.H. (2009), 'Alcohol and drug screening of occupational drivers for preventing injury', *Cochrane Database of Systematic Reviews*, (2), art. CD006566, doi:10.1002/1465 1858.CD006566.pub2.

Center for Behavioral Health Statistics and Quality (2017), 'Results from the 2016 National Survey on Drug Use and Health: detailed tables', 7 September, Substance Abuse and Mental Health Services Administration, Rockville, MD, accessed 28 August 2018 at https://www.samhsa.gov/data/sites/default/files/NSD UH-DetTabs-2016/NSDUH-DetTabs-2016.pdf.

Centers for Disease Control and Prevention (2018), '2018 annual surveillance report of drug-related risks and outcomes: United States', Surveillance Special Report, US Department of Health and Human Services, 31 August, accessed 9 September 2018 at https://www.cdc.gov/drugoverdose/pdf/pubs/2018-cdc-drug-surveillance-report.pdf.

Corral, A., Duran, J. and Isusi, I. (2012), 'Use of alcohol and drugs at the workplace', report, European Foundation for the Improvement of Living and Working Conditions, Dublin, accessed 3 September 2018 at https://www.eurofound.europa .eu/publications/report/2012/use-of-alcohol-and-drugs-at-the-workplace.

Daurat, A., Sagaspe, P., Motak, L., Taillard, J., Bayssac, L., Huet, N. et al.. (2013), 'Lorazepam impairs highway driving performance more than heavy alcohol consumption', *Accident Analysis & Prevention*, **60** (November), 31–4.

Degenhardt, L. and Hall, W. (2012), 'Extent of illicit drug use and dependence, and their contribution to the global burden of disease', *Lancet*, **379** (9810), 55–70.

Degenhardt, L., Whiteford, H.A., Ferrari, A.J., Baxter, A.J., Charlson, F.J., Hall, W.D. et al. (2013), 'Global burden of disease attributable to illicit drug use and dependence: findings from the Global Burden of Disease Study 2010', *Lancet*, **384** (9904), 1564–1574, doi:10.1016/S0140-6736(13)61530-5.

Doumas, D.M. and Hannah, E. (2008), 'Preventing high-risk drinking in youth in the workplace: a web-based normative feedback program', *Journal of Substance Abuse Treatment*, **34** (3), 263–71.

Drummer, O.H. (2002), 'Benzodiazepines – effects on human performance and behavior', *Forensic Science Review*, **14** (1–2), 1–14.

Fell, J.C. and Voas, R.B. (2014), 'The effectiveness of a 0.05 blood alcohol concentration (BAC) limit for driving in the United States', *Addiction*, **109** (6), 869–74.

Ferreira, D.H., Boland, J.W., Phillips, J.L., Lam, L. and Currow, D.C. (2018), 'The impact of therapeutic opioid agonists on driving-related psychomotor skills assessed by a driving simulator or an on-road driving task: a systematic review', *Palliative Medicine*, **32** (4), 786–803.

Frone, M.R. (2004), 'Alcohol, drugs, and workplace safety outcomes: a view from a general model of employee substance use and productivity', in J. Barling and M.R. Frone (eds), *The Psychology of Workplace Safety*, Washington, DC: American Psychological Association, pp. 127–56.

Frone, M.R. (2008), 'Are work stressors related to employee substance use? The importance of temporal context in assessments of alcohol and illicit drug use', *Journal of Applied Psychology*, **93** (1), 199–206.

Frone, M.R. (2013), *Workplace Interventions I: Drug Testing Job Applicants and Employees*, Washington, DC: American Psychological Association.

Goetzel, R.Z., Henke, R.M., Tabrizi, M., Pelletier, K.R., Loeppke, R., Ballard,

D.W. et al. (2014), 'Do workplace health promotion (wellness) programs work?', *Journal of Occupational and Environmental Medicine*, **56** (9), 927–34.

Gunn, C., Mackus, M., Griffin, C., Munafo, M.R. and Adams, S. (2018), 'A systematic review of the next-day effects of heavy alcohol consumption on cognitive performance', *Addiction*, **113** (12), 2182–93.

Hartman, R.L. and Huestis, M.A. (2013), 'Cannabis effects on driving skills', *Clinical Chemistry*, **59** (3), 478–92.

Hayley, A.C., Downey, L.A., Shiferaw, B. and Stough, C. (2016), 'Amphetamine-type stimulant use and the risk of injury or death as a result of a road-traffic accident: a systematic review of observational studies', *European Neuropsychopharmacology*, **26** (6), 901–22.

Heirich, M. and Sieck, C.J. (2000), 'Worksite cardiovascular wellness programs as a route to substance abuse prevention', *Journal of Occupational and Environmental Medicine*, **42** (1), 47–56.

Hermansson, U., Helander, A., Brandt, L., Huss, A. and Rönnberg, S. (2010), 'Screening and brief intervention for risky alcohol consumption in the workplace: results of a 1-year randomized controlled study', *Alcohol and Alcoholism*, **45** (3), 252–7.

International Labour Organization (ILO) (2003), *Alcohol and Drug Problems at Work: The Shift to Prevention*, Geneva: ILO.

Irwin, C., Iudakhina, E., Desbrow, B. and McCartney, D. (2017), 'Effects of acute alcohol consumption on measures of simulated driving: a systematic review and meta-analysis', *Accident Analysis & Prevention*, **102** (May), 248–66.

Kelly, E., Darke, S. and Ross, J. (2004), 'A review of drug use and driving: epidemiology, impairment, risk factors and risk perceptions', *Drug and Alcohol Review*, **23** (3), 319–44.

Kuoppala, J., Lamminpaa, A. and Husman, P. (2008), 'Work health promotion, job well-being, and sickness absences – a systematic review and meta-analysis', *Journal of Occupational and Environmental Medicine*, **50** (11), 1216–27.

Lenne, M.G., Dietze, P., Rumbold, G., Redman, J.R. and Triggs, T.J. (2000), 'Opioid dependence and driving ability: a review in the context of proposed legislative change in Victoria', *Drug and Alcohol Review*, **19** (4), 427–39.

Li, G., Baker, S.P., Zhao, Q., Brady, J.E., Lang, B.H., Rebok, G.W. et al. (2011), 'Drug violations and aviation accidents: findings from the US mandatory drug testing programs', *Addiction*, **106** (7), 1287–92, doi:10.1111/j.1360-0443.2011 .03388.x.

Lindberg, L., Brauer, S., Wollmer, P., Goldberg, L., Jones, A. and Olsson, S. (2007), 'Breath alcohol concentration determined with a new analyzer using free exhalation predicts almost precisely the arterial blood alcohol concentration', *Forensic Science International*, **168** (2–3), 200–207.

Logan, B.K. (2002), 'Methamphetamine – effects on human performance and behavior', *Forensic Science Review*, **14** (1–2), 133–51.

Macdonald, S., Hall, W., Roman, P., Stockwell, T., Coghlan, M. and Nesvaag, S. (2010), 'Testing for cannabis in the work-place: a review of the evidence', *Addiction*, **105** (3), 408–16, doi:10.1111/j.1360-0443.2009.02808.x.

Marques, P.H., Jesus, V., Olea, S.A., Vairinhos, V. and Jacinto, C. (2014), 'The effect of alcohol and drug testing at the workplace on individual's occupational accident risk', *Safety Science*, **68** (October), 108–20, doi:10.1016/j.ssci.2014.03.007.

Martin, T.L., Solbeck, P.A., Mayers, D.J., Langille, R.M., Buczek, Y. and Pelletier, M.R. (2013), 'A review of alcohol-impaired driving: the role of blood alcohol concentration and complexity of the driving task', *Journal of Forensic Sciences*, **58** (5), 1238–50.

McNeilly, B., Ibrahim, J.E., Bugeja, L. and Ozanne-Smith, J. (2010), 'The prevalence of work-related deaths associated with alcohol and drugs in Victoria, Australia, 2001–6', *Injury Prevention*, **16** (6), 423–8, doi:10.1136/ip.2010.027052.

Miller, T.R., Zaloshnja, E. and Spicer, R.S. (2007), 'Effectiveness and benefit–cost of peer-based workplace substance abuse prevention coupled with random testing', *Accident Analysis & Prevention*, **39** (3), 565–73.

Moore, C. (2011), 'Oral fluid and hair in workplace drug testing programs: new technology for immunoassays', *Drug Testing & Analysis*, **3** (3), 166–8, doi:http://dx.doi.org/10.1002/dta.140.

Morantz, A. and Mas, A. (2008), 'Does post-accident drug testing reduce injuries? Evidence from a large retail chain', *American Law and Economics Review*, **10** (2), 246–302.

Murray, J.B. (1998), 'Psychophysiological aspects of amphetamine-methamphetamine abuse', *Journal of Psychology*, **132** (2), 227–37.

National Centre for Education and Training on Addiction (NCETA) (2015), 'National Alcohol and Drug Knowledgebase: short-term risk of alcohol-related injury and employment status', NCETA, Flinders University, Adelaide, accessed July 2018 at http://nadk.flinders.edu.au/kb/alcohol/alcohol-employment/short-term-risk-of-alcohol-related-injury-and-employment-status/.

Parish, M., Rohlman, D., Elliot, D. and Lasarev, M. (2016), 'Factors associated with occupational injuries in seasonal young workers', *Occupational Medicine*, **66** (2), 164–7.

Phan, H.M., Yoshizuka, K., Murry, D.J. and Perry, P.J. (2012), 'Drug testing in the workplace', *Pharmacotherapy: The Journal of Human Pharmacology & Drug Therapy*, **32** (7), 649–56, doi:http://dx.doi.org/10.1002/j.1875-9114.2011.01089.x.

Pidd, K. and Roche, A. (2008), 'Changing workplace cultures: an integrated model for the prevention and treatment of alcohol-related problems', in D. Moore and P. Dietze (eds), *Drugs and Public Health: Australian Perspectives on Policy and Practice*, Melbourne: Oxford University Press, pp. 49–59.

Pidd, K. and Roche, A.M. (2014), 'How effective is drug testing as a workplace safety strategy? A systematic review of the evidence', *Accident Analysis & Prevention*, **71** (October), 154–65.

Pidd, K., Berry, J., Roche, A. and Harrison, J. (2006), 'Estimating the cost of alcohol-related absenteeism in the Australian workforce: the importance of consumption patterns', *Medical Journal of Australia*, **185** (11–12), 637–41.

Pidd, K., Cameron, J., Roche, A., Lee, N., Jenner, L. and Duraisingam, V. (2018), 'Workplace alchol harm reduction intervention in Australia: cluster non-randomised controlled trial', *Drug and Alcohol Review*, **37** (4), 502–13.

Pidd, K., Kostadinov, V. and Roche, A. (2016), 'Do workplace policies work? An examination of the relationship between alcohol and other drug policies and workers' substance use', *International Journal of Drug Policy*, **28** (February), 48–54.

Pidd, K., Roche, A.M. and Buisman-Pijlman, F. (2011), 'Intoxicated workers: findings from a national Australian survey', *Addiction*, **106** (9), 1623–33, doi:10.1111/j.1360-0443.2011.03462.x.

Pidd, K., Roche, A. and Fischer, J. (2015), 'A recipe for good mental health: a pilot randomised controlled trial of a psychological wellbeing and substance use intervention targeting young chefs', *Drugs: Education, Prevention & Policy*, **22** (4), 352–61.

Pidd, K., Roche, A. and Kostadinov, V. (2014), 'Trainee chefs' experiences of alcohol, tobacco and drug use', *Journal of Hospitality and Tourism Management*, **21** (December), 108–15.
Pidd, K., Shtangey, V. and Roche, A. (2008), *Alcohol Use in the Australian Workforce: Prevalence, Patterns, and Implications*, Adelaide: National Centre for Education and Training on Addiction.
Pierce, A. (2007), 'Workplace drug testing outside the U.S.', in S.B. Karch (ed.), *Drug Abuse Handbook*, New York: CRC Press, pp. 765–75.
Price, J. (2014), 'Marijuana and workplace safety: an examination of urine drug tests', *Journal of Addictive Diseases*, **33** (1), 24–7.
Quest Diagnostics (2017), 'Quest Diagnostics Drug Testing index: full year 2016 tables', Secaucus, NJ, accessed 6 September 2018 at http://www.questdiagnostics.com/dms/Documents/Employer-Solutions/Brochures/drug-testing-index-2017-report--tables/drug-testing-index-2017-report-tables.pdf.
Ramaekers, J.G., Berghaus, G., van Laar, M. and Drummer, O.H. (2004), 'Dose related risk of motor vehicle crashes after cannabis use', *Drug & Alcohol Dependence*, **73** (2), 109–19.
Ramaekers, J.G., Kauert, G., van Ruitenbeek, P., Theunissen, E.L., Schneider, E. and Moeller, M.R. (2006), 'High-potency marijuana impairs executive function and inhibitory motor control', *Neuropsychopharmacology*, **31** (10), 2296–303.
Ramchand, R., Pomeroy, A. and Arkes, J. (2009), 'The effects of substance use on workplace injuries', Occassional Paper No. 247, RAND Corporation, Santa Monica, CA, accessed 16 August 2018 at https://www.rand.org/pubs/occasional_papers/OP247.html.
Ramirez, M., Bedford, R., Sullivan, R., Anthony, T.R., Kraemer, J., Faine, B. et al. (2013), 'Toxicology testing in fatally injured workers: a review of five years of Iowa FACE cases', *International Journal of Environmental Research and Public Health*, **10** (11), 6154–68, doi:10.3390/ijerph10116154.
Rapoport, M.J., Lanctot, K.L., Streiner, D.L., Bedard, M., Vingilis, E., Murray, B. et al. (2009), 'Benzodiazepine use and driving: a meta-analysis', *Journal of Clinical Psychiatry*, **70** (5), 663–73.
Rivera, M.S. and Nangle, D.W. (2008), 'Peer intervention', in M. Hersen and A.M. Gross (eds), *Handbook of Clincial Psychology*, Hoboken, NJ: John Wiley and Sons, pp. 759–85.
Roche, A., Pidd, K., Berry, J. and Harrison, J. (2008), 'Workers' drinking patterns: the impact on absenteeism in the Australian workplace', *Addiction*, **103** (5), 738–48.
Sachs, J., McGlade, E. and Yurgelun-Todd, D. (2015), 'Safety and toxicology of cannabinoids', *Neurotherapeutics*, **12** (4), 735–46.
Schofield, K.E., Alexander, B.H., Gerberich, S.G. and Ryan, A.D. (2013), 'Injury rates, severity, and drug testing programs in small construction companies', *Journal of Safety Research*, **44** (February), 97–104.
Sheridan, J., Bennett, S., Coggan, C., Wheeler, A. and McMillan, K. (2006), 'Injury associated with methamphetamine use: a review of the literature', *Harm Reduction Journal*, **3** (1), 14, doi:10.1186/1477-7517-3-14.
Smith, A., Wadsworth, E., Moss, S. and Simpson, S. (2004), 'The scale and impact of illegal drug use by workers', Research Report No. 193, Health and Safety Executive, Norwich, accessed 30 August 2018 at www.hse.gov.uk/research/rrpdf/rr193.pdf.
Spicer, R.S. and Miller, T.R. (2005), 'Impact of a workplace peer-focused substance abuse preventionand early intervention program', *Alcoholism: Clinical and Experimental Research*, **29** (4), 609–11.

Stallones, L. and Kraus, J. (1993), 'The occurrence and epidemiological features of alcohol-related occupational injuries', *Addiction*, **88** (7), 945–51.

Stephens, R., Ling, J., Heffernan, T.M., Heather, N. and Jones, K. (2008), 'A review of the literature on the cognitive effects of alcohol hangover', *Alcohol & Alcoholism*, **43** (2), 163–70.

Stout, P.R. and Farrell, L.J. (2003), 'Opioids – effects on human performance and behavior', *Forensic Science Review*, **15** (1), 29–59.

United Nations Office on Drugs and Crime (UNODC) (2016), *World Drug Report 2016*, Vienna: UNODC, accessed 21 August 2018 at http://www.unodc.org/wdr 2016/.

Waehrer, G.M., Miller, T.R., Hendrie, D. and Galvin, D.M. (2016), 'Employee assistance programs, drug testing, and workplace injury', *Journal of Safety Research*, **57** (June), 53–60.

Walsh, J.M. (2008), 'New technology and new initiatives in U.S. workplace testing', *Forensic Science International*, **174** (2–3), 120–24.

Webb, G., Shakeshaft, A., Sanson-Fisher, R. and Havard, A. (2009), 'A systematic review of work-place interventions for alcohol-related problems', *Addiction*, **104** (3), 365–77.

World Health Organization (2018), *Global Status Report on Alcohol And Health 2018*, Geneva: World Health Organization, accessed 6 September 2018 at http://www.who.int/substance_abuse/publications/global_alcohol_report/en/.

5. Understanding domestic violence as a workplace problem*

Barb MacQuarrie, Katreena Scott, Danielle Lim, Laura Olszowy, Michael D. Saxton, Jen MacGregor and Nadine Wathen

Domestic violence (DV) is part of a broader category of human-centric problems including harassment and violence that have recently been identified as workplace hazards. Understanding DV as both a workplace issue and an occupational health and safety (OHS) hazard has emerged from the work of: feminists working to shift perceptions of DV as a private matter to viewing it as a societal problem (Johnson and Dawson 2011), OHS specialists developing new conceptualizations of psychosocial stresses as workplace hazards (European Agency for Safety and Health at Work 2007) and an international network of women's advocates, trade unionists and researchers who have mobilized knowledge about the impact of DV on workers and the workplace (Domestic Violence at Work Network 2018a). While the work is global in scope, this chapter explores the Canadian experience and how it has reflected, paralleled, differed from and intersected with international initiatives.

American employers were the first to construct DV as a workplace concern, and by the mid-1990s related research in the United States (US) began to emerge (Mighty 1997; Solomon 1998). Surveys with human resource (HR) professionals and security directors showed DV becoming an increasingly common problem in many companies (Personnel Journal 1994; Kinney 1995). In these surveys, DV's negative impacts were framed in one of three ways: as an HR issue impacting productivity; as an anti-discrimination and victims' rights issue emphasizing the need for job protections and supports for victims; and/or as a corporate social responsibility issue emphasizing the response to a societal problem that harms workers and families.

Researchers have also linked employer responsibilities to address DV to family-friendly policies developed to help employees balance work and family obligations as the proportion of women in the workforce increased, particularly in developed economies (de Jonge 2018). Regardless of framing, keeping workers safe was seen as the responsibility of chief executive officers, HR professionals, security personnel and/or employee assistance programmes. Occupational health and safety legislation was silent on the issue and responsibilities for employers were not legislated; OHS experts were rarely involved in crafting prevention and response measures (Duda 1997; McFarlane et al. 2000; Malecha and Wachs 2003; Swanberg et al. 2005; Reeves and O'Leary-Kelly 2007; Lindquist et al. 2010; Pollack et al. 2010).

EMERGING PSYCHOSOCIAL RISKS

Concurrent with the above research, European efforts were identifying emerging psychosocial risks in the OHS field. Cox and Griffiths (2005, p. 786) define psychosocial hazards as 'those aspects of work design and the organization and management of work, and their social and environmental context, which may have the potential to cause psychological or physical harm'. Through these efforts, workplace violence was acknowledged as a psychosocial risk and European researchers and OHS experts prompted research and discussion on the topic.

An International Labour Organization (ILO) report (Lippel 2018) notes an emerging European consensus to adopt instruments that provide practical guidance on how to address violence and harassment in the workplace. As a result, '[t]he national legislation of EU Members has been heavily influenced by the transposition of EU OSH[1] and equal treatment Directives, which contain provisions on the prevention and prohibition of various forms of violence and harassment' (Lippel 2017, p. 42).

In 2005, citing a need to identify and anticipate emerging risks, the European Union (EU) mandated the European Agency for Safety and Health at Work (EU-OSHA) to set up a risk observatory to identify important emerging psychosocial risks (EU-OSHA 2018). In 2007, the EU-OSHA released an expert forecast on emerging psychosocial risks which included a chapter on violence and bullying. The EU-OSHA recognized that workplace violence and harassment are psychosocial risks and major challenges to OHS and, more broadly, to public health. Domestic violence, however, remained an obscure concern, meriting just a brief mention as one potential source of violence in the workplace in the European report (EU-OSHA 2007).

CANADIAN RECOGNITION OF DOMESTIC VIOLENCE AS A WORKPLACE CONCERN

In Canada, the women's shelter movement has long worked to bring increased attention to the issue of DV more generally (Johnson and Dawson 2011; Goodhand 2017). Early efforts focused on how shelters and other social services could address this problem (Department of Justice Canada 2015), with efforts in the criminal justice system leading to mandatory charging and 'no-drop' prosecution policies being in place federally and in most provincial jurisdictions by 1985 (Brown 2002). As the social dimensions of DV became clearer, Canada recognized it as a public health problem, giving the Family Violence Initiative, led by Health Canada, a mandate to promote public awareness and public involvement in to the problem in 1997 (Health Canada 2002). The Public Health Agency of Canada (2016) reiterated the concern that gender-based violence, including DV, is a serious public health problem. Several tragic DV deaths in Canadian workplaces spotlighted DV as a workplace issue. In 2000, Tony McNaughton, manager of a Starbucks café in Vancouver, was killed defending a female colleague. The woman's ex-husband tried to attack her with a butcher's knife in the café. Mr McNaughton intervened and told the woman to run; she would have likely died had he not intervened (Alphonso 2018). In 2004, Aysegul Candir, an English as a second language (ESL) teacher in Brampton, Ontario, was shot and killed by the husband she had just left as she got out of her car in the parking lot of her school (Brampton Guardian 2007). A year later, Lori Dupont, a nurse in Windsor, Ontario, was stabbed and killed in the Emergency Department by her ex-partner, a doctor who worked at the same hospital (CBC News 2007). The recognition of DV as a problem that belongs in the public sphere has generated and is continuing to generate a series of legal and policy changes in Canada, and internationally.

DOMESTIC VIOLENCE AT WORK (DV@WORK) NETWORK

In 2014, the Centre for Research and Education on Violence Against Women and Children received funding from the Social Sciences and Humanities Council of Canada to form an international network of domestic violence researchers, experts, social and labour organizations, and employers. The Domestic Violence at Work Network (DV@Work Net) was inspired by the work of the Australian programme, 'Safe at home, safe at work' (McFerran 2011), funded by the Federal Labor Government in Australia from 2010 to 2013. This support enabled a national Australian survey (the largest of its kind at the time) on the impacts of DV on workers and co-workers (McFerran 2011). Responding to the findings of this survey, Australian trade unions adopted collective bargaining as a widespread strategy to introduce standardized rights for workers affected by DV (Aeberhard-Hodges and McFerran 2018).

The primary goal of DV@Work Net was and is to accelerate the development of an emerging knowledge base and mobilize knowledge about DV in the workplace. In conducting research on the impacts of victimization across member countries, DV@Work Net sought to understand the scope and impact of the problem in Canada and internationally. A 'basis of unity' statement outlined a vision for change, listing strategies for addressing DV at work (Domestic Violence at Work Network 2018b).

DV@Work Net has generated new data from a series of surveys asking about employees' experiences of DV, how this has affected their work and workplace, and employer responses. At the time of writing, victimization surveys have been conducted in Australia (McFerran 2011), New Zealand (Rayner-Thomas 2013), the United Kingdom (Trade Union Congress 2014), Canada (Wathen et al. 2014), Turkey (Ararat et al. 2014), the Philippines (International Trade Union Confederation 2015), Mongolia (International Trade Union Confederation 2017a), Taiwan (International Trade Union Confederation 2017b) and Belgium (Institute for the Equality of Women and Men 2017).

In addition to promoting international cooperation in documenting the impact of DV victimization in the workplace, DV@Work Net has also focused on DV perpetration, an issue that has less often been the focus of research, policy and practice development. A chief reason for this decision is that focusing all (or most) workplace attention on responding to victimization creates an ethical shortcoming; it ignores half of the problem – stopping the violence. As emphasized in movements such as #MeToo and #Time'sUp, downplaying, ignoring, dismissing and ultimately failing to take action against abuse is a way of condoning it and is itself part of the problem. Workers and workplaces need to be able to recognize controlling, degrading, emotionally and physically abusive behaviours, have skills to speak out against these behaviours, and have programmes and policies in place that can prevent and respond to those perpetrating abuse. DV@Work Net partners were also concerned about unintentional harm that can result from workplace responses focused primarily on victimization, which can send a subtle message that DV is primarily a problem for (mostly women) victims, rather than a problem caused by those who perpetrate DV. Adding to emergent literature exploring the intersection of DV perpetration and workplaces (Lim et al. 2004; Rothman and Perry 2004; Rothman and Corso 2008; Galvez et al. 2011; Schmidt and Barnett 2011; Mankowski et al. 2013), DV@Work Net also included a survey of DV offenders conducted in Ontario, Canada (Scott et al. 2017). Together these data, along with a growing research literature (MacGregor et al. forthcoming), provide evidence that DV is a problem that crosses the boundary from home to work and poses risks to productivity and safety. To better understand what we know about the OHS hazards of DV in the Canadian workplace, we next explore the results of the two Canadian studies; detailed results are available elsewhere (Wathen et al. 2015; Scott et al. 2017; MacGregor et al., 2016 and 2017).

CANADIAN STUDY ON THE IMPACT OF DOMESTIC VIOLENCE VICTIMIZATION ON WORKERS AND WORKPLACES

In 2014, Canadian researchers at Western University in partnership with the Canadian Labour Congress conducted a pan-Canadian survey that examined the impacts of DV on workers and the workplace (Wathen et al. 2015). Of the 8429 people completing the survey, a third (34 per cent) reported experiencing DV from an intimate partner. Rates were higher for women (38 per cent) than men (17 per cent), but highest of all were for trans people (65 per cent). Over half (54 per cent) of those reporting DV experiences indicated that at least one type of abusive act occurred at or near the workplace. Of these, the most common were abusive telephone calls or text messages (41 per cent) and stalking or harassment near the workplace (21 per cent). Other behaviours that create OHS hazards included the abuser physically coming to the workplace (18 per cent), abusive email messages (16 per cent) and the abuser contacting co-workers or the employer (15 per cent). In addition to the risks of physical harm, these behaviours create clear psychosocial hazards including stress and fatigue for both the victims (hereafter alternately referred to as victims and survivors) and their co-workers.

Workers described in open-ended survey responses the many ways that DV followed them to work. For some, DV resulted in distraction: 'For myself, when I am at work, I am wondering what is happening with my partner, what type of mood he will be in that night, this will impact productivity at work' and 'If I have had a particularly bad time with him, I am not productive at all, I just can't concentrate, can't think, not motivated, I think of ways to leave and not have my belongings smashed'.

Some were closely monitored: 'Would phone my workplace to see what time I had left, and phoned when I arrived to make sure I was actually going to work'. While others experienced physical violence: 'He pretended to be security and dragged me out of work.'

Other workers experienced work-related threats from the abuser, including threats that they would come to the workplace. Not surprisingly, these victims were often distracted and anxious at the prospect that their abusive partner or ex-partner could invade their working life: 'My former spouse would threaten to wait for me outside work at the end of the day if I didn't answer him quickly enough or didn't give him the answer he wanted.'

Some victims (8 per cent) reported working in the same workplace as the abusive partner or ex-partner. In these cases, victims reported ongoing abuse that undermined their psychological safety and their ability to focus on their work.

Survivors also described the harm they experienced when employers and co-workers do not understand the dynamics of abusive relationships and the impact on the health and safety: 'Victims may be further victimized at work (decisions made for them without consent, i.e. forced stress leave); Negative labels created gossip'.

For some, the harm of experiencing DV was compounded by bullying, resulting in even more negative health outcomes:

> Domestic violence combined with abusive bosses can be a horrible place to be. I did not have workplace support for my situation due to being a temporary worker at the time and to add to that, my manager routinely bullied and marginalized me in the workplace. It was awful and my health was permanently damaged as a result.

Others saw work as a place of safety and refuge:

> My work place was my safety zone. I knew that he would not show up there to abuse me in any way because then his behaviour would have become known to others outside our home. He worked very hard to show others 'what a good guy he was', but behind closed doors . . .

When DV threatened to spill over into the workplace, the health and safety of the co-workers of victims was also at risk. Some co-workers (10 per cent) had to deal with frequent phone calls, messages or emails from the abusive person and many (29 per cent) were stressed or concerned about their co-worker's DV. Although less common, co-workers were also reported to be harmed or threatened by the perpetrator: 'I could see how my situation could place others in danger and was lucky that none of the threats were brought forth or followed up.'

Co-workers suffered along with victims when abusers did come to work: 'It is very frightening for everyone at work when any spouse comes to the workplace and rages in front of everyone. Everyone, including staff, victim, and patients/clients/residents are at risk of both physical and psychological harm.'

As might be expected, the abuse had a ripple effect on work relationships: 'The domestic violence caused unease between me and my co-workers because I had to miss work or sometimes cried. Also, some people felt helpless; they would have liked to intercede, but did not dare for fear of endangering me or themselves.'

Workers themselves linked the psychosocial hazards of stress and distraction to a risk to safety:

> Domestic violence affects the personality of the people abused. It affects their concentration as they are continuously looking for an escape plan from where they are emotionally. They cannot focus safely on the task at hand due to stress and worry and fear of what is going on at home, the fear of returning home after a shift . . .

Some admitted to using substances to try to cope with the DV. Along with stress and fatigue, substance use can also create health and safety risks: 'Living a double life, trying to "act normal", pain, sorrow, fatigue, along with lowered concentration and effectiveness, feeling alienated, alcohol or substance abuse to self-medicate, shame etc. And a fear of losing maybe the only thing that gives meaning and purpose or some form of independence/pride . . . one's job.'

Overall, our survey confirmed that workplaces were seen as a place that could, but did not always, provide a safe and supportive response to DV survivors. Most respondents believed that workplace supports such as paid leave and safety policies for DV can reduce the impact of DV on the work lives of workers.

One respondent explained the benefit of viewing DV as an OHS issue:

> [W]e bring to work everything that happens at home. We can't compartmentalize or mentally separate these different aspects of our lives. While it might not technically be the responsibility of the employer or union to provide shelter or assistance for employees being victimized by abusers at home, the workplace is a logical place to provide help, support, and resources for victims of violence.

CANADIAN STUDY ON THE IMPACT OF DOMESTIC VIOLENCE PERPETRATION ON WORKERS AND WORKPLACES

Adding to the picture provided by the pan-Canadian survey of those experiencing DV, members of the DV@Work Net research group carried out an investigation of the impact of DV perpetration on workers and work-places in Ontario, Canada's most populous province (Scott et al. 2017). For this study, a geographically representative sample of participants was recruited from the Ontario's Partner Assault Response programmes; part of the province's criminal justice response to DV. The majority of the 501 respondents to this survey were heterosexual men (88 per cent). Participants completed survey questions measuring the degree to which DV perpetration occurred in the workplace, lost work productivity and time due to DV, as well as workplace response to DV perpetration issues.

Consistent with findings from victims, examination of responses from perpetrators found that DV persists into the workplace and that it is associated with substantial negative impacts on workplace safety and productivity. About one-third of respondents (34 per cent) reported being in contact with their (ex-)partner during work hours to engage in behaviours that were emotionally abusive (that is, to be hurtful or degrading, intimidate, threaten, or scare the (ex-)partner) or to monitor her actions or whereabouts). Of those who reported emotionally abusing or monitoring their partners during work hours, the vast majority used messages (that is, calls, emails and/or texts), though around a quarter reported going by their (ex)partner's workplace or home. Further, of the men who admitted to engaging in abusive behaviour within the workplace, almost 20 per cent had someone at work 'cover' for them while they engaged in these activities. As one man explained: 'All I could think about was what was going on at home and if my wife was even going to be there.' Of additional concern was the impact of DV perpetration on workplace safety and productivity as reported by respondents who had engaged in these behaviours: about one-tenth (9 per cent) reported that they caused or almost caused a work accident as a result of being distracted or preoccupied by DV issues. For example, one man explained that: 'I forgot that I was assigned six patients on day shift, so I missed one of them when it came time to administering medication. The incident caused me to be reprimanded and questioned by my manager.' Another reported 'accidentally dropping pieces of stone . . . off a forklift while trying to fight off tears'.

With regard to decreased work productivity, close to half of respondents (45 per cent) reported that DV issues at least sometimes affected their job performance, with around one in ten reporting that their productivity was impacted very often and to a very large degree. One man described that 'Anxiety, depression resulted from conflicts and I was unable to focus/concentrate on my work. When I did work, much of my work was substandard.' Another described high levels of irritability that led to loss of enjoyment in his work and to making those around him feel miserable. Respondents also wrote about texting and fighting with their partners 'all the time' or 'while running machines'. Absenteeism was another impact; around one-quarter to one-third of respondents indicated that DV issues led to difficulties getting to and staying at work, and similar proportions reported taking time off as a result of DV.

Domestic violence perpetration also impacts employee recruitment and retention. In the Canadian survey, over a quarter of respondents indicated that they had lost their job either directly (in half the cases) or indirectly for example, as a result of too many missed days or low productivity) as a result of DV perpetration. Those who did lose their jobs, many of whom had specific trades, skills or qualifications, reported substantial difficulties getting new work; they sometimes attributed these difficulties to having a charge on their record, but many other times reported that the informal spread of information through the community about their DV perpetration was a significant barrier to them. As one respondent explained: 'I'm looking to slow my career progression and get a day job. With charges against me, [it] will become harder to find new job opportunities, career paths I would like to pursue, but with a tarnished background check.'

The Canadian study, similar to the few previous studies on DV perpetration in the workplace (Lim et al. 2004; Rothman and Perry 2004; Galvez et al. 2011; Schmidt and Barnett 2011; Mankowski et al. 2013), used a sample of respondents from an intervention programme for DV perpetrators (often, as a result of referral from the justice system). Given this, one concern was whether reported workplace impacts were associated with DV perpetration itself, or whether they occurred as a result of having been charged with DV (for example, perhaps missed days were associated with court appearances). To explore this, men were asked to report on DV-related impacts both prior to and after the DV incident that led to their referral to intervention. Results found that DV issues were equally disruptive to workplaces across both time frames: an equal proportion of men indicated that DV affected their work productivity (approximately 20 per cent) and performance (approximately 30 per cent) both before and after the incident, and around one-quarter of men reported taking paid and/ or unpaid time off, with a median of around one to two weeks taken off before and after the DV incident that led to their referral to intervention.

Finally, respondents reported mixed views on the extent to which they felt workplaces should be involved in addressing DV issues, with some (40 per cent) expressing the view that workplaces should be more involved, but others either unsure (29 per cent) or unsupportive (24 per cent). A number of respondents commented that workplaces needed to be more informed about the nature of DV and the possibility of change. As expressed by one respondent: 'Domestic violence happens more and more these days and people are just in denial. Employees and employers need to keep an open mind. We're all human and make mistakes. It's what we do with those mistakes and learn from them.'

UNION RESPONSES TO DOMESTIC VIOLENCE IN THE WORKPLACE

Bolstered by findings from studies such as the Canadian surveys of the impact of DV victimization and perpetration on workers and workplaces, national and global union federations have played a key role in motivating change. As members of the DV@Work Net, various unions have collaborated with researchers to conduct national surveys, drawing upon their large memberships to solicit information about how DV impacts workers and their workplaces. Through this involvement, unions have started to make DV in the workplace a priority for training and education, for collective bargaining and for lobbying efforts (International Trade Union Confederation 2015, 2017a, 2017b; Yussuff and Byers 2015; Pillinger 2017; New Zealand Public Service Association 2018; Wibberley et al. 2018). In an increasing number of countries, the trade union movement, which had not previously played a leadership role in the protection of workers and members affected by DV, has successfully established DV as a national workplace issue that requires practical support, such as paid leave and safety planning for all workers (Aeberhard-Hodges and McFerran 2018).

The Australian Council of Trade Unions (ACTU) led this work, adopting collective bargaining in 2010 to introduce standardized rights for workers affected by DV. By 2016, they had won access to paid leave for over a third of all workers through collective bargaining, and launched a multiyear campaign to have the Fair Work Commission (FWC) implement a Modern Award that entitles employees experiencing family and domestic violence to ten paid days of leave (Aeberhard-Hodges and McFerran 2018). Although they were not successful in the first bid for paid leave, the ACTU did eventually win national access to family and DV leave with the FWC granting five days of unpaid leave in 2018 'if the employee needs to do something to deal with the impact of that violence and it is impractical for them to do it outside their ordinary hours of work' (FWC 2018, item 5).

Unions in Canada, Australia, New Zealand and countries in Europe and Asia Pacific also have plans to educate members, stewards and leaders about DV. For some, lobbying for change to labour legislation to provide paid leave and for OHS legislation to have DV recognized as a workplace hazard is also a priority. Many unions are invested in collective bargaining as a strategy to address workplace DV (Yussuff and Byers 2015; Pillinger 2017; Wibberley et al. 2018). The Canadian Labour Congress has a

Domestic Violence Resource Centre with a section specifically for leaders and negotiators that states, 'Collective bargaining can play an important role in keeping people safe and supported at work. It provides unions with a powerful tool to secure workplace support and policies and to hold employers accountable' (Canadian Labour Congress 2015). Unions in many other countries have also developed online resources about DV and the workplace (International Trade Union Confederation 2015, 2017a, 2017b; New Zealand Public Service Association 2018).

NATIONAL RESPONSES TO DOMESTIC VIOLENCE IN THE WORKPLACE

Augmenting the work of unions, federal and regional governments are increasingly requiring employers to take steps to protect employee-victims when DV spills over into the workplace (Lippel 2017; Pillinger in press). These obligations are integrated into laws on DV, gender equality law, labour law and labour codes and OHS law. They might contain provisions for prohibition of discrimination or retaliation against employees based on their status as a DV victim, paid or unpaid DV leave, security of employment and guarantee of a job following paid or unpaid leave, support for victims in the workplace and/or obligations for employers to take steps to ensure women's safety in the workplace (Pillinger in press). Argentina, France, Italy, the Philippines, Spain, Australia, New Zealand, some provinces in Canada and many states in the US provide legislated leave of absence from work. The length of the allowed leave varies, and in some cases it is paid and in others not. New Zealand has recently received a great deal of media attention for passing national legislation that not only gives victims of DV ten paid days away from work, but also allows victims to request flexible working arrangements and gives them protection from discrimination. In addition, in many countries, government regulations and Acts are being passed that require employers to identify and manage risks to health and safety. With increasing frequency, wording of these documents includes psychological risks stemming from another person's behaviour, including that of a current or an ex-partner or spouse, whether or not that is spelled out in the legislation (Clayton Utz 2018; Graham-McLay 2018; Pillinger in press). For example, in the US several states have passed Workplace Protection Order Statutes, enabling employers to apply for workplace protection orders as part of the duty of care to their workforce (Widiss, 2008).

Of all countries to date, Canada has made the most extensive use of OHS legislation to explicitly outline employer responsibilities for preventing and responding to DV in the workplace. Canadian governments' moves to introduce OHS legislation that addresses harassment, violence and domestic violence in the workplace has been supported by trade unions. It has also been influenced by the long-time work of women's advocates, international developments framing violence and harassment as OHS hazards, and tragedies such as the death of Lori Dupont and the associated inquests. In 2010 Bill 168 'An Act to amend the Occupational Health and Safety Act with respect to violence and harassment in the workplace and other matters' was passed into law in the province of Ontario (Ontario Ministry of Labour 2010). It specifically spelled out obligations for employers to prevent and respond to DV when it could cause physical harm in the workplace.

Recognizing the emerging trend, also in 2010, the Canadian Centre for Occupational Health and Safety (CCOHS), which provides guidance for jurisdictions across Canada, published the 3rd edition of its *Violence in the Workplace Prevention Guide*. This guide clearly identifies DV as a form of workplace violence and an OHS risk, outlining measures to recognize and respond to it (CCOHS 2010).

Shortly afterwards, in 2012, the British Columbia governmental agency responsible for OHS (WorkSafeBC 2018) launched a new education programme to help people address DV in the workplace (CBC News 2012). British Columbia OHS regulations also clearly spell out the responsibilities of employers to assess and eliminate risks, instruct workers and respond to incidents of DV (WorkSafeBC 2012).

Taken together, these laws and initiatives have influenced the interpretation of the OHS legislation in other jurisdictions in Canada. Many legal experts are of the opinion that although OHS laws in some Canadian jurisdictions do not specifically mention DV, 'those that set out general violence prevention obligations typically define "violence" broadly enough to cover domestic violence' (Jurczak and Mueller 2015).

Change is happening rapidly in Canada; in June 2018 Alberta passed new OHS legislation that imposes a positive duty on employers and supervisors to ensure that no workers are subject to or participate in harassment or violence at work. The new definitions of violence and harassment address physical, psychological and social well-being, including sexual and domestic violence (Alberta OHS 2018). New Brunswick is the most recent jurisdiction to announce new OHS regulatory changes that define harassment and violence, including DV, as workplace hazards, clearly signalling a new direction in the scope of Canadian OHS legislation (Canadian Occupational Safety Magazine 2018).

At the time of writing, the federal government is consulting on proposed changes to the Canada Occupational Health and Safety Regulations which would include measures requiring employers to outline in their prevention policies how they will respond to family violence. Proposed changes would have policies include obligations to provide risk assessments, put preventative measures in place, provide training to employees, identify a designated contact person and provide individual support to those experiencing DV (Government of Canada 2018).

INTERNATIONAL RESPONSES TO DOMESTIC VIOLENCE AT WORK

With countries increasingly acknowledging the role of the workplace as an entry point for addressing DV, the ILO has adopted a new international convention on violence and harassment in the world of work that includes reference to DV's workplace impacts. As a tripartite body of the United Nations, representing governments, employers and workers, the ILO exists to develop policies, devise programmes and set labour standards; that is, legal instruments drawn up by governments, employers and workers to promote opportunities for women and men to obtain decent and productive work, in conditions of freedom, equity, security and dignity (ILO 2018a).

The ILO first acknowledged gender-based violence at work as a concern in its occupational safety and health research and policy advice in 2002. Sexual harassment was the first issue identified (Di Martino et al. 2002; McCann 2005). Domestic violence appeared in 2006 in the groundbreaking book, *Violence at Work*. Domestic violence is described as a '"hidden issue" that has rapidly become a public one' with the 'potential to have a negative "spillover" impact on the workplace' (Chappell and Di Martino 2006, p. 17). In late 2012, the ILO started considering whether to place gender-based violence at work on the agenda of a future session of the International Labour Conference (ILC). Three years later, in November 2015, the Governing Body of the International Labour Office placed a standard-setting item on 'Violence against women and men in the world of work' on the agenda of the one hundred and seventh session (ILO 2018b) of the ILC. In preparation for this, a tripartite meeting of experts was held in Geneva in October 2016. Largely owing to the influence of members of DV@Work Net, DV and its impact on workers and the workplace was extensively discussed. The resultant meeting report recognized that, 'Domestic violence and other forms of violence and harassment are relevant to the world of work when they impact the workplace' (ILO 2016, p. 35). The conclusions also note that, 'Collective agreements could address the effects of domestic violence. The workplace provides an entry point to mitigate the effects, and employers could be allies to address such violence, though they are not responsible for it' (ILO 2016, p. 35).

In the two-year process of adopting a new labour standard, the ILC agreed that the standard would take the form of a convention and recommendation, which provides the strongest possible obligations for countries that ratify the new standard. Countries that ratify the new international treaty will have to develop laws that prohibit workplace violence and take preventative actions, such as addressing violence and harassment through their occupational health and safety management, and requiring companies to have policies on violence. The convention sets out standards for how governments should mitigate the effects of domestic violence in the world of work, including by having flexible working arrangements and leave for domestic violence survivors (International Labour Conference, 2019).

THE WAY FORWARD

Since US employers and researchers first identified DV as a threat to workplace safety, the pace of change has been relatively rapid. Governments, unions, employers, academics and advocates have worked independently and together to raise awareness, advance prevention and workplace education initiatives, develop policies, craft new legislation, interpret existing legislation in new ways and secure entitlements for workers through collective bargaining. Promising as these changes are, there is still a considerable amount of work to do.

Despite the undisputed success of Australian unions in gaining entitlements for workers experiencing DV through collective bargaining, and the burgeoning success of unions in other countries, some advocates have concluded that this strategy has not met the goal of introducing minimum terms and conditions (Aeberhard-Hodges and McFerran 2018). They suggest that concentration on the issue of leave has been to the detriment of the other equally important conditions, including flexible work arrangements, protection from discrimination, counselling support, safety planning, broader workplace education, and confidentiality. This neglect of conditions other than paid leave and the varied nature of the conditions being negotiated points to the need for minimum standards in national instruments such as employment law and OHS legislation (Aeberhard-Hodges and McFerran 2018). In addition, it is worth noting that new legislation, regulation and policies often remain silent on employer responsibilities to hold offenders accountable for their behaviour and to implement education and supports designed to prevent DV.

To be effective, laws need to be complemented by comprehensive policies and practical measures that tackle the root causes of violence and harassment and that strive to change social norms in the world of work. Workplaces have a key role to play in making laws relevant and applicable for workers, but this is a challenging task.

Workplace policies addressing DV need to take into consideration the unique aspects of the problem. Although DV shares qualities with other forms of workplace violence, the differences are important: DV involves persistent behaviours and may result in a chronic pattern of crises where co-workers and management blame the victim for having brought the violence into the workplace. The still prevalent perception that DV is primarily a personal problem may result in victims, or co-workers, being ambivalent or hesitant to disclose safety concerns. This suggests the need for workplace education programmes that can help supervisors and co-workers understand the dynamics of DV, to recognize warning signs and risk factors, as well as prepare them to provide supportive response to DV disclosures. Our Canadian survey showed that despite ongoing attitudes that DV is a personal problem, 43 per cent of survivors discussed it with someone at work, with almost half telling more than one person. Those who did not disclose were ashamed, embarrassed, afraid of being judged, believed it was a private matter or did not want to get others involved. Unfortunately, some of these fears around disclosure are not unfounded and 19 per cent of those who talked to someone said they experienced judgement or a lack of confidentiality. Survivors of DV are much more likely to talk to co-workers (82 per cent) than to supervisors or managers (45 per cent) or to HR representatives (11 per cent), underlining the importance of providing training to everyone in the workplace. Employers and unions can benefit from the experience and expertise of specialized DV services and women's advocates as they navigate these often complex and potentially dangerous situations (MacGregor et al. 2016, 2017; Saxton et al. 2018).

Ongoing work is also needed to develop and improve policy and practices for addressing DV perpetration. In the Canadian DV perpetration survey, men reported minimal workplace knowledge of their DV situation despite the fact that all the men surveyed were attending an intervention programme and the majority had been arrested for DV perpetration. Specifically, almost two-thirds of men (61 per cent) did not talk about issues relating to their DV situation with people at their workplace. Of those respondents who did, the majority spoke with a co-worker (79 per cent) or supervisor (59 per cent); very few talked to a union, worker association or to the HR, or personnel, department. Men were frank about why they kept this information private – they felt embarassed or ashamed, wanted privacy, felt that that is was 'none of their workplace's business' and they feared judgement and job loss. 'You can't talk to anyone, you are stressed, . . . depression [is] high, no resources, you are completely alone.' Moreover, almost half of the men surveyed in the Canadian study felt that their workplace was closed to these kind of discussions; that a fairly clear message was conveyed that such issues should not be 'brought to work'.

Policies that embrace zero tolerance can also be problematic, and some experts are already raising alarm about the potential use of DV policies that lead to dismissal as the only possible outcome. As outlined in a thought-provoking article on the *Make It Our Business* blog site:

> Zero tolerance policies can allow leaders to wash their hands of what they see as the 'problem' individual, without appreciating or exploring how problematic individual behaviour is rooted in attitudes and beliefs that also exist in our workplaces . . . When the 'problem' person is . . . fired, the organization hasn't done anything to address abusive behaviour in intimate relationships . . . The 'one bad apple' approach to domestic violence obscures the social context in which these behaviours thrive. (MacPherson 2018)

Another problem with this approach is that it can, ironically, place victims at heightened risk of victimization. In the context of other risk factors, unemployment among DV perpetrators is a recognized risk factor for serious and lethal DV (Domestic Violence Death Review Committee 2017). Victims who do not want to be the cause of their partner being fired can be put in the untenable and unsafe position of hiding the violence to save his job. In the Canadian survey, 13 per cent of respondents reported losing their job as a direct result of their DV perpetration. An additional 13 per cent reported job loss as an indirect (for example, too many missed days or poor productivity) result. In addition, many men, including those who were self-employed or contractually employed, commented that DV issues have made it more difficult to get new work.

While job loss as a result of DV perpetration is a concern, there are many, many stories from DV victims being ignored with no actions taken: clearly, changes are needed. Workplaces and community agencies with expertise in this area need to work collaboratively to chart a better path of response that supports victims and creates multiple paths to accountability.

In conclusion, the past 20 years have seen extremely rapid change in the way that DV is viewed by and within workplaces. Research evidence has established that DV is an issue for safety and productivity in the workplace and, more recently, OHS legislation is beginning to identify DV as a hazard. The challenge that lies ahead is where to focus efforts and how to work collaboratively to make the best possible use of opportunities to reduce DV, and its impacts, in workplaces and beyond.

NOTES

* We would like to acknowledge Melanie Stone, PhD candidate, Women's Studies and Feminist Research, Western University, and Natalia Musielak, MA Counselling Psychology candidate, 2018, Western University, for the research work they conducted for this chapter.
1. Different acronyms are used in North America and Europe. In North America, OHS (occupational health and safety) is used, whereas in Europe, OSH (occupational safety and health) is used.

REFERENCES

Aeberhard-Hodges, J. and McFerran, L. (2018), 'An international labour organization instrument on violence against women and men at work: the Australian influence', *Journal of Industrial Relations*, **60** (2), 246–65.

Alberta Occupational Health and Safety (Alberta OHS) (2018), 'Occupational health and safety changes', Government of Alberta, Edmonton, accessed 31 August 2018 at https://www.alberta.ca/ohs-changes.aspx.

Alphonso, C. (2018), 'Trial opens in Vancouver in Starbucks hero's death', *The Globe and Mail*, 10 April, accessed 18 September 2018 at https://www.theglobeandmail.com/news/national/trial-opens-in-vancouver-in-starbucks-heros-death/article4146024/.

Ararat, M., Alkan, S., Bayazit, M., Yuksel, A. and Budan, P. (2014), 'Domestic violence against white-collar working women in Turkey', Sabancı University, Istanbul, accessed 16 September 2018 at http://dvatworknet.org/sites/dvatworknet.org/files/Turkey_survey_report_%202014.pdf

Brampton Guardian (2007), 'Husband guilty of murder; sentenced to life in prison', *Brampton Guardian*, 2 April, accessed 21 September 2018 at https://www.bramptonguardian.com/news-story/3073681-husband-guilty-of-murder-sentenced-to-life-in-prison/.

Brown, T. (2002), 'Charging and prosecution policies in cases of spousal assault: a synthesis of research, academic, and judicial responses', Department of Justice Canada, Ottawa, accessed 29 April 2019 at http://www.justice.gc.ca/eng/rp-pr/csj-sjc/jsp-sjp/rr01_5/rr01_5.pdf.

Canadian Centre for Occupational Health and Safety (CCOHS) (2010), *Violence in the Workplace Prevention Guide*, 3rd edn, Hamilton: CCOHS, accessed 29 April 2019 at https://www.ccohs.ca/products/publications/pdf/samples/violence.pdf.

Canadian Labour Congress (2015), 'Leaders and negotiators', Canadian Labour Congress, Ottawa, accessed 3 October 2018 at http://canadianlabour.ca/leaders-and-negotiators.

Canadian Occupational Safety Magazine (2018), 'New Brunswick passes workplace violence, harassment legislation', *Canadian Occupational Safety Magazine*, 17 September, accessed 30 April 2019 at https://www.cos-mag.com/psychological-health-safety/37991-new-brunswick-passes-workplace-violence-harassment-legislation/.

CBC News (2007), 'Slain Windsor nurse often harassed by doctor, colleague testifies', *CBC News*, 27 September, accessed 16 September 2018 at https://www.cbc.ca/news/

canada/toronto/slain-windsor-nurse-often-harrassed-by-doctor-colleague-testi
fies-1.664590.

CBC News (2012), 'Starbucks' hero motivates violence prevention toolkit', *CBC News*,
22 March, accessed 7 September 2018 at https://www.cbc.ca/news/canada/british-
columbia/starbucks-hero-motivates-violence-prevention-toolkit-1.1287973.

Chappell, D. and Di Martino, V. (2006), *Violence at Work*, 3rd edn, Geneva:
International Labour Office, accessed 29 April 2019 at https://www.ilo.org/global/
publications/ilo-bookstore/order-online/books/WCMS_PUBL_9221108406_EN/
lang--en/index.htm.

Clayton Utz (2018), 'Compulsory family and domestic violence leave is on the
way for Modern Awards', *Clayton Utz*, 27 March, accessed 4 September 2018
at https://www.claytonutz.com/knowledge/2018/march/compulsory-family-and-
domestic-violence-leave-is-on-the-way-for-modern-awards.

Cox, T. and Griffiths, A. (2005), 'The nature and measurement of work stress:
theory and practice', in J.R. Wilson and Ni.E. Corlett (eds), *Evaluation of
Human Work*, 3rd edn, London: CRS Press, pp. 783–803.

De Jonge, A. (2018), 'Corporate social responsibility through a feminist lens:
domestic violence and the workplace in the 21st century', *Journal of Business
Ethics*, **148** (3), 471–87.

Department of Justice Canada (2015), 'Final report of the Ad Hoc Federal-
Provincial-Territorial Working Group reviewing spousal abuse policies and
legislation', Department of Justice Canada, Ottawa, accessed 18 September 2018
at http://www.justice.gc.ca/eng/rp-pr/cj-jp/fv-vf/pol/spo_e-con_a.pdf.

Di Martino, V., Gold, D. and Schaap, A. (2002), 'Managing emerging health-
related problems at work: SOLVE training package', International Labour
Organization, Geneva.

Domestic Violence at Work Network (2018a), 'Dv@Work Net: generating knowledge on the impacts of domestic violence in the workplace', accessed 4
September 2018 at http://dvatworknet.org/.

Domestic Violence at Work Network (2018b), DV@Work Net: basis of unity,
accessed 7 September 2018 at http://dvatworknet.org/about/basis-of-unity.

Domestic Violence Death Review Committee (2017), 'Domestic Violence Death
Review Committee: 2016 annual report', Office of the Chief Coroner, Toronto,
accessed 10 October 2018 at https://www.mcscs.jus.gov.on.ca/sites/default/files/con
tent/mcscs/docs/2016%20DVDRC%20Annual%20Report%20Accessible%20%2
8ENGLISH%29.pdf.

Duda, R.A. (1997), 'Workplace domestic violence: Intervention through program
and policy development', *AAOHN Journal*, **45** (12), 619–24.

European Agency for Safety and Health at Work (EU-OSHA) (2007), *European
Risk Observatory Report: Expert Forecast on Emerging Psychosocial Risks
Related to Occupational Safety and Health*, Luxembourg: European Agency for
Safety and Health at Work, accessed 31 August 2018 at https://osha.europa.eu/
en/tools-and-publications/publications/reports/7807118.

European Agency for Safety and Health at Work (EU-OSHA) (2018), 'European
Agency for Safety and Health at Work', accessed 17 October 2018 at https://osha.
europa.eu/en.

Fair Work Commission (FWC) (2018), 'Decision: 4 yearly review of modern awards
– family and domestic violence leave', Melbourne, 6 July, accessed 3 October 2018
at https://www.fwc.gov.au/documents/decisionssigned/html/2018fwcfb3936.htm.

Galvez, G., Mankowski, E.S., McGlade, M.S., Ruiz, M.E. and Glass, N. (2011),

'Work-related intimate partner violence among employed immigrants from Mexico', *Psychology of Men & Masculinity*, **12** (3), 230–46.

Goodhand, M. (2017), *Runaway Wives and Rogue Feminists: The Origins of the Women's Shelter Movement in Canada*, Halifax, NS: Fernwood.

Government of Canada (2018), 'Labour Program Forward Regulatory Plan from 2018 to 2020', Government of Canada, Ottawa, accessed 11 September 2018 at https://www.canada.ca/en/employment-social-development/programs/laws-regulations/labour/forward-regulatory-plan/labour-forward-regulatory-plan.html.

Graham-McLay, C. (2018), 'New Zealand grants domestic violence victims paid leave', *New York Times*, 26 July, accessed 24 September 2018 at https://www.nytimes.com/2018/07/26/world/asia/new-zealand-domestic-violence-leave.html.

Health Canada (2002), *The Family Violence Initiative Year Five Report*, Ottawa: Health Canada, accessed 22 May 2019 at http://publications.gc.ca/collections/Collection/H72-2-2004E.pdf.

Institute for the Equality of Women and Men (2017), 'National survey results on the impact of domestic violence on work, workers and workplaces in Belgium', Domestic Violence at Work Network and Belgian Institute for Equality of Women and Men, accessed 18 September 2018 at http://dvatworknet.org/sites/dvatworknet.org/files/dvatwork-belgium-surveyresults-Sept18-2017.pdf.

International Labour Conference (2019), 'Convention 190. Concerning the Elimination of Violence and Harassment in the World of Work', Geneva: International Labour Organization, accessed 1 July 2019 at https://www.ilo.org/wcmsp5/groups/public/---ed_norm/---relconf/documents/meetingdocument/wcms_711570.pdf

International Labour Organization (ILO) (2016), 'Meeting of experts on violence against women and men in the world of work', International Labour Office, Conditions of Work and Equality Department, Geneva, accessed 11 September 2018 at https://www.ilo.org/wcmsp5/groups/public/---dgreports/---gender/documents/meetingdocument/wcms_546303.pdf.

International Labour Organization (ILO) (2018a), 'Introduction to International Labour Standards', accessed 19 September 2018 at https://www.ilo.org/global/standards/introduction-to-international-labour-standards/lang--en/index.htm.

International Labour Organization (2018b), 'Violence and harassment against women and men in the world of work', 8 June, provisional record at International Labour Conference, Geneva, accessed 13 September 2018 at https://www.ilo.org/wcmsp5/groups/public/---ed_norm/---relconf/documents/meetingdocument/wcms_631787.pdf.

International Trade Union Confederation (2015), 'Key findings of national survey on the impact of domestic violence on workers and in workplaces in the Philippines', International Trade Union Confederation – Asia Pacific and Philippine Affiliates, 23 September, accessed 21 September 2018 at https://www.ituc-csi.org/IMG/pdf/philippine_domestic_violence_survey_key_findings_23_september_2015.pdf.

International Trade Union Confederation (2017a), 'National survey results on the impact of domestic violence on workers and workplaces in Mongolia', International Trade Union Confederation – Asia Pacific and Confederation of Mongolian Trade Unions, accessed 27 September 2018 at https://www.ituc-csi.org/IMG/pdf/dvatwork-mongolian-surveyresults.pdf.

International Trade Union Confederation (2017b), 'National survey results on the impact of domestic violence on work, workers and workplaces in Taiwan', International Trade Union Confederation – Asia Pacific and Chinese Federation of Labour, accessed 3 October 2018 at https://www.ituc-csi.org/IMG/pdf/dvat work-taiwanese-surveyresults.pdf.

International Transport Workers' Federation (2018), 'International law on violence: ITF pushes work forward', 26 June, accessed 24 September 2018 at http://www. itfglobal.org/en/news-events/news/2018/june/international-law-on-violence-itf-pushes-work-forward/.

Johnson, H. and Dawson, M. (2011), *Violence Against Women: Research and Policy Perspectives*, Don Mills, ON: Oxford University Press.

Jurczak, J. and Mueller, P. (2015), 'Domestic violence is a workplace issue', *OHS Insider*, 25 February, accessed 7 September 2018 at https://ohsinsider.com/topics/ workplace-violence/domestic-violence-workplace-issue.

Kinney, J.A. (1995), 'When domestic violence strikes the workplace', *HR Magazine*, **40** (8), 74–8.

Lim, K.C., Rioux, J. and Ridley, E. (2004), 'Impact of domestic violence offenders on occupational safety and health: a pilot study', Maine Department of Labour, Augusta, ME, accessed 17 October 2018 at https://www.maine.gov/labor/ labor_stats/publications/dvreports/domesticoffendersreport.pdf.

Lindquist, C.H., McKay, T., Clinton-Sherrod, A.M., Pollack, K.M., Lasater, B.M. and Walters, J.L.H. (2010), 'The role of employee assistance programs in workplace-based intimate partner violence intervention and prevention activities', *Journal of Workplace Behavioral Health*, **25** (1), 46–64.

Lippel, K. (2017), 'Ending violence and harassment against women and men in the world of work', report submitted to International Labour Conference, Geneva, 12 May, accessed 4 September 2018 at https://www.ilo.org/wcmsp5/groups/public/---ed_norm/---relconf/documents/meetingdocument/wcms_553577.pdf.

Lippel, K. (2018), 'Ending violence and harassment in the world of work', report submitted to International Labour Conference, Geneva, 7 March, accessed 27 September 2018 at https://www.ilo.org/wcmsp5/groups/public/---ed_norm/---relconf/documents/meetingdocument/wcms_619730.pdf.

MacGregor, J.C.D., Oliver, C., MacQuarrie, B.J. and Wathen, C.N. (forthcoming), 'Intimate partner violence and work: a scoping review of published research', manuscript under review.

MacGregor, J.C.D., Wathen, C.N. and MacQuarrie, B.J. (2017), 'Resources for domestic violence in the Canadian workplace: results of a pan-Canadian survey', *Journal of Workplace Behavioral Health*, **32** (3), 190–205.

MacGregor, J.C.D., Wathen, C.N., Olszowy, L., Saxton, M. and MacQuarrie, B.J. (2016), 'Gender differences in workplace disclosure and supports for domestic violence: Results of a pan-Canadian survey', *Violence and Victims*, **31** (6), 1135–54.

MacPherson, M. (2018), 'Zero tolerance complicit', *Make it Our Business* blog, 14 August, accessed 30 April 2019 at http://makeitourbusiness.ca/blog/zero-tolerance-complicit.

Malecha, A. and Wachs, J.E. (2003), 'Screening for and treating intimate partner violence in the workplace', *AAOHN Journal*, **51** (7), 310–16.

Mankowski, E.S., Galvez, G., Perrin, N. A., Hanson, G.C. and Glass, N. (2013), 'Patterns of work-related intimate partner violence and job performance among abusive men', *Journal of Interpersonal Violence*, **28** (15), 3041–58.

McCann, D. (2005), 'Sexual harassment at work: national and international responses', International Labour Organization, accessed at https://www.ilo.org/ wcmsp5/groups/public/---ed_protect/---protrav/---travail/documents/publication/ wcms_travail_pub_2.pdf.

McFarlane, J., Malecha, A., Gist, J., Schultz, P., Willson, P. and Fredland, N. (2000), 'Indicators of intimate partner violence in women's employment: Implications for workplace action', *AAOHN Journal*, **48** (5), 215–20.
McFerran, L. (2011), 'Safe at home, safe at work? National domestic violence and the workplace survey', University of New South Wales, Sydney, accessed 11 September 2018 at https://www.arts.unsw.edu.au/media/FASSFile/National_Domestic_Vio lence_and_the_Workplace_Survey_2011_Full_Report.pdf.
Mighty, E.J. (1997), 'Conceptualizing family violence as a workplace issue: A framework for research and practice', *Employee Responsibilities and Rights Journal*, **10** (4), 249–62.
New Zealand Public Service Association (2018), 'Family violence and workplaces', accessed 13 September 2018 at https://www.psa.org.nz/media/campaigns/family-violence-and-new-zealand-workplaces.
Ontario Ministry of Labour (2010), *Bill 168, Occupational Health and Safety Amendment Act (Violence and Harassment in the Workplace) 2009*, Legislative Assembly of Ontario, accessed 1 September 2018 at https://www.ola.org/en/legisla tive-business/bills/parliament-39/session-1/bill-168.
Personnel Journal (1994), 'What you thought – domestic violence is a workplace issue', *Personnel Journal*, **73** (11), 14.
Pillinger, J. (2017), 'Safe at home, safe at work: trade union strategies to prevent, manage, and eliminate workplace harassment and violence against women', European Trade Union Confederation, Brussels, accessed 27 September 2018 at https://www.etuc.org/sites/default/files/document/files/en_-_brochure_-_safe_at_ home_1.pdf.
Pillinger, J., Carlson, E., Proios Torras, I. and Deligiorgis, D. (2019), *Handbook: Addressing Violence and Harassment against Women in the World of Work*, New York: UN Women, accessed 19 June 2019 at http://www.world-psi.org/sites/ default/files/documents/research/work-handbook-web.pdf.
Pollack, K.M., Cummiskey, C., Krotki, K., Salomon, M., Dickin, A., Gray, W.A. et al. (2010), 'Reasons women experiencing intimate partner violence seek assistance from employee assistance programs', *Journal of Workplace Behavioral Health*, **25** (3), 181–94.
Public Health Agency of Canada (2016), *The Chief Public Health Officer's Report on the State of Public Heath in Canada 2016: A Focus on Family Violence in Canada*, Ottawa, ON: Public Health Agency of Canada, accessed 19 September 2018 at http://www.healthycanadians.gc.ca/publications/department-ministere/state-public-health-family-violence-2016-etat-sante-publique-violence-familiale/alt/pdf -eng.pdf.
Rayner-Thomas, M.M. (2013), 'The impacts of domestic violence on workers and the workplace', Master's thesis, University of Auckland, accessed 16 September 2018 at http://dvatworknet.org/sites/dvatworknet.org/files/New%20Zealand_survey _report_2013.pdf.
Reeves, C. and O'Leary-Kelly, A.M. (2007), 'The effects and costs of intimate partner violence for work organizations', *Journal of Interpersonal Violence*, **22** (3), 327–44.
Rothman, E.F. and Corso, P.S. (2008), 'Propensity for intimate partner abuse and workplace productivity: why employers should care', *Violence Against Women*, **14** (9), 1054–64.
Rothman, E.F. and Perry, M.J. (2004), 'Intimate partner abuse perpetrated by employees', *Journal of Occupational Health Psychology*, **9** (3), 238–46.

Saxton, M.D., Olszowy, L., MacGregor, J.C.D., MacQuarrie, B.J. and Wathen, C.N. (2018), 'Experiences of intimate partner violence victims with police and the justice system in Canada', *Journal of Interpersonal Violence*, February, 1–27, doi:10.1177/0886260518758330.

Schmidt, M.C. and Barnett, A. (2011), 'How does domestic violence affect the Vermont workplace? A survey of male offenders enrolled in batterer intervention programs in Vermont', Vermont Council on Domestic Violence, Montpelier, VT, accessed 16 September 2018 at https://www.uvm.edu/crs/reports/2012/VTDV_WorkplaceStudy2012.pdf.

Scott, K.L., Lim, D.B., Kelly, T., Holmes, M., MacQuarrie, B.J., Wathen, C.N. et al. (2017), 'Domestic violence at the workplace: investigating the impact of domestic violence perpetration on workers and workplaces', University of Toronto, accessed 31 August 2018 at http://dvatworknet.org/sites/dvatworknet.org/files/PAR_Partner_report-Oct-23-2017dl.pdf.

Solomon, C.M. (1998), 'Picture this: a safer workplace', *Workforce*, **77** (2), 82–6.

Swanberg, J.E., Logan, T.K. and Macke, C. (2005), 'Intimate partner violence, employment, and the workplace: consequences and future directions', *Trauma, Violence, and Abuse: A Review Journal*, **6** (4), 286–312.

Trade Union Congress (2014), 'Domestic violence and the workplace: a TUC survey report', Trade Union Congress, London, accessed 13 September 2018 at http://dvatworknet.org/sites/dvatworknet.org/files/Britain_survey_report_2014.pdf.

Wathen, C.N., MacGregor, J.C.D. and MacQuarrie, B.J. (2014), 'Can work be safe, when home isn't? Initial findings of a pan-Canadian survey on domestic violence and the workplace', Centre for Research and Education on Violence Against Women and Children and Canadian Labour Congress, London, ON, accessed 12 September 2018 at http://dvatworknet.org/sites/dvatworknet.org/files/Canada_survey_report_2014_EN_0.pdf.

Wathen, C.N., MacGregor, J.C.D. and MacQuarrie, B.J. (2015), 'The impact of domestic violence in the workplace: results from a pan-Canadian survey', *Journal of Occupational and Environmental Medicine*, **57** (7), 65–71.

Wibberley, G., Bennett, T., Jones, C. and Hollinrake, A. (2018), 'The role of trade unions in supporting victims of domestic violence in the workplace', *Industrial Relations Journal*, **49** (1), 69–85.

Widiss, D.A. (2008), 'Domestic violence and the workplace: the explosion of state legislation and the need for a comprehensive strategy', *Florida State University Law Review*, **35** (3), 669–728.

WorkSafeBC (2012), 'Addressing domestic violence in the workplace: a handbook for employers outside of B.C.', Workers' Compensation Board of British Columbia, accessed 11 September 2018 at https://www.worksafebc.com/en/resources/health-safety/books-guides/addressing-domestic-violence-in-the-workplace-a-handbook-for-employers-outside-of-bc?lang=en.

WorkSafeBC (2018), WorkSafeBC website, accessed 27 September 2018 at https://www.worksafebc.com/en.

Yussuff, H. and Byers, B. (2015), 'Follow-up to domestic violence at work survey', Canadian Labour Congress, briefing note to Members of the Canadian Council, Women's Advisory Committee, 1 October.

6. Job resources and outcomes in the process of bullying: a study in a Norwegian healthcare setting

Espen Olsen, Maria Therese Jensen, Gunhild Bjaalid and Aslaug Mikkelsen

INTRODUCTION

Over the past 20 years, there has been an extensive amount of research on bullying. However, theoretical explanations and frameworks for explaining predictors of bullying have been relatively neglected (Hewett et al. 2018). We know that bullying is associated with several negative outcomes such as reduced mental health, and negative physiological stress responses (Hogh et al. 2012). A meta-analytic review concluded that bullying is associated with both job-related and health- and well-being-related outcomes, such as mental and physical health problems, symptoms of post-traumatic stress, burnout, increased intentions to leave, and reduced job satisfaction and organizational commitment (Nielsen and Einarsen 2012).

Based on a review, Branch et al. concluded that,

> as a phenomenon, workplace bullying is now better understood with reasonably consistent research findings in relation to its prevalence; its negative effects on targets, bystanders and organizational effectiveness; and some of its likely antecedents. However [. . .] many challenges remain, particularly in relation to its theoretical foundations and efficacy of prevention and management strategies. (Branch et al. 2013, p. 280)

The aim of the current study is to contribute to the development of theory that can strengthen the understanding of both antecedents and outcomes of bullying. An essential question is whether reducing job demands or increasing job resources is the most rewarding in preventing bullying. Schaufeli (2015) recommends the latter and argues that increasing job resources is normally a more feasible strategy compared with reducing job demands, such as reducing treatment needs in populations. Thus, in this study we primarily focus on job resources when investigating antecedents

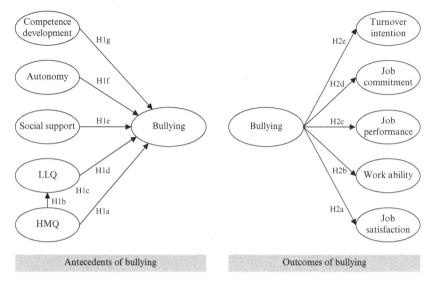

Note: LLQ = local leadership quality; HMQ = hospital management quality.

Figure 6.1 Theoretical models underlying the current study

of bullying. Studies on job resources often lack management and leadership factors (Schaufeli 2015). Hence, in the current study we include two new job resources which might be related to bullying. These are labeled hospital management quality (HMQ) and local leadership quality (LLQ). Both are integrated in our theoretical model together with three job resources: competence development, autonomy and social support. We expect to see that the factors included in our study model have the potential to explain workplace bullying, and give new direction to management strategies at different levels in the healthcare sector. Finally, we validate the bullying indicator applied, by testing a separate bullying outcome model. The underlying theoretical models of the current study are presented in Figure 6.1.

THEORETICAL BACKGROUND

Bullying can be seen as repeated and persistent negative acts towards one or more individual(s), which involve a perceived power imbalance and create a hostile work environment (Einarsen and Skogstad 1996; Zapf et al. 1996). A more precise definition of bullying is:

Bullying at work means harassing, offending, or socially excluding someone or negatively affecting someone's work. In order for the label bullying (or mobbing) to be applied to a particular activity, interaction or process, the bullying behavior has to occur repeatedly and regularly (e.g. weekly) and over a period of time (e.g. about six months). Bullying is an escalating process in the course of which the person confronted ends up in an inferior position and becomes the target of systematic negative social acts. A conflict cannot be called bullying if the incident is an isolated event or if two parties of approximately equal strength are in conflict. (Einarsen et al. 2010, p. 22)

Research on the causes and antecedents of workplace bullying has in general been approached from two perspectives. In the first, emphasis has been put on individual characteristics of targets and perpetrators. For instance, as regards the target, bullying has been related to personal characteristics such as low social skills (Zapf et al. 1996), shyness (Einarsen et al. 1994) and neuroticism (Mikkelsen and Einarsen 2002). Regarding the perpetrator, bullying has been related to various personality types as authoritarian and tyrannical (for example, Ashforth 1994). The second and most dominant perspective is inspired by the work environment hypothesis (Hauge et al. 2007), where bullying is attributed to organizational and workplace characteristics and a stressful work environment (Hauge et al. 2007). In line with research on the work environment hypothesis, it has been recognized that organizational factors play a vital role in the development of bullying in organizations. The basis of this view is that when organizations are exposed to internal or external pressures, it will create good conditions for conflicts and aggressive behavior (for example, Einarsen et al. 1994; Zapf 1999; Salin, 2003).

One of the worst psychosocial work climates an organization can offer employees consists of high levels of conflicts and bullying (Einarsen 2000). In a meta-analysis based on 90 studies, Bowling and Behr (2006, p. 1005) revealed that 'characteristic of the work environment might strongly contribute to workplace harassment'.

Baillien et al. (2009) suggest that work characteristics can indirectly and directly stimulate bullying, by allowing, or even promoting, bullying behavior (Baillien et al. 2009).

Psychosocial work environment can be defined as the socio-structural range of opportunities that are available to an individual employee to meet his or her needs of well-being, productivity and positive self- experience in the work environment (Peter et al. 2002). Agervold (2009) found that a poor psychosocial work environment correlated positively with workplace bullying.

Risk factors that might contribute to a poor psychosocial work environment, and thereby an increased risk of bullying, include high psychological job demands (high workload and work pressure), work decision latitude (lack of control over the work tasks and how to perform them) and lack of social support from colleagues and supervisors. Other risk factors include an experienced imbalance between high effort spent at work and low reward received, in addition to lack of meaningful work tasks. Whether the workers perceive if they are treated in a fair, polite, transparent and considerate manner by supervisors and co-workers, may also be a determinant of a poor psychosocial work environment (Agervold and Mikkelsen 2004).

The major difference between normal conflict expected to occur in every organization at times, and bullying, is not necessarily what and how it is done, but rather the frequency and longevity of what is done (Salin 2003). Einarsen and Skogstad (1996) stress that bullying is a repeated, persistent and continuous behavior. The structural and organizational reasons behind bullying can seldom be explained by one factor alone, but is instead a multi-causal phenomenon (Zapf 1999). Generally, leaders and managers have little control over, and ability to change workers' individual personality traits and characteristics. Thus, from a management perspective it is important to emphasize the organizational, contextual and work environment factors, and to some extent group-level factors, when trying to prevent bullying. Work environment factors are to higher degree under the control of management, who may exert considerable influence on organizational structures, workflow, reward systems and job design. Also, leaders are able to decide with what types of sanctions and consequences the organization meets negative behavior and bullying (Salin 2003).

Prospective studies have shown that work characteristics and work environment factors influence bullying levels and result in health problems among employees in healthcare (Reknes et al. 2014; Rosta and Aasland 2018). Physicians and nurses berating less experienced physicians and nurses, and supervisors' publicly belittling staff, are all common examples of negative behavior reported regularly in the hospital sector (Martin 2008). In a study of hospital workers in the USA including 1500 nurses and physicians in 12 different states, 68 percent of the nurses and 47 percent of the physicians reported witnessing disruptive behavior in which fellow hospital workers targeted other hospital workers (Rosenstein 2002).

Over time, exposure to bullying relates negatively to mental health. Furthermore, the relationship between bullying and mental health are reciprocally related, meaning that mental health problem over time may also increase the risk of experiencing bullying (Reknes et al. 2014). The challenge is therefore to develop strategies to break this negative cycle. Changing individual characteristics might not be very beneficial since individual characteristics, similar to personality traits (Cobb-Clark and Schurer 2012) and cognitive abilities (Johnson et al. 2011), tend to be stable throughout the lifespan. Furthermore, in addition to focusing on job resources, Schaufeli (2015) recommends more studies exploring direct and indirect influence from leadership. Based on these recommendations, this study integrates factors related to management, leadership and job resources when studying bullying. Hopefully, these combination of factors can contribute to an increased understanding of antecedents of bullying.

Management and Leadership Quality

Management and leadership can be considered both a resource and a threat when studying bullying. For instance, in a large-scale study, Hoel and Cooper (2000) reported that employees who experienced being bullied perceived pressure of work as high, and that management and the social climate were poorer compared with colleagues' who were not bullied. However, research on relationships between leadership styles and bullying has been relatively limited (Hoel et al. 2010). Still, in the few studies conducted, results have shown that various leadership styles can potentially influence bullying. Specifically, Hoel et al. (2010) found that autocratic leadership, which is characterized by a willingness to apply force to achieve organizational goals, was most strongly associated with bullying. Other studies have shown that authentic (Avolio and Gardner, 2005) and transformational leadership (Yukl 1999) decrease the risk of exposure to bullying (Laschinger and Fida 2014; Nielsen et al. 2014).

Managers and leaders are part of, or can influence, work characteristics and the psychosocial work environment, which again might be related to bullying. Among nurses and physicians, it has been documented that work characteristics and psychosocial work environment factors influence the risk of bullying (Reknes et al. 2014; Rosta and Aasland 2018). We therefore expect that management and leaders, either directly or indirectly, can influence both job resources and exposure to bullying. Thus, as previously noted, we included HMQ and LLQ in our study model. The development of these concepts was based on earlier research (Karasek et al. 1982; Aasland 1994; Krogstad et al. 2004). In this study, HMQ is defined as hospital managerial skills related to (1) conducting correct prioritization, (2) obtaining adequate knowledge at the institutional levels and (3) having knowledge at the regional level which constitutes the hospital context. At the lower level, we assume that other types of leadership skills are required. Hence, we suggest that three leadership elements are the most important for LLQ, including (1) relational skills to retain workers, (2) having overview and knowledge at the local hospital level and (3) having the ability to develop employees.

Specifically, we aimed to investigate whether HMQ is directly associated with bullying. Further, we expect to see that HMQ relates indirectly to bullying through LLQ and, finally, that LLQ relates directly to bullying. Accordingly, we propose the following hypotheses:

Hypothesis 1a: HMQ is negatively related to bullying.

Hypothesis 1b: HMQ is positively related to LLQ.

Hypothesis 1c: LLQ will mediate the association between HMQ and self-reported bullying.

Hypothesis 1d: LLQ is negatively related to bullying.

Job Resources

As previous research found that when basic psychological needs are met in the workplace, this will result in several positive organizational outcomes (Meyer and Maltin 2010; Deci and Ryan 2012; Van den Broeck et al. 2016) we expect that employees' satisfaction with competence development, autonomy and social support from colleagues will prevent and relate negatively to bullying. Autonomy can be defined as the capacity of an agent to determine its own actions through independent choices within a system of principles and laws to which the agent is dedicated (Ballou 1998). Previous studies demonstrated that low autonomy and influence over work were related to bullying (Vartia, 1996; Agervold and Mikkelsen 2004; Trepanier et al. 2013).

Workers have an inherent desire to feel effective in interacting with the environment, which defines the need for competence (Deci and Ryan 2000). Similar to social support and autonomy, we consider competence to be a job resource. This further implies that we expect to see that employees who experience that their competence is stimulated and approved of, will also experience less bullying.

Social support can be defined as the desire of an individual to experience a sense of social connectedness or to be valued, helped or appreciated by other individuals (Van der Heijden 1998). Previous studies have demonstrated that social support may function as a buffering role between victimization and internalizing distress from bullying (Davidson and Michelle 2007), and that bullied respondents experience lower social support compared with co-workers (Hansen et al. 2006). Accordingly, we propose the following hypotheses:

Hypothesis 1e: Social support from colleagues is negatively related to bullying.

Hypothesis 1f: Autonomy is negatively related to bullying.

Hypothesis 1g: Competence development is negatively related to bullying.

Outcomes of Bullying in the Current Study

In addition to having detrimental health effects on the individual employee's health and well-being, bullying may also have negative effects on organizational outcomes (Nielsen and Einarsen 2012). For instance, previous research found that bullying associates positively with turnover intentions among nurses (Hogh et al. 2011; Blackstock et al. 2015). Moreover, negative associations between bullying and job satisfaction have been revealed (for a meta-study, see Bowling and Behr 2006). Further, studies conducted among nurses demonstrated that targets of bullying reported significantly lower levels of job satisfaction (Einarsen et al. 1998). Bullying has also been found to relate negatively to organizational commitment (for meta-studies, see Bowling and Behr 2006; Nielsen and Einarsen 2012). Employee performance is a critical factor in order to ensure efficiency and quality of care in hospitals. However, bullying may affect job performance negatively (for meta-study see Bowling and Behr 2006). Yildirim (2009) conducted a study among nurses where workplace bullying related to decreased motivation, decreased ability to concentrate, poor productivity, lack of commitment to work, and poor relationships with patients, managers and colleagues. Based on the above-mentioned research, we expect bullying to be negatively associated with several organizational outcomes and to increase turnover intentions. Therefore, the following hypotheses were developed:

Hypothesis 2a: Bullying is negatively related to job satisfaction.

Hypothesis 2b: Bullying is negatively related to work ability.

Hypothesis 2c: Bullying is negatively related to job performance.

Hypothesis 2d: Bullying is negatively related to job commitment.

Hypothesis 2e: Bullying is positively related to turnover intentions.

Table 6.1   Sample characteristics

Variables	N	%
Gender		
Female	7186	78.4
Male	1976	21.6
Age		
< 31 years	1417	15.5
31–40 years	2290	25.0
41–50 years	2419	26.4
51–60 years	2293	25.0
> 60 years	743	8.1
Employment type		
Fixed position	7310	87.4
Other	1057	12.6

METHOD

Data Collection and Sampling

A self-completion questionnaire was distributed electronically via e mail to all healthcare employees employed in the Western Norway Regional Health Authority. The Western Norway Regional Health Authority is comprised of four public health enterprises and a pharmacy trust, and has overall responsibility for the specialist health service in the region. The response rate was 40 percent (N = 9162). In the sample 78.4 percent were female, 40.5 percent were less than 40 years, and 87.4 percent had a fixed position (Table 6.1).

Measures

Hospital management quality and LLQ were based on earlier studies using the Work Research and Quality Improvement Questionnaire (Karasek et al. 1982; Aasland 1994; Krogstad et al. 2004). Some small adjustments of HMQ and LLQ were conducted.

Hospital management quality was measured with three items measured on a five-point scale (1 = not correct, 5 = totally correct). The items assessed employee's perceptions of hospital management skills: (1) corrects prioritization skill, (2) having knowledge at the regional level and (3) having knowledge at the institutional level. One example of the items is, 'The hospital management has good knowledge about the work in the different departments'. Cronbach's alpha was 0.84.

Local leadership quality was measured using three items measured on a five-point scale (1 = not correct, 5 = totally correct). The items assessed employee's perceptions of leaders' skills: (1) relational skills to retain workers, (2) having overview and knowledge and (3) having the ability to develop employees. A sample item is, 'My closest leader has good knowledge about my working situation'. Cronbach's alpha was 0.84.

Competence development was assessed with four items (Kristensen 2000) measured on a five-point scale (1 = to a very small extent, 5 = to a great extent). An example item being, 'Do you have the opportunity to learn new things through your work?' Cronbach's alpha was 0.79.

Autonomy was measured by applying four items, using a five-point scale (1 = to a very small extent, 5 = to a great extent) from the Organization Assessment Survey (Dye 1996). An example item is, 'In my department, we work together to influence the standards that constitute good work'. Cronbach's alpha was 0.92.

Social support from colleagues (van der Heijden 1998) was assessed with three items measured on a five-point scale (1 = never, 5 = very often). An example item is, 'Are your colleagues able to appreciate the value of your work and see the results of it?' Cronbach's alpha was 0.75.

Bullying was measured using a 12-item trimmed version of the Negative Acts Questionnaire–Revised (NAQ–R) instrument (Einarsen et al. 2009). The items assess exposure to negative acts within the last six months using five response alternatives (1 = never, 2 = now and then, 3 = monthly, 4 = weekly, 5 = daily). All items are formulated in behavioral terms, with no reference to the term bullying, and the items are referring to both direct (for example, verbal abuse) and indirect behavior (for example, withdrawal of information). Cronbach's alpha was 0.92.

Turnover intention was measured with the use of three items from Allen and Meyer, (1990). Five response alternatives (1 = never, 5 = every day) are used and items assess whether employees have considered quitting during the last year, as well as considered moving to another type of profession, or considered changing job to another hospital. Cronbach's alpha was 0.77.

Job commitment was measured with three items (Allen and Meyer 1990) using a five point scale (1 = to a very low degree, 5 = to a very high degree). Items assess belonging to the organization, personal attachment to the organization, as well as being proud of working in the organization. Cronbach's alpha was 0.89.

Job performance was measured with four items (Elo et al. 2000) using a five-point scale (1 = never/seldom, 5 = always/very often). Items include employees' self-assessment of their job performance and level of mastering. Topics include the quantity and quality of work performance, the ability to solve problems at work, and the satisfaction with their own capacity to develop and maintain good work relationships with colleagues. Cronbach's alpha was 0.79.

Work ability was measured with two items using a scale ranging from 0 (not capable of working) to 10 (optimal work ability). One item measured self-rated current work ability while the other item measured estimated work ability in the forthcoming six months. Items of the work ability index (Tuomi et al. 1998) originally included additional items that used different scales, which is not optimal in structural equation modelling (SEM). Hence, this measure was adapted for the purpose of the study. Cronbach's alpha was 0.84.

Job satisfaction was measured with four items (Kristensen 2000) using a four-point scale (1 = very unsatisfied, 4 = very satisfied). The items measure job satisfaction related to job prospects, physical working conditions, the use of skills and overall satisfaction with the job. Cronbach's alpha was 0.76.

Data Analyses

To test the hypothesized research models, we used SEM. A two-step analytical approach was used. First, confirmatory factor analyses (CFA) with the use of maximum likelihood estimation was performed to test the validity of constructs. All the latent variables and observed variables were entered simultaneously to assess the construct validity. Second, SEM with the use of maximum likelihood estimation was performed to test the two hypothetical models developed. Moreover, to reduce the complexity of the model estimation, the theoretical model was divided into one antecedents' model and one outcome model, as presented in Figure 6.2. This corresponds well with hypotheses being investigated, without exaggerating and challenging too much the complexity of the model estimations, which is not recommended (Hair et al. 2010). Hence, the measurement model was also divided and performed separately for the antecedents' model and the outcome model. To evaluate fit we applied normed fit index (NFI), comparative fit index (CFI) and root mean square error of approximation (RMSEA). The cut-off for CFI and NFI was 0.90, and RMSEA should not exceed 0.08 (Hair et al., 2010). Composite reliability (CR > 0.7) was performed to assess the reliability of constructs (Hair et al. 2010). Average variance extracted (AVE) was used to estimate the convergent validity. Average variance extracted is sometimes referred to as a conservative measure (Malhotra and Dash 2011). The AVE 0.50 threshold was not applied very strictly in the current study. However, AVE was expected to be in the area of 0.50 or above.

To estimate the indirect effects of HMQ on bullying via LLQ, bootstrap analyses was performed for the antecedents' model (Hayes 2013). We

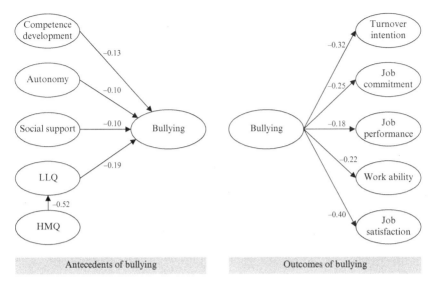

Figure 6.2   Final structural modeling with standardized beta coefficients

followed Hayes's (2013) recommendation of 5000 bootstrapped resamples and 95 per cent bias-corrected confidence intervals (CIs) estimation.

RESULTS

Descriptive Statistics, Cronbach's Alpha and Correlations

Table 6.2 contains descriptive statistics, Cronbach's alpha and correlations between constructs. Generally, gender and age had very low correlations with the constructs, indicating that gender and age are not confounding variables. Further, bullying was negatively correlated with the five job resources included in our study. In addition, and according to expectations, bullying was positively correlated with turnover intention, and negatively correlated with the other outcome indicators. The reliability of constructs is generally considered adequate with Cronbach's alpha within the range 0.75–0.92. Thus, the properties of the measurement constructs are satisfactory. The correlations between constructs are adequate and indicate satisfactory discriminant validity of measures.

Table 6.2   Descriptive statistics, Cronbach's alphas (diagonal) and correlations

Constructs	M	SD	1	2	3	4	5	6	7	8	9	10	11	12	13
1. Gender	1.22	0.41	–												
2. Age	2.85	1.19	.08**	–											
3. HMQ	3.28	0.83	.08**	-.09**	(.85)										
4. LMQ	3.79	0.96	.05**	.00**	.45**	(.84)									
5. Competence development	4.37	0.62	.03**	.17**	.14**	.30**	(.79)								
6. Autonomy	3.29	0.84	.05**	.01**	.39**	.58**	.32**	(.92)							
7. Social support from colleagues	3.42	0.72	-.12**	.08**	.12**	.24**	.29**	.28**	(.75)						
8. Bullying	1.25	0.40	.03**	-.03**	-.15**	-.30**	-.22**	-.28**	-.22**	(.92)					
9. Job performance	4.10	0.49	.02**	-.09**	.19**	.20**	.16**	.19**	.25**	-.18**	(.79)				
10. Turnover intention	1.84	0.79	.00**	.03**	-.23**	-.33**	-.22**	-.30**	-.13**	.29**	-.17**	(.77)			
11. Job commitment	3.89	0.94	.07**	-.02**	.37**	.42**	.37**	.41**	.24**	-.22**	.26**	-.37**	(.89)		
12. Work ability	9.01	1.28	-.04**	.02**	.12**	.14**	.16**	.14**	.13**	-.19**	.24**	-.18**	.17**	(.84)	
13. Job satisfaction	2.96	0.50	.06**	.01**	.32**	.47**	.39**	.46**	.25**	-.32**	.26**	-.43**	.44**	.20**	(.76)

Notes:
* p < .05, ** p < .05.
Gender: female = 1, male = 2.
LLQ = local leadership quality, HMQ = hospital management quality.

Confirmatory Factor Analyses

The psychometric properties of the latent factors were assessed using CFA. Corresponding with the two theoretical models applied in the study, two estimates of CFA were performed; one for the antecedents' model and one for the outcome model. As indicated in Table 6.3, CFA indicated acceptable fit for the six-factor model used to measure the constructs in the antecedents' model: NFI = 0.93; CFI = 0.94; RMSEA = 0.05. The constructs in the antecedents' model showed satisfactory score on CR (0.76–0.92), but competence development scored 0.48 on AVE which is marginally below 0.50 (Table 6.4). For the other constructs in the antecedents' model AVE ranged from 0.52 to 0.74. Since AVE is referred to as relatively strict compared with CR (Malhotra and Dash 2011), and based on the total impression of factor to item relations, the psychometric properties of the measurements construct were considered adequate for the antecedents' model.

The fit statistics of the measurement constructs included in the outcome model are shown in Table 6.3. The CFA showed a similar and acceptable fit for the six-factor hypothesized measurement outcome model: NFI = 0.94; CFI = 0.94; RMSEA = 0.05. For the outcome measurement model, CR ranged from 0.78 to 0.93 and AVE ranged from 0.47 to 0.73 (Table 6.4). Job satisfaction (0.47) had AVE scores marginally below 0.50 (Table 6.4), while the remaining AVE scores ranged from 0.50 to 0.73.

Path Analyses: Antecedents of Bullying

Structural equation modelling was applied to examine the two hypothesized models. The results (Figure 6.2) supported that the hypothesized

Table 6.3   Model fit

Model	X2	DF	NFI	CFI	RMSEA
Measurement model: antecedents	9459.62	362.00	0.93	0.94	0.05
Measurement model: outcomes	7356.45	335.00	0.94	0.94	0.05
Structural model of bullying antecedents	13801.07	368.00	0.90	0.91	0.06
Adjusted structural model of bullying antecedents; path from hospital and bullying is removed	13802.98	369.00	0.90	0.91	0.06
Structural model: outcomes	11962.24	345.00	0.90	0.90	0.06

Notes:
X2 values for the measurement and structural model are significant at $p > 0.001$.
The antecedents' structural models are estimated using 5000 bootstrap samples.

Table 6.4   CFA results

Constructs	CR	AVE
Measurement model: antecedent		
Competence development	0.79	0.48
Autonomy	0.92	0.74
Social support from colleagues	0.76	0.52
LLQ	0.85	0.65
HMQ	0.86	0.67
Bullying	0.89	0.58
Measurement model: outcomes		
Bullying	0.93	0.51
Job satisfaction	0.78	0.47
Well-being	0.84	0.73
Job performance	0.80	0.50
Job commitment	0.89	0.73
Turnover intention	0.80	0.59

Note: LLQ = local leadership quality; HMQ = hospital management quality.

antecedents' model fitted the data within recommended thresholds (NFI = 0.90; CFI = 0.91; RMSEA = 0.06). All of the hypotheses in the first test of the antecedents' model were significant, with exception of the direct relation between HMQ and bullying. Hence, the antecedents' model was modified removing the path between HMQ and bullying. This did not reduce the fit of the model (NFI = 0.90; CFI = 0.91; RMSEA = 0.06).

Hospital management quality was not significantly related to bullying. Thus, hypothesis 1a (hereafter H1a, and similar for the other hypotheses) was rejected. Furthermore, HMQ was significantly related to the mediator variable LLQ (H1b: β = 0.52, p < 0.001). A strong relation between the independent variable and the mediator is considered an important assumption that indicates a mediation effect is taken place (Baron and Kenny 1986). Bootstrap analyses showed that the indirect effect was significant in support of H1c: HMQ $\rightarrow$ LLQ $\rightarrow$ bullying (standardized indirect effect = − 0.10; 95 percent CI = − 0.08, − 0.12).

Moreover, LLQ (H1d: β = − 0.19, p < 0.001), social support (H1e: β = − 0.10, p < 0.001), autonomy H1f: (β = − 0.10, p < 0.001) and competence development (H1g: β = − 0.13, p < 0.001) were all negatively related to bullying.

Path Analyses: Outcomes of Bullying

Results from SEM supported that the hypothesized bullying outcome model (Figure 6.2) fitted the data (Table 6.3) within recommended thresholds (NFI = 0.90; CFI = 0.91; RMSEA = 0.06). All paths in the bullying model were significant. Specifically we found that bullying relates negatively to job satisfaction (H1a: $\beta = -0.40$, $p < 0.001$), work ability (H2b: $\beta = -0.22$, $p < 0.001$), job performance (H2c: $\beta = -0.18$, $p < 0.001$), job commitment (H2d: $\beta = -0.25$, $p < 0.001$) and positively to turnover intention (H2e: $\beta = 0.32$, $p < 0.001$).

Furthermore, the results confirm the validity of the Negative Act Questionnaire (NAQ) scale used in the current study, both to function as an outcome scale in relation to job resources, as well as being an explanatory factor in relation to organizational outcomes.

DISCUSSION

In the current study we explored antecedents and outcomes of bullying among hospital workers in a Norwegian healthcare setting. Results provide empirical support for the notion that hospital managers and local leaders play an important role in relation to bullying. Moreover, findings illustrate that HMQ has an indirect role in preventing bullying, via the strengthening of LLQ. Further, our findings revealed that bullying was associated with several organizational outcomes. Specifically, bullying was related to higher turnover intentions. Healthcare personnel exposed to bullying had lower job satisfaction, lower job commitment, and lower work ability and job performance. Results illustrate the importance of adjusting the work environment and the need to increase job characteristics to include competence development, autonomy in relation to work tasks and reinforcing social support from co-workers. The results and implications of our findings are discussed in the following subsection.

Job Resources and Prevention of Bullying

In line with previous research (for example, Hansen et al. 2006), negative associations were found between competence development and bullying, and social support and bullying. The negative association found between autonomy and bullying is according to previous research findings (Vartia 1996; Agervold and Mikkelsen 2004; Trepanier et al. 2013), and supports our theoretical model. Autonomy, competence and social support are all described in work motivational theories as basic, innate, universal human psychological needs (for example, Deci and Ryan 2012). Our findings clearly support this notion, as the job resources included in our study were negatively related to bullying.

A work environment characterized by high possibilities to thrive and develop competence, where employees have a sense of belonging and feeling of social support from colleagues, and the opportunity to influence work tasks, is crucial to avoid hostility and bullying among healthcare workers. Bullying can be seen as an accumulation of minor acts that amounts to a pattern of systematic maltreatment (for example, Einarsen and Skogstad 1996; Salin 2003). Bullying acts may arise from individual, situational, organizational and societal characteristics (for example, Hoel and Salin 2003; Salin 2003). Furthermore, individual and organizational factors interact towards influencing bullying in organizational settings (Branch et al. 2013). Focus of the first part of the current study relates primarily to how bullying associates with organizational factors; specifically job resources. In summary, results demonstrated that when employees perceived their job resources to be high, including top management quality and local leadership quality, the extensiveness of bullying was lower. This is an important finding, which hospital managers need to take into account in the daily running and management of hospitals.

The power imbalance which exists between subordinates and leaders can be a risk factor for bullying towards subordinates. Thus, early bullying research took an interest in the leader in relation to bullying (Hodson et al. 2006), and it is now acknowledged that managers play a significant role when it comes to bullying in organizations (Einarsen et al. 1994; Hoel and Cooper 2001). Bullying research has also emphasized the importance of managers' and leaders' function as role models, as well as communicating expectations for a collaborative and healthy environment (Cleary et al. 2009). The findings of our study give further support to the significant role leaders have in the prevention of bullying. This applies to leaders both at lower and higher levels in the organization. Our findings demonstrated that whereas local leadership relates directly to bullying, top management also plays an indirect role in the prevention of bulling through LLQ.

Results from our study suggest that hospital managers need to establish relevant knowledge about their hospital and further ensure correct prioritizing in decision making, as these are factors which were found to associate positively with LLQ. The aspects of LLQ in the current study related to having appropriate knowledge, as well as ensuring retention and development of employees. These aspects were found to be important with regard to the prevention of bullying in hospitals.

In addition, our study shows that it is very important that employees are given opportunities to develop their competence, are given autonomy in their work and that they experience social support from their colleagues, as these are aspects which most likely reduce the occurrence of bullying acts in the hospital context. Researchers have previously focused on conditions that may provide a good starting point for a healthy psychosocial work climate leading to engaged and satisfied employees performing meaningful and well-organized work tasks without high levels of conflicts and bullying. These conditions include an organizational willingness to give employees a certain autonomy over their work tasks, in addition to having a transparent resource distribution to prematurely avoid potential conflicts (Einarsen et al. 1994; Zapf et al. 1996). Other factors include having an organizational leadership structure able to define and work after clear strategic goals and values, and working to avoid organizational structures and job positions with a high level of role conflicts and ambiguity (Einarsen and Nielsen 2015; Salin 2003).

Outcomes of Bullying

A second aim of our study was to investigate the outcomes of bullying. Based on earlier research (for example, Bowling and Behr 2006; Nielsen and Einarsen 2012), we expected to find that bullying was negatively related to job commitment and positively related to turnover intentions. Our findings supported these hypothesizes, and therefore illustrate the importance of avoiding bullying in hospitals, thereby ensuring employee commitment, and preventing them from quitting their job.

Further, results demonstrated that bullying was negatively related to job performance, job satisfaction and work ability. Work ability and employee health conditions are indisputably important (for example, Ilmarinen 2009; van Rijn et al. 2014), as is the significant area of developing job satisfaction and job performance (for example, Harter et al. 2002; Diener and Seligman 2004; Bowling and Beehr 2006; Saks 2006; Wang et al. 2011) among hospital employees. Generally, our findings confirmed all the negative outcomes of bullying, emphasizing the importance of avoiding and preventing bullying in hospital settings.

LIMITATIONS AND FUTURE RESEARCH

This study uses a cross-sectional design which limits conclusion related to causality of research models. To compensate for this, several analytical steps have been conducted to establish discriminant and convergent validity of measures, thereby reducing the likelihood of common method bias and type I error. Furthermore, the likelihood of type II error is substantially reduced based on the use of a large sample consisting of 9162 responses. Hence, the study design is relatively strong with regard to the external validity of findings. It is likely that study findings are representative in similar hospitals' regions, especially in Norway, and thereafter in other countries with comparable organizational characteristics of hospitals. Nevertheless, we suggest that future research applies longitudinal designs and potentially multiple data sources.

REFERENCES

Aasland, O. (1994), 'Study on physicians' conditions in Norway', *Nordisk Medicin*, **110** (2), 65–7.

Agervold, M. (2009), 'The significance of organizational factors for the incidence of bullying', *Scandinavian Journal of Psychology*, **50** (3), 267–76.

Agervold, M. and Mikkelsen, E.G. (2004), 'Relationships between bullying, psychosocial work environment and individual stress reactions', *Work & Stress*, **18** (4), 336–51.

Allen, N.J. and Meyer, J.P. (1990), 'The measurement and antecedents of affective continuance and normative commitment to the organization', *Journal of Occupational Psychology*, **63** (1), 1–18.

Ashforth, B. (1994), 'Petty tyranny in organizations', *Human Relations*, **47** (7), 755–78.

Avolio, B.J. and Gardner, W.L. (2005), 'Authentic leadership development: getting to the root of positive forms of leadership', *Leadership Quarterly*, **16** (3), 315–38.

Baillien, E., Neyens, I., De Witte, H. and De Cuyper, N. (2009), 'A qualitative study on the development of workplace bullying: towards a three way model', *Journal of Community and Applied Social Psychology*, **19** (1), 1–16.

Ballou, K.A. (1998), 'A concept analysis of autonomy', *Journal of Professional Nursing*, **14** (2), 102–10.

Baron, R.M. and Kenny, D.A. (1986), 'The moderator–mediator variable distinction in social psychological research: conceptual, strategic, and statistical considerations', *Journal of Personality and Social Psychology*, **51** (6), 1173–82.

Blackstock, S., Harlos K., Macleod M.L.P. and Hardy C.L. (2015), 'The impact of organisational factors on horizontal bullying and turnover intentions in the nursing workplace', *Journal of Nursing Management*, **23** (8), 1106–14.

Bowling, N.A. and Behr, T.A. (2006), 'Workplace harassment from the victim's perspective: a theoretical model and meta-analysis', *Journal of Applied Psychology*, **91** (5), 998–1012.

Branch, S., Ramsay, S. and Barker, M. (2013), 'Workplace bullying, mobbing and general harassment: a review', *International Journal of Management Reviews*, **15** (3), 280–99.

Cleary, M., Hunt, G.E., Walter, G. and Robertson, M. (2009), 'Dealing with bullying in the workplace: towards zero tolerance', *Journal of Psychosocial Nursing & Mental Health Services*, **47** (12), 34–41.

Cobb-Clark, D.A. and Schurer, S. (2012), 'The stability of big-five personality traits', *Economics Letters*, **115** (1), 11–15.

Davidson, L.M. and Michelle, K.D. (2007), 'Social support as a moderator between victimization and internalizing-externalizing distress from bullying', *School Psychology Review*, **36** (3), 383–405.

Deci, E.L. and Ryan, R.M. (2000), 'The "what" and "why" of goal pursuits: human needs and the self-determination of behaviour', *Psychological Inquiry*, **11** (4), 227–68.

Deci, E.L. and Ryan, R.M. (2012), 'Self-determination theory', in P.A.M.V. Lange, A.W. Kruglanski and E.T. Higgins (eds), *Handbook of Theories of Social Psychology*, vol. 1, Thousand Oaks, CA: Sage, pp. 416–37.

Diener, E. and Seligman, M.E. (2004), 'Beyond money: toward an economy of well-being', *Psychological Science in the Public Interest*, **5** (1), 1–31.

Dye, D.A. (1996), *Organizational Assessment Survey*, Washington, DC: US Office of Personnel Management.

Einarsen, S. (2000), 'Harassment and bullying at work: a review of the Scandinavian approach', *Aggression and Violent Behavior*, **5** (4), 379–401.

Einarsen, S. and Nielsen, M.B. (2015), 'Workplace bullying as an antecedent of mental health problems: a five-year prospective and representative study', *International Archives of Occupational and Environmental Health*, **88** (2), 131–42.

Einarsen, S. and Skogstad, A. (1996), 'Bullying at work: epidemiological findings in public and private organizations', *European Journal of Work and Organizational Psychology*, **5** (2), 185–201.

Einarsen, S., Hoel, H. and Notelaers, G. (2009), 'Measuring exposure to bullying and harassment at work: validity, factor structure and psychometric properties of the Negative Acts Questionnaire-Revised', *Work & Stress*, **23** (1), 24–44.

Einarsen, S., Hoel, H., Zapf, D. and Cooper, C.L. (eds) (2010), *Bullying and Harassment in the Workplace. Developments in Theory, Research, and Practice*, Boca Raton, FL: CRC Press.

Einarsen, S., Matthiesen, S.B. and Skogstad, A. (1998), 'Bullying, burnout and well-being among assistant nurses', *Journal of Occupational Health and Safety – Australia and New Zealand*, **14** (6), 563–8.

Einarsen, S., Raknes, B.R.L. and Matthiesen, S.B. (1994), 'Bullying and harassment at work and their relationships to work environment quality: an exploratory study', *European Journal of Work and Organizational Psychology*, **4** (4), 381–401.

Elo, A.-L., Skogstad, A., Dallner, M., Gamberale, F., Hottinen, V. and Knardahl, S. (2000), 'User's guide for the QPSNordic: general Nordic questionnaire for psychological and social factors at work', Copenhagen: Nordic Council of Ministers.

Hair, J.F., Black, W.C., Babin, B.J. and Anderson, R.E. (2010), *Multivariate Data Analysis*, 7th edn, Upper Saddle River, NJ: Prentice Hall.

Hansen, Å.M., Hogh, A., Persson, R., Karlson, B., Garde, A.H. and Palle, Ø. (2006), 'Bullying at work, health outcomes and physiological stress response', *Journal of Psychosomatic Research*, **60** (1), 63–72.

Harter, J.K., Schmidt, F.L. and Hayes, T.L. (2002), 'Business-unit-level relationship between employee satisfaction, employee engagement, and business outcomes: a meta-analysis', *Journal of Applied Psychology*, **87** (2), 268–79.

Hauge, L.J., Skogstad, A. and Einarsen, S. (2007), 'Relationships between stressful work environments and bullying: results of a large representative study', *Work & Stress*, **21** (3), 220–42.

Hayes, A. (2013), *Introduction to Mediation, Moderation, and Conditional Process Analysis: A Regression-Based Approach*, New York: Guilford Press.

Hewett, R., Liefooghe, A., Visockaite, G. and Roongrerngsuke, S.J. (2018),

'Bullying at work: cognitive appraisal of negative acts, coping, wellbeing, and performance', *Journal of Occupational Health Psychology*, **23**, (1), 71–84.

Hodson, R., Roscigno, V.J. and Steven, H.L. (2006), 'Chaos and the abuse of power: workplace bullying in organizational and interactional context', *Work and Occupations*, **33** (4), 382–416.

Hoel, H. and Cooper. C.L. (2000), 'Destructive conflict and bullying at work', November, Manchester School of Management, University of Manchester Institute Science and Technology.

Hoel, H. and Cooper, C.L. (2001), 'Origins of bullying: theoretical frameworks for explaining workplace bullying', in N. Tehrani (ed.), *Building a Culture of Respect: Managing Bullying at Work*, London: Taylor & Francis, pp. 3–19.

Hoel, H. and Salin, D. (2003), 'Organizational antecedents of workplace bullying', in S. Einarsen, H. Hoel, D. Zapf and C.L. Cooper (eds), *Bullying and Emotional Abuse in the Workplace: International Perspectives in Research and Practice*, London: Taylor & Francis, pp. 203–18.

Hoel, H., Glasø, L., Hetland. J., Cooper. C.L. and Einarsen. S. (2010), 'Leadership styles as predictors of self-reported and observed workplace bullying', *British Journal of Management*, **21** (2), 453–68.

Hogh, A., Hansen, Å.M., Mikkelsen, E.G. and Persson, R. (2012), 'Exposure to negative acts at work, psychological stress reactions and physiological stress response', *Journal of Psychosomatic Research*, **73** (1), 47–52.

Hogh, A., Hoel, H. and Carneiro, I.G. (2011), 'Bullying and employee turnover among healthcare workers: a three-wave prospective study', *Journal of Nursing Management*, **19** (6), 742–51.

Ilmarinen, J. (2009), 'Work ability – a comprehensive concept for occupational health research and prevention', *Scandinavian Journal of Work, Environment & Health*, **35** (1), 1–5.

Johnson, W., Corley, J., Starr, J.M. and Deary, I.J. (2011), 'Psychological and physical health at age 70 in the Lothian Birth Cohort 1936: links with early life IQ, SES, and current cognitive function and neighborhood environment', *Health Psychology*, **30** (1), 1–11.

Karasek, R.A., Theorell, T., Schwartz, J., Pieper, C. and Alfredsson, L. (1982), 'Job, psychological factors and coronary heart disease', in H. Denolin (ed.), *Psychological Problems before and after Myocardial Infarction*, Basel: Karger, pp. 62–7.

Kristensen, T.S. (2000), *A New Tool for Assessing Psychological Factors at Work: The Copenhagen Psychosocial Questionnaire*, Copenhagen: National Institute of Health.

Krogstad, U., Hofoss, D. and Hjortdahl, P. (2004), 'Doctor and nurse perception of inter-professional co-operation in hospitals', *International Journal for Quality in Health Care*, **16** (6), 491–7.

Laschinger, H.K.S. and Fida. R. (2014), 'New nurses burnout and workplace well-being: the influence of authentic leadership and psychological capital', *Burnout Research*, **1** (1), 19–28.

Malhotra N.K. and Dash S. (2011), *Marketing Research an Applied Orientation*, London: Pearson.

Martin, W.F. (2008), 'Is your hospital safe? Disruptive behavior and workplace bullying', *Hospital Topics*, **89** (3), 21–8.

Meyer, J.P. and Maltin, E.R. (2010), 'Employee commitment and well-being: a critical review, theoretical framework and research agenda', *Journal of Vocational Behavior*, **77** (2), 323–37.

Mikkelsen. E.G.E. and Einarsen, S. (2002), 'Basic assumptions and symptoms of post–traumatic stress among victims of bullying at work', *European Journal of Work and Organizational Psychology*, **11** (1), 87–111.

Nielsen, M.B. and Einarsen, S. (2012), 'Outcomes of exposure to workplace bullying: a meta-analytic review', *Work & Stress*, **26** (4), 309–332.

Nielsen. M.B., Magerøy. N., Gjerstad. J. and Einarsen. S. (2014), 'Workplace bullying and subsequent health problems', *Tidsskrift for den Norske Laegeforening*, **134** (12–13), 1233–8.

Peter, R., Siegrist, J., Hallqvist, J., Reuterwall, C. and Theorell, T. (2002), 'Psychosocial work environment and myocardial infarction: improving risk estimation by combining two complementary job stress models in the SHEEP Study', *Journal of Epidemiology & Community Health*, **56** (4), 294–300.

Reknes, I., Pallesen, S., Magerøy, N., Moen, B.E., Bjorvatn, B. and Einarsen, S. (2014), 'Exposure to bullying behaviors as a predictor of mental health problems among Norwegian nurses: results from the prospective SUSSH-survey', *International Journal of Nursing Studies*, **51** (3), 479–87.

Rosenstein, A.H. (2002), 'Nurse–physician relationships: impact on nurse satisfaction and retention', *American Journal of Nursing*, **102** (6), 26–34.

Rosta, J. and Aasland, O.G. (2018), 'Perceived bullying among Norwegian doctors in 1993, 2004 and 2014–2015: a study based on cross-sectional and repeated surveys', *Epidemiology Research*, **8** (2), e018161, doi:10.1136/bmjopen-2017-018161.

Saks, A.M. (2006), 'Antecedents and consequences of employee engagement', *Journal of Managerial Psychology*, **21** (7), 600–619.

Salin, D. (2003), 'Ways of explaining workplace bullying: a review of enabling, motivating and precipitating structures and processes in the work environment', *Human Relations*, **56** (10), 1213–32.

Schaufeli, W.B. (2015), 'Engaging leadership in the job demands-resources model', *Career Development International*, **20** (5), 446–63.

Trepanier, S.G., Fernet, C. and Austin, S. (2013), 'Workplace bullying and psychological health at work: the mediating role of satisfaction of needs for autonomy, competence and relatedness', *Work & Stress*, **27** (2), 123–40.

Tuomi, K., Ilmarinen, J., Jahkola, A., Katajarinne, L. and Tulkki, A. (1998) *Work Ability Index*, 2nd edn, Helsinki: Finnish Institute of Occupational Health.

Van den Broeck, A., Ferris, D.L., Chang, C.H. and Rosen, C.C. (2016), 'A review of self-determination theory's basic psychological needs at work', *Journal of Management*, **42** (5), 1195–229.

Van der Heijden, B.I. (1998), 'The measurement and development of professional expertise throughout the career. A retrospective study among higher level Dutch professionals', PhD thesis, Universitet Twente, Enschede.

Van Rijn, R.M., Robroek, S.J., Brouwer, S. and Burdorf, A. (2014), 'Influence of poor health on exit from paid employment: a systematic review', *Occupational Environmental Medicine*, **71** (4), 295–301.

Vartia, M. (1996), 'The sources of bullying–psychological work environment and organizational climate', *European Journal of Work and Organizational Psychology*, **5** (2), 203–14.

Wang, G., Oh, I.S., Courtright, S.H. and Colbert, A.E. (2011), 'Transformational leadership and performance across criteria and levels: a meta-analytic review of 25 years of research', *Group & Organization Management*, **36** (2), 223–70.

Yildirim, D. (2009), 'Bullying among nurses and its effects', *International Nursing Review*, **56** (4), 504–11.

Yukl, G. (1999), 'An evaluation of conceptual weaknesses in transformational and charismatic leadership theories', *Leadership Quarterly*, **10** (2), 285–305.

Zapf, D. (1999), 'Organisational, work group related and personal causes of mobbing/bullying at work', *International Journal of Manpower*, **20** (1–2), 70–85.

Zapf, D., Knorz. C. and Kulla. M. (1996), 'On the relationship between mobbing factors, and job content, social work environment, and health outcomes', *European Journal of Work and Organizational Psychology*, **5** (2), 215–37.

7. Safety, health and climate: taking the temperature on nurses' work health and safety

Valerie O'Keeffe

1. INTRODUCTION

Nurses' work health and safety (WHS) is shaped by social interactions within their working environment. Nurses' working conditions are inextricably linked to the quality of care they can provide to their patients, and to their own health, safety and well-being (Dekker 2011; Bae and Fabry 2014; Adriaenssens et al. 2015). Healthcare workers face a diverse range of hazards leading to high rates of injury and illness (Geiger-Brown and Lipscomb 2010; Laschinger, 2014; Weiner et al. 2015; Nayak et al, 2016). In seeking to address this problem; managers, researchers, practitioners and regulators have searched beyond traditional technical and management system approaches to managing risks. Rather they are recognizing the role of social behaviour and have turned to safety climate and culture as the next age in the evolution of health and safety management (Borys et al. 2009). Health and safety is a social enterprise, a concept which the climate age or culture age now recognizes. The construction of WHS is influenced by the social environment in which work takes place (Hasle et al. 2009; Gherardi 2018). Thus, safety climate is an aggregate measure of group perceptions toward health and safety, providing an indicator of motivations and behaviours across work groups or organizations (Flin 2007), while culture is seen as more associated with safety values, assumptions and norms (Myers et al. 2014; Reiman and Rollenhagen 2014).

Safety climate is known to predict incidents and injuries in the workplace (Griffin and Neal 2000; Clarke 2006a, 2006b; Nahrgang et al. 2011; Leitão and Greiner 2016), thereby providing a diagnosis on which to base the treatment of health and safety dysfunction. Over recent decades, studies examining relationships between safety climate and safety outcomes have been prolific. Despite differences in definitions and questionnaire instruments used, overall results reveal positive, although in some cases modest, relationships between safety climate, safety behaviours and injury and incident occurrences (Neal and Griffin 2002; Katz-Navon et al. 2005; Beus et al. 2010; Clarke 2010; Singh et al. 2013; Steyrer et al. 2013). Clarke (2006a) found a stronger negative relationship between worker involvement in health and safety and accident occurrence, than between adherence to procedures and accident occurrence. These findings support the role of social engagement in prevention of work-related incidents. Modest relationships may be symptomatic of the underlying causes of injuries and incidents. However, just as symptoms may indicate the presence of many different maladies, many causes contribute to the outcome of incidents and injuries. Safety climate is typically measured using questionnaires, although they do not account for the contextual specificity of the work or the environment in which it is performed (Guldenmund 2007; Tholén et al. 2013). Therefore, measures of safety climate alone are unlikely to be sufficient to provide a definitive diagnosis of a group's state of WHS, although they efficiently identify the salient symptoms. Alternatively, qualitative methods emphasize context and in doing so, have value in extending understanding of how group safety perceptions develop (Zohar 2010). Qualitative methods not only reveal the significant components of WHS, but point to how they manifest, and the relationships between them, allowing hypothesis development and testing with quantitative methods.

2.   BACKGROUND

2.1   The Nature of Safety Climate

Safety climate is a specific form of organizational atmosphere that draws on and combines individual perceptions of the importance of safety in the work environment (Hofmann et al. 2017). Specifically, safety climate is considered to be the 'individual perceptions of policies, procedures and practices relating to safety in the workplace' (Neal and Griffin 2006 pp. 946–7). Contemporary WHS conceptualizes the work environment to include both physical and psychosocial factors, such as those that arise from personal interaction (Safe Work Australia 2011; Dollard et al. 2012). Climate is a collective phenomenon that focuses on data aggregated at the group level to reveal relationships between clusters of perceptions regarding the workplace. For the nurses in this study, those perceptions are towards health and safety (Schneider and Reichers 1983). Since people at work are exposed to myriad activities, policies, procedures and practices, they make sense of them as 'related sets', suggesting that climates are 'specific to particular referents' (Schneider and Reichers 1983 p. 21) such as patient care, safety and quality, or WHS.

Safety climate is regarded as the surface manifestation of an organization's underlying safety culture, through which it reflects its deeper and more enduring values (Flin et al. 2006). Schein (1990, 2004) supports this view suggesting that climate, as a product of attitudes and espoused values, is a surface expression of culture and that culture springs from deeper levels of unconscious assumptions (Flin 2007). There is much debate about the nature of culture, whether it is a product or characteristic of social interaction (for example, Silbey 2009; Bellot 2011) and the differences, if any, between safety culture and safety climate (Neal and Griffin 2006; Flin 2007; Guldenmund 2010; Petitta et al. 2017). Despite the features that distinguish safety climate and safety culture remaining contested, safety climate is the terminology used when attitudes and perceptions are measured using questionnaires (Flin et al. 2000; Guldenmund 2000, 2007); therefore, safety climate is the terminology that is used throughout this chapter.

Budworth (1997) describes safety climate as taking the safety temperature of an organization, suggesting it is an indicator of the state of health of the organization's overall safety management. However, research efforts on safety climate have lacked practical application. Notwithstanding the desire to contribute to the pragmatic goal of improving the health and safety outcomes of nurses and patients, a great deal of safety climate research has been preoccupied with measurement rather than conceptual and practical development (Zohar 2010; Schneider et al. 2017). Consequently, safety climate in healthcare has become a burgeoning area of research, driven by its relationship with unacceptable levels of patient and staff injury (McCaughey et al. 2013; Pousette et al. 2017; Zadow et al. 2017; Mohr et al. 2018), but with questionable tangible impacts on improving nurses' lived WHS experience, or injury reductions.

Research on organizational safety climate has primarily focused on systems-level structures and processes, with much less emphasis on the micro-level social processes fundamental to achieving health and safety at work (Barton and Sutcliffe 2009). It is with the micro-level social processes that this study is most concerned. While it is generally accepted that climate encompasses the sharing of attitudes, values and perceptions, safety research has not yet explained why and by what processes this sharing occurs. During social interactions, individuals test their perceptions, exploring, checking and refining them in the context of others' interpretations, thereby learning about their sociocultural environment (Waring 2009). Over time, it is expected that participants' perceptions merge, forming a group perspective (Zohar 2010), although this assumption remains largely unexplained. Through comparing qualitative and quantitative data about nurses' health and safety perceptions, this study adds weight to the facilitative role that social processes play in shaping safety climate.

Questionnaires remain an economical and efficient means of measuring safety climate, although they lack the richness, depth and contextual specificity of the qualitative data available through observation and interview (Flin et al. 2006). There is an unmet need to test the validity of questionnaire methods through qualitative data (Tholén et al. 2013), thereby extending the understanding of social factors and group-level dynamics on WHS performance. Safety and risk are inextricably linked, in that they are culturally based and derived from social organization (Dekker 2011). Consequently, individuals' perceptions of risk have implications for how they assess risks, recognize hazardous situations and respond. How people perceive risk is influenced by their sense of control, their knowledge of the risk, how risks and benefits are shared, and the immediacy or delay of harm, should it eventuate (Bluff 2011). Thus, safety climate perceptions should be differentiated from other organizational perceptions.

In seeking to understand the relationship between safety climate perceptions, behaviours and WHS outcomes, this study addresses the question: how do safety climate measures inform nurses' interpretations of their WHS? In examining this question, part A describes the nature of nurses' safety climate, identifying the key features. Part B compares the elements comprising nurses' safety climate with their self-reported perceptions derived from interview and observational data. Integrating these findings, we discuss how workers' interpretations of their WHS are most effectively assessed, and the implications of safety climate measures on WHS policy and practice.

3.   METHOD

This study examined safety climate in nurses working in acute care settings in five hospital units (including medical, surgical, intensive care and emergency) in an Australian capital city. The analysis drew on data obtained from administering the Safety Climate Questionnaire (SCQ) (Glendon and Litherland 2001). The SCQ examines the interface of policies, procedures and practices at the operational level. At the operational level, workers evaluate the relative priority of health and safety, as reflected by the degree to which perceptions are shared, against the precedence assigned to other organizational priorities, such as patient safety and quality (Zohar 2010). The SCQ includes 40 items (see Table 7.1) and was distributed to 320

Table 7.1 Nurses' six-factor solution from the Safety Climate Questionnaire

Oblimin rotated factor loadings for six-factor solution of the Safety Climate Questionnaire for nurses

Questionnaire item	Loading
Factor 1 – Consultation and communication: 49% variance, α = 0.92, eigen value 19.61	
4. Workers can express their views about work policy	0.52
5. Workers can discuss important policy issues	0.74
6. Workers are spoken to when changes in working practices are suggested	0.74
7. Work problems are openly discussed between workers and supervisors	0.64
8. Changes in working procedures and their effects on safety are effectively communicated to workers	0.57
9. Workers are told when changes are made to the working environment on the work site	0.52
17. Hospital policy is effectively communicated to workers	0.49
Factor 2 – Work pressure: 6% variance, α = 0.94, eigen value 2.58	
23. Workers have enough time to carry out their tasks	0.77
24. There are enough workers to carry out the required work	0.71
25. There is sufficient 'thinking time' to enable workers to plan and carry out their work to an adequate standard	0.78
27. Time schedules for completing work are realistic	0.70
28. Workload is reasonably balanced	0.74
29. Changes in workload which have been made at short notice can be accommodated without negatively affecting safety	0.49
Factor 3: – Best practice and procedures: 4.8% variance, α = 0.93, eigen value 1.92	
30. Personal protective equipment use is enforced	0.46
31. Relevant workers are specifically trained in the use of emergency personal protective equipment	0.51 0.68
35. Workers can easily identify the relevant procedure for each job	
36. An effective documentation management system ensures the availability of procedures	0.70
37. Work procedures are technically accurate	0.76
38. Work procedures are complete and comprehensive	0.84
39. Work procedures are clearly written	0.71
40. Written work procedures match the way tasks are done in practice	0.54

Table 7.1 (continued)

Oblimin rotated factor loadings for six-factor solution of the Safety Climate Questionnaire for nurses

Questionnaire item	Loading
Factor 4 – Management of PPE: 4.5% variance, α = 0.90, eigen value 1.81	
32. PPE users are consulted for suggested design improvements	0.73
33. PPE use is monitored to identify problem areas	0.78
34. Findings from PPE monitoring are acted upon	0.79
Factor 5 – Practicality of safety rules: 3.7% variance, α = 0.79, eigen value 1.46	
1. Safety rules are followed even when a job is rushed	0.50
2. Safety rules can be followed without conflicting with work practices	0.70
3. Safety rules are always practical	0.50
Factor 6 – Trust and relationships: 2.7% variance, α = 0.91, eigen value 1.07	
16. Workers are encouraged to support and look out for each other	0.43
18. Workers trust the management in this hospital	0.66
19. Management trust the workers in this hospital	0.59
20. Workers are confident about their future in this hospital	0.46
21. Good working relationships exist in this hospital	0.76
22. Morale is good	0.64

Note: PPE = personal protective equipment.

nurses; with 152 responses received, representing a response rate of 48 percent. Exploratory factor analysis (principal axis factoring) was conducted in preference to confirmatory factor analysis because although a factor structure had been established by Glendon and Litherland (2001), with road construction workers, safety climate in nursing may differ owing to the tension between nurses' own safety and that of their patients. Oblimin rotation was performed in the expectation that a degree of correlation would exist between the variables (Field 2005), since they examined different aspects of WHS. No cases were removed owing to missing data greater than 10 percent. The five cases with missing data less than 10 percent were replaced with mean values for the relevant items.

In addition, data from interviews with nurses recalling a salient safety-related incident occurring during their normal work was used to access nurses' interpretation of WHS in practice. Observations of nurses' work practices were also recorded in field notes, transcribed and analysed thematically together with the interview text, consistent with the processes for text-based analysis described by Saldana (2009). The analysis aimed to identify overarching themes relevant to nurses' understanding of WHS, contextualized to their work environment. The thematic analysis proceeded to the level at which overarching themes were distinguished that were comparable with the broad themes evident in the SCQ factor analysis. Weick's (1995) model of sensemaking was used to inform the aggregation process through which repeating concepts and phrases were organized into categories and themes.

To reveal the depth and breadth of nurses' interpretations of WHS, we integrated the quantitative and qualitative analyses. In part A, we identified the factor structure and means of the SCQ relevant to nursing. In part B, we compared the content of the six factors resulting from the nurses' SCQ to the content of each of the eight themes identified from the thematic analysis of nurses' interviews and observations. We mapped the questionnaire factors against the interview themes to identify content that was not accounted for in the questionnaire results. The mapping was achieved by transposing the content of each factor over each theme, and identifying the similarities and differences in content at the item and category level respectively.

4.   RESULTS AND DISCUSSION

4.1   Part A: Quantitative Data

4.1.1   Factor structure of safety climate within the nursing sample

The adequacy of sample sizes for exploratory factor analysis has been subject to long-standing debate. The interaction of sample size, variable-to-factor ratio and level of communality most significantly affect the rep-licability of factor patterns (Pearson and Mundform 2010). This sample size was 152, with variable-to-factor ratio of 6:1 and criterion fit of 0.92. Based on empirical research, Mundfrom et al. (2005) recommend sample sizes between 70 (for low level of communality) and 120 (for high level of communality). The sample size exceeded these criteria.

The analysis produced six factors representing what participating nurses considered to be the salient features of health and safety, accounting for 71 per cent of the variance in the sample responses. The six factors each yielded eigen values greater than 1.0, (see Table 7.1). Factor 1 was 'consultation and communication', factor 2 was 'work pressure' and factor 3 was 'best practice and procedures'. 'Management of PPE' emerged as factor 4, 'practicality of safety rules' was factor 5 and factor 6 comprised 'trust and relationships'. Table 7.1 presents the factor loadings following the oblimin rotation. Items with factor loadings of less than 0.4 were suppressed (Field 2005). Cronbach's alpha was calculated for each factor and ranged between 0.79 and 0.94, as shown in Table 7.1, indicating a high degree of consistency in how items were answered within each factor. Factor 1, addressing worker consultation and communication, explained 49 per cent of the variance.

Table 7.2 displays the correlations between factors. The matrix indicates that all of the factors are interrelated to some degree, with the use of personal protective equipment (PPE) showing the weakest relationships with other factors. The degree of correlation suggests that the data are not independent and that the oblimin rotated solution provides an appropriate base for interpretation (Field 2005).

In summary, the factor structure of nurses' safety climate, revealed six factors that describe nurses' perceptions of their WHS. The SCQ was selected for its designed focus on operationally derived components of WHS. The extracted factors reflect aspects of the enactment of WHS relevant to patient care, being central to the work of nurses and depicting how they understand salient aspects of WHS. Safety is built through shared perceptions, arguably transmitted and shaped through social interactions. Four of the six factors revealing nurses' perceptions, namely factor 1 'consultation and communication', factor 3 'best practice and procedures', factor 5 'the practicality of safety rules' and factor 6 'trust and relationships', are characteristics of health and safety practice that are highly mediated through social interaction on the job.

Table 7.2   Nurses' factor correlation matrix

Factor	Consultation and communication	Work pressure	Best practice and procedures	PPE	Relationships and trust	Practicality of rules
Consultation and communication	1.00	0.30	0.53	0.34	0.40	0.56
Work pressure	0.30	1.00	0.47	0.27	0.33	0.45
Best practice and procedures	0.53	0.47	1.00	0.36	0.38	0.46
PPE	0.34	0.27	0.36	1.00	0.25	0.33
Relationships and trust	0.40	0.33	0.38	0.25	1.00	0.35
Practicality of rules	0.56	0.45	0.46	0.33	0.35	1.00

Note: PPE = personal protective equipment.

In part B, we extend our understanding of nurses' WHS by examining their perceptions discovered in interviews. Interviews add nuance and context to the factors derived from questionnaire methods, enriching comprehension of WHS from the nurses' perspective.

4.2 Part B: Qualitative Data

In contrasting qualitative and quantitative findings, we illustrate how qualitative data can extend our capacity to understand the influence of social interactions on the state of WHS. Interviews have the potential to reveal nuances and details, producing greater specificity and a more diverse range of symptoms on which to base a diagnosis and treatment of health and safety. In examining the qualitative data, we delve into the meaning that nurses ascribe to their WHS.

Tables 7.3 and 7.4 depict the alignment between the qualitative themes and the factors derived from the SCQ. Table 7.3 summarizes the

Table 7.3   Alignment of text-derived themes with safety climate factors

Text-derived theme	Factor					
	1	2	3	4	5	6
Clinical practice						
Audit review, handover, clinical policies and procedures	✓		✓			
Communication and support						
Consultation, leadership, meetings, networks, relationships	✓				✓	
Safety information						
Patient status, equipment, general policies and procedures			✓	✓		
Safe practice	✓		✓	✓	✓	✓
Knowledge and skill, practice standards, teamwork, development, compliance, just culture, non-compliance, resistance, rules, duty of care, corrective action						
WHS management system						
Hazards, physical and psychological environment, shift work, injuries and incidents	✓		✓	✓		
Work pressure		✓				
Efficiency, planning, anticipating, patient comes first, workload, staffing, priorities						

Note: WHS = work health and safety.

relationship between the qualitative themes that were accounted for in the factors derived from the SCQ, while Table 7.4 summarizes the additional themes and the categories that comprise them, derived from the qualitative data. The themes and categories detailed in Table 7.4 were not accounted for in the quantitative data obtained by analysing the SCQ. However, they provide additional rich information about nurses' perceptions and understandings of WHS and the meaning it holds for them.

Underpinning the quality of nurses' practice is their sense of identity. The theme 'nurses' identity' embraced features of their professional persona, including their experience, self-efficacy and task-orientation. It was reflected in nurses' accounts and observations of using experience to anticipate and manage the unexpected:

> I usually find if I am on top of things then they don't develop into a crisis. After a while, you develop experience, a sense of how things are looking and signals to look out for that there might be a crisis and you avert it before it happens – it's prescience. (Nurse H5-IV10)[1]

Another nurse reflected the sentiments of many interviewed nurses who focus on continuing their work, despite impediments, to benefit the patient. Often such occasions suggested the possibility of bypassing safe work practices: 'I think nurses are in-bred to go "I just want to do it; I just want to get my job done. I just want to do it and get on with it"' (Nurse H5-IV05).

Nurses' sense of identity originates in caring for patients and relatives. Nurses' interpretations of the dynamics of working to achieve patient care were captured in the theme 'clinical practice'. Nurses related the following examples of how clinical practice is challenging and presents risks:

> I really love my job but I'd like to put a big sign on the front door saying that we don't deal with alcohol and drug-related incidences. I think it is not an emergency if you've worked really hard to get to the state you are in before you come here. We have people come in here drunk, they're unco-operative . . . we give them a bed to stay in for the night and a bowl to puke in but they should not be here . . . where there are legitimate emergencies. (Nurse H4-IV01)

> If you've been with an aggressive patient for hours and hours and you're just beginning to lose your patience, maybe you are going to respond in a way that might arc them up. So you just need to say to the co-ordinator can I swap with someone for a while. (Nurse H4-IV10)

The theme 'decision processes' incorporated aspects affecting decision-making, including attention, errors, interruptions and problem-solving. Nurses provided the following examples of how decision processes affected their work, particularly in situations of work pressure:

Table 7.4   Additional text-derived themes not accounted for by safety climate factors

Text-derived theme	Categories comprising theme	Sample quote
Clinical practice	Caring for relatives Patient presentation Patient rights	The patient died, however to me the therapeutic nature of resuscitating a patient in front of the next of kin is really important and his wife left here knowing that we had done everything in our power to save his life. We were both comfortable with that (Nurse H4-IV09) Patients have a right to privacy and respect. Their dependence or lack of power should not be used against them (Nurse H4-IV11)
Decision processes	Attention Debrief Errors Problem-solving	It's so busy I feel really fatigued and just worn out and sort of a bit dangerous 'cause you're non-competent, and that's when errors start to happen and you think oh I'll take a shortcut here because that's alright. (Nurse H3-IV02) Usually it is specific clinical situations that need problem-solving, like how to best move a patient with a certain problem, like a stroke and a knee replacement on the opposite side (Nurse H2-IV15)
Identity	Experience Reward and recognition Self-efficacy Task focused	Making decisions? I use my experience really, that's even better than the policies. And I suppose there's 30 years nursing. I've seen a fair bit. Ultimately it goes back to safety for the staff and safety for the patients. (Nurse H2-IV03). I probably don't do the right thing. I probably will pull my patient off the bed rather than ask for help sometimes. I know some people think they're invincible when sometimes I think I just get on with it and I do those things and I know that I shouldn't. (Nurse H5-IV05)

The workload is really heavy and mentally draining. In terms of health and safety it's really fatiguing and you're just worn out and a bit dangerous 'cause you're non-competent, and that's when errors start to happen and you think I'll take a shortcut here because that's alright. (Nurse H3-IV02)

If you are having a busy day you can accidentally miss things. Your mind may be thinking about something else when someone asks you to do something simple and routine and you probably know in the back of your head that you need to take appropriate precautions but you just have a lapsing memory. (Nurse H5-IV08)

The theme 'safe practice' embraced patient and nurse safety and featured competency, compliance and consequences, and duty of care. Nurses often raised competency and training as central to safe practice:

Anything that would make my job easier would make it safer. Honestly, it would just come down to me to having more and better trained staff readily available to assist when you need assistance. It's not about having special equipment, or new interventions or things that can do stuff for you, it's about being able to have adequate knowledge to carry out your job properly and knowing that your colleagues have that same knowledge and are able to use it. (Nurse H5-IV04)

Manual handling updates help us work safer – they focus on problem solving situations, like a patient collapsed on the floor. We have to work out what lifter to use and how to get patients up. We have to work through a problem, play a role, assess the situation – it's practical, better than watching a video. (Nurse H1-IV04)

The theme 'work pressure' was concerned with needs for efficiency and productivity, work pace, priorities, staffing and workloads. When working under pressure, nurses were more concerned about making errors or failing to follow safe work practices:

Sometimes I think the pace is too busy. They put patients through [surgery] too quickly and I think sometimes mistakes can be made and injuries can happen because the pace is too fast, and nurses have a horrible reputation of trying to please. They're putting more on our plate and we can't possibly do it. (Nurse H1-IV12)

Often I hear nurses say to each other 'oh just get on with it' or 'I didn't have time to get that piece of equipment' or 'I was too busy to worry about that'. Like 'give me a hand moving a patient up the bed', 'should we get whatever piece of equipment, nah just give me a hand, we'll pull them up by the sheet'. (Nurse H4-IV12)

4.2.1 Summary Part B

Interview and observational data extend our understanding of the social cues that influence decision-making and behaviour. Since patients are at the heart of nurses' work, they are expected to be central to how risks manifest and how nurses' experience those risks. The patient-care imperative for nurses is inherent within their occupational identity and is expected to drive much of their decision-making, ultimately affecting their own health and safety and the climate in which they work. Intimately connected to patient care was nurses' own sense of professional identity, which embraces their desire for and satisfaction in helping others. The core features of safety climate emerging in this study were 'communication and support', 'relationships', 'best practice and procedures' and 'work pressure', which exhibited high levels of consistency when comparing qualitative and quantitative data sources.

5. GENERAL DISCUSSION

This study focused on how safety climate contributes to the social context of nurses' interpretations of WHS, which inform their subsequent behaviours, influencing health and safety outcomes. Nurses' distinction between the informative value of communication and practical and emotional support is consistent with nurses' sense of professional identity. Professional identity embodies altruism and seeking personal satisfaction through imparting care and developing relationships with patients (Kenny 2015; Steege and Rainbow 2017). Identity defines the sense of self, providing the lens through which participants view their world (Weick et al. 2005). Through this perspective, participants determine what is significant for attention and action (McDonald et al. 2006). Identity also frames how organizational participants learn about their work, through both their cognitive knowledge and its embodiment through physical action and acquisition of technical skills (Harquail and Wilcox King 2010).

Safety climate measures provide an indicator of the overall perceived state of health and safety within a group by revealing perceptions of the different components of work organization. However, safety climate data may present a narrow view of WHS when considered diagnostic and used in isolation from other indicators that reflect a broader, dynamic conceptualization of the workplace. While questionnaires are useful for distinguishing differences in the degree to which perceptions are held, they are limited in examining the meaning that workers attach to their WHS (Colley and Neal 2012). Questionnaires do not delve into the fundamental values, assumptions and beliefs of the participants (Guldenmund 2007) nor provide an understanding of the social context (Antonsen 2009). Qualitative data allow exploration of the context and meaning attached to work. Nurses' work is predicated on providing care and compassion. In contemporary healthcare, these must coexist with increasing demands for skills and efficiency (Kohlen 2018). Consequently, nurses must rely on each other to provide practical, technical and emotional support (Hofmann et al. 2009). Nurses are deeply motivated to care for their patients, although working together to achieve quality care also requires trust. Nurses need to trust each other to be competent, and not to make errors or to harm the patients, either through their acts or omissions.

The insights available through the combination of qualitative and quantitative data suggest that the temperature of nurses' safety climate is within a healthy range, although with scope for improvement, particularly related to the use of PPE and conditions leading to work pressure. The results obtained from the qualitative data were consistent with but extended beyond the factors produced from the questionnaire. Nurses' identity, decision processes, patient-related care, and training and education were considered important in nurses' own accounts of interpreting and locating WHS in their daily work. These occupational-specific factors add weight to the assertion that safety climate informs nurses' interpretations and behaviours toward WHS, acting through social processes to reconcile patient, organizational and personal goals.

5.1 Beyond Safety Climate

Significantly, this study revealed aspects of nurses' work crucial to their ongoing practice of health and safety as a dynamic and situated activity within a complex system. Nurses' actions were directed by continuously updating information to maintain situation awareness of current events within a specific context. This interaction was facilitated by verbal communication to share information (Hunter et al. 2008) and conversation facilitated problem-solving to overcome obstacles impeding continuing care (Dunford and Perrigino 2018). Through these localized social interactions, nurses acted on their understandings, testing and refining them.

Nurses' identity came to the fore when experiencing conflict between providing optimal patient service that epitomized their values of caring, and following procedures that governed quality and WHS standards. Time pressures forced choices between high productivity and working conscientiously in accordance with rules and procedures. In interviews, nurses emphasized the influence of work pressure and inadequate staffing on their own risk-taking and level of compliance, particularly with manual handling and infection control procedures. Nurses rationalized these trade-offs by devoting effort to those activities that supported their sense of professionalism, often favouring patient outcomes.

Nurses saw themselves as task-focused and efficient problem-solvers with the ability to acquire resources. Problem-solving to maintain task continuity is an adaptive strategy to make nursing work functional in an environment that is hierarchical and can be oppressive, constraining communication, collaboration and efficiency (Garon 2012). Nurses constantly process cues from their patients, their colleagues and the environment, interpreting these cues through their experience. Experience enabled nurses to plan and to anticipate problems and generate solutions, enhancing their sense of competency and self-efficacy, thereby reinforcing their sense of identity (Tucker 2004).

Capability development as reflected through training and education supported nurses' understandings of WHS. In interviews, nurses interpreted training as a primary mechanism for obtaining formal health and safety information and as the basis for building competency in nursing skills. Through training, nurses were able to learn how health and safety interfaced with nursing processes. In the questionnaire, training did not emerge as a distinct factor; rather it loaded on three other factors, consistent with nurses' interpretations of training expressed in interviews. First, nurses discussed training as a way of promoting and transferring the content inherent in best practices and procedures. Second, training was seen as a way of communicating, particularly related to risks and normal work. Third, training was viewed as a way of disseminating safety rules, such as how to move and position patients.

The results of this study suggest that interview and observational methods reveal how WHS is enacted in greater depth than questionnaire methods and are instructive for more effective prevention efforts. Questionnaire findings for communication, work pressure, best practices and procedures, the practicality of safety rules and trust and relationships were consistent with the findings from interviews, highlighting these as critical targets for prevention efforts. However, questionnaire findings did not illuminate the interfaces between these important components of health and safety or the potential mechanisms by which nurses make WHS decisions, where prevention efforts are likely to be most successful.

5.2 Implications for Policy and Practice

Our findings emphasize the value of contextual information obtained through qualitative data for extending understanding of the meaning of WHS for occupational groups. Occupational and group identity, the development of capability, competency and proficiency, the meaningfulness of tasks and the capacity to make decisions added new dimensions to making sense of nurses' perceptions of WHS. Contextual information allows explanations for safety behaviours to be proposed and investigated, thereby aiding in developing tailor-made interventions targeted at improving specific aspects of WHS.

To promote improvements in WHS, managers need to understand and respond to sociocultural factors that enhance or impair performance, and then target these to facilitate change. For example, via factor loadings, quantitative findings assist in identifying the specific areas of WHS that warrant improvement. Understanding sources of power, conflict and meaning, within and between teams, enables managers and practitioners to focus on how they will implement the specific interventions (what they will do) necessary to achieve sustainable WHS performance. For regulators, quantitative findings enable comparisons between organizations and highlight areas of performance that require improvement and possibly compliance action. As part of a comprehensive diagnostic approach, when combined with injury and incident data, audit or interview findings, quantitative safety climate data make a significant contribution to diagnosing the state of WHS and selecting options for treatment (cf Turcanu et al. 2015). Interview and observational data (often part of a comprehensive audit process) extend diagnostic information by highlighting the relationships between workers' perceptions and behaviours, identifying the reasons behind them.

This study also raises implications for WHS researchers. As noted by Zohar (2010) much research effort over recent decades has focused on methodological and definitional aspects of safety climate or safety culture research. Several authors (Flin et al. 2006; Tholén et al. 2013; Turcanu et al. 2015) have highlighted the need to invest greater effort in applied aspects of research to better understand real world influences and impact of safety climate. Extending our understanding of the underlying mechanisms, how it manifests and the meaning safety climate holds for organizations, are areas to which qualitative research can significantly contribute. In particular, there is need for longitudinal studies that combine qualitative and quantitative measures to reveal processes operating over time, within and between groups, since safety climate or safety culture is a sociocultural process that shapes attitudes, values, beliefs and behaviours across time.

5.3   Limitations of the Study

This study aimed to examine the influences on nurses' understanding of WHS using multiple methods, although methodological limitations restrict application beyond this sample. First, the SCQ is not widely used and lacks the validity and reliability available from questionnaires more commonly used in healthcare settings, such as the Safety Attitudes Questionnaire (SAQ) (Sexton et al. 2006). The SAQ is heavily focused on climate related to patient safety and quality, and consequently was not used. Second, since the SCQ was not specific to the healthcare setting, it may have brought into sharp focus the contextual idiosyncrasies of the hospital environment.

The small sample size also limits the conclusions that can be drawn from this study. Although the conditions for performing factor analysis recommended by several authors were met (see section 4.1.1), Field (2005) suggests that a sample size 20 times that of the number of questionnaire items as ideal. For this study that would have required 800 respondents, a number greater than the population of all nurses working in the study sites.

6. CONCLUSION

This study demonstrates that nurses' safety climate measures are largely consistent with nurses' self-reported interpretations of WHS used to guide their decisions and actions, although with subtle but important differences. Adequately explaining the mechanisms through which these similarities and differences occur requires further research. Also, the findings support the notion that safety climate is like taking the health and safety temperature of the organization to indicate its general state of health (Budworth 1997). Similar to a litmus test, safety climate provides a point-in-time assessment of health indicators (Flin et al. 2000), but it is neither comprehensive nor diagnostic. Safety climate highlights the places to look in the organization worthy of further examination, to deepen the understanding of the complex interactions within socially enacted groups. Qualitative methods offer greater scope for identifying and understanding the complex issues inherent in health and safety, risk perception and management, rather than merely quantifying their existence. This study indicates the value in applying multiple methods to examine a phenomenon such as WHS from different perspectives, particularly as a basis for implementing and evaluating tailored interventions.

NOTE

1. The codes identifying participants denote hospital unit number (H) and interview number (IV).

REFERENCES

Adriaenssens, J., De Gucht, V. and Maes, S. (2015), 'Determinants and prevalence of burnout in emergency nurses: a systematic review of 25 years of research', *International Journal of Nursing Studies*, **52** (2), 649–61.

Antonsen, S. (2009), *Safety Culture: Theory, Method and Improvement*, Farnham: Ashgate.

Bae, S.-H. and Fabry, D. (2014), 'Assessing the relationships between nurse work hours/overtime and nurse and patient outcomes: systematic literature review', *Nursing Outlook*, **62** (2), 138–56.

Barton, M.A. and Sutcliffe, K.M. (2009), 'Overcoming dysfunctional momentum: organizational safety as a social achievement', *Human Relations*, **62** (9), 1327–56.

Bellot, J. (2011), 'Defining and assessing organizational culture', *Nursing Forum*, **46** (1), 29–37, doi:10.1111/j.1744-6198.2010.00207.x.

Beus, J.M., Payne, S.C., Bergman, M.E. and Arthur Jr, W. (2010),' Safety climate and injuries: an examination of theoretical and empirical relationships', *Journal of Applied Psychology*, **95** (4), 713–27.

Bluff, E. (2011), *Something to Think About – Motivations, Attitudes, Perceptions and Skills in Work, Health and Safety*, Canberra: Safe Work Ausyralia, accessed 26 September 2018 at https://www.safeworkaustralia.gov.au/doc/something-think-ab out-motivations-attitudes-perceptions-and-skills-work-health-and-safety.

Borys, D., Else, D. and Leggett, S. (2009), 'The fifth age of safety: the adaptive age?', *Journal of Health and Safety Practice and Research*, **1** (1), 19–27.

Budworth, N. (1997), 'The development and evaluation of a safety climate measure as a diagnostic tool in safety management', *Journal of the Institution of Occupational Safety and Health*, 1, 19–29.

Clarke, S. (2006a), 'Contrasting perceptual, attitudinal and dispositional approaches to accident involvement in the workplace', *Safety Science*, **44** (6), 537–50, accessed 20 September 2018 at http://dx.doi.org/10.1016/j.ssci.2005.12.001.

Clarke, S. (2006b), 'The relationship between safety climate and safety performance: a meta-analytic review', *Journal of Occupational Health Psychology*, **11** (4), 315–27.

Clarke, S. (2010), 'An integrative model of safety climate: linking psychological climate and work attitudes to individual safety outcomes using meta-analysis', *Journal of Occupational and Organizational Psychology*, **83** (3), 553–78.

Colley, S.K. and Neal, A. (2012), 'Automated text analysis to examine qualitative differences in safety schema among upper managers, supervisors and workers', *Safety Science*, **50** (9), 1775–85, doi:10.1016/j.ssci.2012.04.006.

Dekker, S. (2011), *Patient Safety: A Human Factors Approach*, Boca Raton, FL: CRC Press.

Dollard, M.F., Tuckey, M.R. and Dormann, C. (2012), 'Psychosocial safety climate moderates the job demand-resource interaction in predicting workgroup distress', *Accident Analysis and Prevention*, **45** (March), 694–704, doi:10.1016/j. aap.2011.09.042.

Dunford, B.B. and Perrigino, M.B. (2018), 'The social construction of workarounds', in D. Lewin and P.J. Gollan (eds), *Advances in Industrial and Labor Relations, 2017: Shifts in Workplace Voice, Justice, Negotiation and Conflict Resolution in Contemporary Workplaces*, Bingley: Emerald, pp. 7–28.

Field, A. (2005), *Discovering Statistics Using SPSS*, 2nd edn, London: Sage.

Flin, R. (2007), 'Measuring safety culture in healthcare: a case for accurate diagnosis', *Safety Science*, **45** (6), 653–67.

Flin, R., Burns, C., Mearns, K., Yule, S. and Robertson, E.M. (2006), 'Measuring safety climate in health care', *Quality and Safety in Health Care*, **15** (2), 109–15.

Flin, R., Mearns, K., O'Connor, P. and Bryden, R. (2000), 'Measuring safety climate: identifying the common features', *Safety Science*, **34** (1–3), 177–92.

Garon, M. (2012), 'Speaking up, being heard: registered nurses' perceptions of workplace communication', *Journal of Nursing Management*, **20** (3), 361–71.

Geiger-Brown, J. and Lipscomb, J. (2010), 'The health care work environment and adverse health and safety consequences for nurses', *Annual Review of Nursing Research*, **28**, 191–231.

Gherardi, S. (2018), 'A practice-based approach to safety as an emergent competence', in C. Bieder, C. Gilbert, B. Journé and H. Laroche (eds), *Beyond Safety Training*, Cham: Springer, pp. 11–21.

Glendon, A.I. and Litherland, D.K. (2001), 'Safety climate factors, group differences and safety behaviour in road construction', *Safety Science*, **39** (3), 157–88.

Griffin, M.A. and Neal, A. (2000), 'Perceptions of safety at work: a framework for linking safety climate to safety performance, knowledge and motivation', *Journal of Occupational Health Psychology*, **5** (3), 347–58, doi:10.1037/1076-8998.5.3.347.

Guldenmund, F.W. (2000), 'The nature of safety culture: a review of theory and research', *Safety Science*, **34** (1), 215–57.

Guldenmund, F.W. (2007), 'The use of questionnaires in safety culture research: an evaluation', *Safety Science*, **45** (6), 723–43.

Guldenmund, F.W. (2010), '(Mis)understanding safety culture and its relationship to safety management', *Risk Analysis*, **30** (10), 1466–80, doi:10.1111/j.1539-6924.2010.01452.x.

Harquail, C.V. and Wilcox King, A. (2010), 'Construing organizational identity: the role of embodied cognition', *Organization Studies*, **31** (12), 1619–48, doi:10.1177/0170840610376143.

Hasle, P., Kines, P. and Andersen, L.P. (2009), 'Small enterprise owners' accident causation attribution and prevention', *Safety Science*, **47** (1), 9–19.

Hofmann, D.A., Burke, M.J. and Zohar, D. (2017), '100 years of occupational safety research: from basic protections and work analysis to a multilevel view of workplace safety and risk', *Journal of Applied Psychology*, **102** (3), 375–88.

Hofmann, D.A., Lei, Z. and Grant, A.M. (2009), 'Seeking help in the shadow of doubt: the sensemaking processes underlying how nurses decide whom to ask for advice', *Journal of Applied Psychology*, **94** (5), 12610–74.

Hunter, C.L., Spence, K., McKenna, K. and Iedema, R. (2008), 'Learning how we learn: an ethnographic study in a neonatal intensive care unit', *Journal of Advanced Nursing*, **62** (6), 657–64, doi:10.1111/j.1365-2648.2008.04632.x.

Katz-Navon, T., Naveh, E. and Stern, Z. (2005), 'Safety climate in health care organizations: a multidimensional approach', *Academy of Management Journal*, **48** (6), 1075–89.

Kenny, D. (2015), 'Vocation, caring and nurse identity', PhD dissertation, Manchester Metropolitan University, accessed 29 September 2018 at https://e-space.mmu.ac.uk/582274/1/Deborah%20Kenny%20Vocation%20caring%20and%20nurse%20identity%20-%20Doctorate%20in%20Education%20June%202015.pdf.

Kohlen, H. (2018), 'Caring about care in the hospital arena and nurses' voices in hospital ethics committees: three decades of experiences', in F. Krause and J. Boldt (eds), *Care in Healthcare: Reflections on Theory and Practice*, Cham: Springer International, pp. 237–63.

Laschinger, H.K.S. (2014), 'Impact of workplace mistreatment on patient safety risk and nurse-assessed patient outcomes', *Journal of Nursing Administration*, **44** (5), 284–90.

Leitão, S. and Greiner, B.A. (2016), 'Organisational safety climate and occupational accidents and injuries: an epidemiology-based systematic review', *Work & Stress*, **30** (1), 71–90.

McCaughey, D., DelliFraine, J.L., McGhan, G. and Bruning, N.S. (2013), 'The negative effects of workplace injury and illness on workplace safety climate perceptions and health care worker outcomes', *Safety Science*, **51** (1), 138–47, doi:10.1016/j.ssci.2012.06.004.

McDonald, R., Waring, J.J. and Harrison, S. (2006), 'Rules, safety and the narrativisation of identity: a hospital operating theatre case study', *Sociology of Health and Illness*, **28** (2), 178–202.

Mohr, D.C., Eaton, J.L., Mcphaul, K.M. and Hodgson, M.J. (2018), 'Does employee safety matter for patients too? Employee safety climate and patient safety culture in health care', *Journal of Patient Safety*, **14** (3), 181–5.

Mundfrom, D.J., Shaw, D.G. and Ke, T.L. (2005), 'Minimum sample size recommendations for conducting factor analyses', *International Journal of Testing*, **5** (2), 159–68.

Myers, D.J., Nyce, J.M. and Dekker, S.W. (2014), 'Setting culture apart: distinguishing culture from behavior and social structure in safety and injury research', *Accident Analysis & Prevention*, **68** (July), 25–9.

Nahrgang, J.D., Morgeson, F.P. and Hofmann, D.A. (2011), 'Safety at work: a meta-analytic investigation of the link between job demands, job resources, burnout, engagement, and safety outcomes', *Journal of Applied Psychology*, **96** (1), 71–94. doi:10.1037/a0021484.

Nayak, S., Mayya, S., Chakravarthy, K., Andrews, T., Goel, K. and Pundir, P. (2016), 'Work-related injuries and stress level in nursing professional', *International Journal of Medical Science and Public Health*, **5** (8), 1693–7.

Neal, A. and Griffin, M. (2002), 'Safety climate and safety behaviour', *Australian Journal of Management*, **27** (September), special issue, 67–75.

Neal, A. and Griffin, M.A. (2006), 'A study of the lagged relationships among safety climate, safety motivation, safety behavior, and accidents at the individual and group levels', *Journal of Applied Psychology*, **91** (4), 946–53, doi:10.1037/0021-9010.91.4.946.

Pearson, R. H. and Mundform, D.J. (2010), 'Recommended sample size for conducting exploratory factor analysis on dichotomous data', *Journal of Modern Applied Statistical Methods*, **9** (2), 359–68.

Petitta, L., Probst, T.M., Barbaranelli, C. and Ghezzi, V. (2017), 'Disentangling the roles of safety climate and safety culture: multi-level effects on the relationship between supervisor enforcement and safety compliance', *Accident Analysis & Prevention*, **99** (February), 77–89.

Pousette, A., Larsman, P., Eklöf, M. and Törner, M. (2017), 'The relationship between patient safety climate and occupational safety climate in healthcare – a multi-level investigation', *Journal of Safety Research*, **61** (June), 187–98.

Reiman, T. and Rollenhagen, C. (2014), 'Does the concept of safety culture help or hinder systems thinking in safety?', *Accident Analysis & Prevention*, **68** (July), 5–15.

Safe Work Australia (2011), 'Work health and safety model bill', Commonwealth of Australia, Canberra.

Saldana, J. (2009), *The Coding Manual for Qualitative Researchers*, London: Sage.

Schein, E.H. (1990), 'Organizational culture', *American Psychologist*, **45** (2), 109–19.

Schein, E.H. (2004), *Organizational Culture and Leadership*, 3rd edn, San Francisco, CA: Jossey-Bass.

Schneider, B. and Reichers, A. (1983), 'On the etiology of climates', *Personnel Psychology*, **36** (1), 19–39.

Schneider, B., González-Romá, V., Ostroff, C. and West, M.A. (2017), 'Organizational climate and culture: reflections on the history of the constructs in the Journal of Applied Psychology', *Journal of Applied Psychology*, **102** (3), 468–82.

Sexton, J., Helmreich, R., Neilands, T., Rowan, K., Vella, K., Boyden, J. et al. (2006), 'The Safety Attitudes Questionnaire: psychometric properties, benchmarking data, and emerging research', *BMC Health Services Research*, **6** (1), 1–10, doi:10.1186/1472-6963-6-44.

Silbey, S. (2009), 'Taming Prometheus: talk about safety and culture', *Annual Review of Sociology*, **35** (August), 341–69.

Singh, B., Winkel, D.E. and Selvarajan, T. (2013), 'Managing diversity at work: does psychological safety hold the key to racial differences in employee performance?', *Journal of Occupational and Organizational Psychology*, **86** (2), 242–63.

Steege, L.M. and Rainbow, J.G. (2017), 'Fatigue in hospital nurses – 'supernurse' culture is a barrier to addressing problems: a qualitative interview study', *International Journal of Nursing Studies*, **67** (February), 20–28.

Steyrer, J., Schiffinger, M., Huber, C., Valentin, A. and Strunk, G. (2013), 'Attitude is everything? The impact of workload, safety climate, and safety tools on medical errors: a study of intensive care units', *Health Care Management Review*, **38** (4), 306–16.

Tholén, S.L., Pousette, A. and Törner, M. (2013), 'Causal relations between psychosocial conditions, safety climate and safety behaviour – a multi-level investigation', *Safety Science*, **55** (June), 62–9.

Tucker, A.L. (2004), 'The impact of operational failures on hospital nurses and their patients', *Journal of Operations Management*, **22** (2), 151–69.

Turcanu, C., Mkrtchyan, L., Nagy, A. and Faure, P. (2015), 'Can belief structures improve our understanding of safety climate survey data?', *International Journal of Approximate Reasoning*, **66** (November), 103–18.

Waring, J.J. (2009), 'Constructing and re-constructing narratives of patient safety', *Social Science and Medicine*, **69** (12), 1722–731.

Weick, K. (1995), *Sensemaking in Organizations*, Thousand Oaks, CA: Sage.

Weick, K.E., Sutcliffe, K.M. and Obstfeld, D. (2005), 'Organizing and the process of sensemaking', *Organization Science*, **16** (4), 409–21, doi:10.1287/orsc.1050.0133.

Weiner, C., Alperovitch-Najenson, D., Ribak, J. and Kalichman, L. (2015), 'Prevention of nurses' work-related musculoskeletal disorders resulting from repositioning patients in bed: comprehensive narrative review', *Workplace Health & Safety*, **63** (5), 226–32, doi:10.1177/2165079915580037.

Zadow, A.J., Dollard, M. F., McLinton, S.S., Lawrence, P. and Tuckey, M.R. (2017), 'Psychosocial safety climate, emotional exhaustion, and work injuries in healthcare workplaces', *Stress and Health*, **33** (5), 558–69.

Zohar, D. (2010), 'Thirty years of safety climate research: reflections and future directions', *Accident Analysis and Prevention*, **42** (5), 1517–22.

8. Antecedents of aggression in nursing: a review

Katharine McMahon, Lauren S. Park and Liu-Qin Yang

INTRODUCTION

Aggression against nurses occurs ubiquitously as they constantly interact with various people. The aggression that nurses face can manifest in multiple forms and originate from multiple sources, both internal and external to the organization. A combination of job-related factors such as ample interpersonal interactions, a tense and emotionally charged environment and a lack of job control result in frequent occurrences of aggression, and qualifies nursing as a high-risk industry for workplace aggression.

Workplace aggression or violence is an umbrella term that includes a range of behaviors with various levels of intensity. The form with the lowest intensity is incivility, described as low intensity and deviant behavior with ambiguous intent to harm a target (Andersson and Pearson 1999). Once intent is known, the violence can be classified as verbal abuse, or verbal behaviors that humiliate, degrade or otherwise indicate a lack of respect for the dignity and worth of another (Cook et al. 2001). In addition, and more broadly, emotional abuse indicates hostile verbal and non-verbal behaviors, not necessarily connected to sexual or racial content, but with the intent of acquiring compliance from others (Keashly 1997). Progressively, the behavior can then be classified as bullying when it becomes repeated inappropriate behavior that undermines an individual's right to dignity at work (Einarsen 2000; Task Force on the Prevention of Workplace Bullying 2001). Finally, words become action, and physical violence or assault become the ultimate offense. Throughout this chapter, we use workplace aggression and workplace violence as interchangeable terms.

The high prevalence of violence in nursing has been documented extensively. In Rowe and Sherlock (2005), 96.4 percent of nurse participants reported having experienced verbal abuse from other nurses, patients and family, doctors and others. One study showed that over 90 percent of nurses experienced physical assault from patients and that all forms of violence were under-reported (Hesketh et al. 2003). In a more recent study, Speroni et al. (2014) reported that 76 percent of nurses experienced verbal abuse and physical violence, often by patients and visitors. Verbal abuse tends to be experienced the most by nurses, with reports ranging from 39 percent to 91 percent (Cameron 1998; Rowe and Sherlock 2005; Celik et al. 2007; Kamchuchat et al. 2008; Pai and Lee 2011; Speroni et al. 2014), followed by bullying at 30 percent to 48 percent (Simons 2008; Pai and Lee 2011; Etienne 2014). Furthermore, physical abuse has been reported by up to 20 percent.

In addition to the frequency by type of aggression, the frequency by source of the violence are important to examine. In general, patients and their families have been the main perpetrators for non-physical and physical violence (Hesketh et al. 2003; Celik et al. 2007; Kamchuchat et al. 2008; AbuAlRub and Al-Asmar 2014), but coworkers and physicians have also been cited as perpetrators of non-physical violence (Celik et al. 2007; Hesketh et al. 2003). Specifically, Rowe and Sherlock (2005) demonstrated that 27 percent reported other nurses as the most frequent source of verbal abuse, followed by 25 percent reporting families of patients as the most frequent source, 22 percent for doctors, 17 percent for patients, 4 percent for residents, 3 percent for other and 2 percent for interns.

The negative impact of workplace violence has been well established; for example, lower job satisfaction (Hesketh et al. 2003; Spence Laschinger et al. 2009; Budin et al. 2013), lower organizational commitment and higher intent to turnover (Simons 2008; Spence Laschinger et al. 2009; Budin et al. 2013), poorer relationships with colleagues (Kamchuchat et al. 2008), and lower autonomy and more unfavorable perceptions of work environment (Budin et al. 2013). It has been profoundly agreed upon that workplace violence leads to negative outcomes for both the individual and their organization.

Precursors of aggression in the nursing profession have received less attention than the outcomes, but are necessary to understand if practitioners hope to prevent or reduce violence against nurses. Knowledge of the preceding factors that lead to violence in nursing can lead to the development of effective interventions and strategies to deter aggression. This chapter focuses on antecedents to violence in nursing at the individual, interpersonal and organizational levels, and provide implications for practitioners hoping to address violence at its outset in their organizations.

To complete this review, we focused on empirical articles that identified one or more antecedents of aggression in nursing. We used Google Scholar, Business Source Premier, PubMed and PsycInfo to search for the following terms in the context of nursing: aggression, violence, bullying, verbal abuse, physical abuse, mistreatment, conflict, hostility, incivility and victimization. We excluded any meta-analyses or review articles; however, we did use some of these articles – namely, Bowling and Beehr (2006); Hershcovis and Barling (2010) and Yang et al. (2014) – to direct us toward empirical articles. We excluded studies focusing on the outcomes of aggression, unless one or more antecedents were also reported. We also excluded articles that sampled nursing students and/or other medical staff in addition to professional nurses, unless results were reported from the subsample of nurses, as we wanted to represent only the experiences of the professional nursing population. We chose to include qualitative studies in which participants shared experiences of any form of aggression, such as aggression in general, incivility, violence, harassment or conflict. Finally, as our research team has completed substantial past work concentrating on the nursing population, we selected remaining articles from our comprehensive filing of nursing literature. After our consideration of exclusion criteria, we included 64 articles from our comprehensive list consisting of the original 182 articles.[1] The variables uncovered in our literature review were divided by level – individual, interpersonal, and organizational – and we structured our review using this division.

LEVELS OF VIOLENCE

Individual

Individual factors may explain the frequency or timing of violence in nursing and can provide cues for the risk of violence. These factors may be attributed to nurses specifically or to individuals they work with, most often the patients. Identifying characteristics of individuals that relate to violence against nurses allows practitioners to be mindful of characteristics that may pose risk.

Victim

Studies have observed certain characteristics of individuals that can be linked to increased exposure to aggression. The vast majority of these characteristics have been simply classified as risk factors or correlates, rather than suggesting causation. For example, in Budin et al. (2013), registered nurses reported more verbal abuse from their colleagues if they were unmarried, although Chen et al. (2010) uncovered the opposite. In another study, core-self evaluations, measured on four dimensions of self-esteem (overall value individuals place on themselves), generalized self-efficacy (ability to cope, perform, and achieve success), locus of control (belief that one has control over events in their lives), and neuroticism (a negativistic cognitive/explanatory style), negatively predicted intra-group conflict in a nursing population (Almost et al. 2010).

A common risk factor of aggression is age or experience. In a study by Pai and Lee (2011), nurses under 30 were over twice as likely as those over 30 to experience verbal abuse. Kamchuchat et al. (2008) also found age to be a personal risk factor, as younger age is related to higher odds of experiencing verbal abuse. In most workplaces, younger age or lack of experience creates a vulnerability for aggression, but in nursing this phenomenon is even more prevalent; the common phrase in the profession 'nurses eat their young' refers to older nurses treating younger nurses with disrespect during the beginning stages of their careers. In the qualitative study done by Simons and Mawn (2010), this phrase was repeatedly used by participants and even more nurses had alluded to their past experiences with this phenomenon. However, some studies have found that older age and longer employment serve as risk factors for experiencing workplace violence (Celik et al. 2007; Campbell et al. 2011).

Other demographic characteristics can also contribute to the likelihood of experiencing violence at work. For nurses who belong to groups that have been historically marginalized, such as racial or gender minorities, incivility may act as a covert, modern manifestation of discrimination, a phenomenon termed selective incivility (Cortina et al. 2001). Although ample evidence for selective incivility exists outside the field of nursing, few studies have examined selective incivility among nurses. In terms of gender, multiple studies have found that male nurses, a minority in the nursing population, are at greater risk for aggression (Campbell et al. 2011), while some have found that females are at greater risk (Chen et al. 2009). In a qualitative study of nurses, Simons and Mawn (2010) found that incivility often occurred based on the target's ethnicity, and Park and Martinez (2018) found that black nurses were more likely than white nurses to experience incivility from nursing peers.

Negative affectivity (NA) was also described as an antecedent for aggression in nursing. Demir and Rodwell (2012) found high NA to be linked to bullying and reports of threat of assault from those outside the organization, such as patients and visitors. Furthermore, NA has been linked to internal emotional abuse and external physical assault (Rodwell et al. 2013; Rodwell and Demir 2014), internal and external referring to the source based on their ties to the organization, meaning that internal represents a person working for the organization as the source of violence and external represents someone who does not. In another study by Rodwell et al. (2015), higher NA related to more bullying, more external and internal emotional abuse, and more external threat of assault among aged care nurses. In addition to NA, anxiety has also been linked to bullying. In a study examining the risk factors of violence among nurses, higher levels of anxiety significantly increased their odds of experiencing verbal abuse and bullying (Pai and Lee 2011). Anxiety has been classified as one of the most common causes of verbal abuse; in one study, 43 percent of nurses reported anxiety as a perceived cause (Kamchuchat et al. 2008). Finally, Bilgin (2009) demonstrated that nurses who are less social and less tolerant were more exposed to physical violence from patients.

Certain qualities of an individual may increase their risk of violence exposure or predispose them to experience higher rates of violence incidents. Similarly, the individual state and traits of the perpetrator may also increase the frequency of violence.

Perpetrator

Perpetrators may mistreat others regularly or episodically depending on their individual characteristics and circumstances. Perpetrators can be any of the various people with whom nurses interact; however, when examining individual differences of the perpetrators, studies have largely focused on patients as the offenders.

Patients who are under the influence of drugs or alcohol tend to be perpetrators of violence against nurses. Approximately one-fourth of nurses perceived that perpetrators of violence against them were under the influence of drugs and alcohol (Crilly et al. 2004). Speroni et al. (2014) examined nurse perceptions of antecedents of aggression and found that 45 percent of nurses believed alcohol led to patient aggression, and 40 percent believed the influence of drugs produced patient aggression. In the study, the authors also found that perpetrators tended to be white men in their late twenties or early thirties who were confused or under the influence. Drugs and alcohol are not only perceived as a common cause of violence in general, but as the most common cause of physical violence (Kamchuchat et al. 2008; Pich et al. 2017). Although all forms of violence are significant, this displays the importance of appropriate training for nurses in order to protect themselves from physical harm committed by aberrant and unruly patients.

Patients may also have an altered state of mind from mental illness. Approximately 38 percent of nurses perceive mental illness to be associated with violent behaviors (Crilly et al. 2004). In addition, nurses have perceived mental illness or cognitive impairment to be the main reason for aggression (Duxbury and Whittington 2005; Pich et al. 2017). When interviewed, most nurses perceived poor physical or mental health of patients and their families to be a factor of violence (Najafi et al. 2018).

More specifically, approximately half of nurses in the study by Speroni et al. (2014) believed that dementia or Alzheimer's disease caused aggressive acts. It is harder to prevent aggression when precursors include interactions with patients who are in altered state of minds, as these interactions are unavoidable owing to the nature of nursing. This underscores the need to provide nurses with an environment and resources which allow them to control potentially dangerous situations.

While characteristics of the perpetrator and victim of violence at the individual level can influence the likelihood of violence, they are only taking into account the lowest level of an organizational system. This focus can lead to blaming the victim for these incidents and neglecting consideration of the organization as a multi-level system with complex dimensions that may have more influence on the likelihood of violence. Furthermore, healthcare requires a high level of interpersonal cooperation among nurses, physicians, staff and patients, and interpersonal factors that impede this cooperation have a substantial impact on the likelihood of violence.

Interpersonal

Aggression is inherently about the interaction between two or more people. On an interpersonal level, aggression may result from the dynamics between the perpetrator and victim, or from an influential third party. Perpetrators may have malicious and self-serving intent for aggression or they may mistreat others as a reaction of dissent and disdain.

Leadership

Supervisors, managers, and other forms of leaders have significant impact on their subordinates. Their attitudes and behaviors, or lack thereof, signal to employees what is appropriate, acceptable or, even, encouraged. Passive leaders neglect the opportunity to establish or enforce zero-tolerance policies until intervention is absolutely necessary. Nurses have criticized their nurse managers for failing to provide supportive structures to respond to aggression and failing to respond aptly themselves to prevent future incidents (Farrell 1997). By contrast, ethical and transformational forms of leadership promote more civil work environments (Kaiser 2017; Islam et al. 2018). In addition, lower levels of incivility were linked to higher positive perceptions of nurse manager's ability, leadership and support of nurses (Smith et al. 2018).

In addition to allowing aggression to occur under their supervision, leaders are also often the enablers and instigators of aggression themselves. A recurring theme among violence in nursing is the misuse of authority.

This can manifest as leaders protecting those who perpetrate violence. Individuals may openly commit unethical behavior, knowing that more senior leaders will give them protection (Hutchinson et al. 2009). Leaders may also contribute to an atmosphere that encourages bullying by adapting formal rules and subverting their obvious purpose (Hutchinson et al. 2009). Ulterior motives also interfere with appropriate action when leaders minimize or suppress events to prevent risking public exposure (Hutchinson et al. 2009). In general, the misuse of authority, process and procedures enables bullying (Hutchinson et al. 2010; Blackstock et al. 2015). Finally, leaders may mistreat nurses through structural bullying, in which nurses perceive their supervisor's actions as unfair and punitive through scheduling, patient assignments or workload, or approval of sick and/or vacation time (Simons and Mawn 2010). Thus, leaders play a significant role as they have the ability to directly or indirectly encourage or thwart workplace violence.

Miscommunication
Another common antecedent to aggression in the workplace is miscommunication. Miscommunication has been identified as the number one cause of aggression, specifically for verbal abuse (Kamchuchat et al. 2008). In addition, 17 percent of respondents in AbuAlRub and Al-Asmar (2014) supported the idea that lack of communication skills between nurses and patients and their families contributed to psychological violence. Patients also perceive miscommunication as the number one precursor to aggression, even though nurses themselves do not necessarily always see their interactions with patients as problematic (Duxbury and Whittington 2005). This discrepancy in perception related to miscommunication, alone, provides a basis for further miscommunication as patients believe their caretakers do not respect or care about them, whereas nurses only acknowledge that communication could be improved when managing aggression (Duxbury and Whittington 2005). This disparity may also provide a basis for patients to react violently if they perceive that nurses are disregarding their needs. However, nurses in other studies have recognized poor communication and misunderstandings as influencing the probability of experiencing violence (Najafi et al. 2018). Sometimes language or dialect differences, or time pressure, also contribute to miscommunication between nurse and patient (Najafi et al. 2018). Kamchuchat et al. (2008) found that training in communication skills helped to clarify patients' needs and identify appropriate responses, reducing the risk of verbal abuse by 40 percent. Other than miscommunication between nurses and patients and their family, miscommunication among staff has also been reported. For example, nurses disclosed that inappropriate verbal, non-verbal, or written communication from physicians in patient status-reports was a

form of humiliating aggression from doctors (Najafi et al. 2018). The capacity to effectively communicate in a fast-paced, high-stress environment, as well as to continuously develop communication skills, is key to preventing violence in nursing. Accordingly, healthcare management should treat this as a priority while designing and implementing programs to train nursing staff and leaders.

Relations and interactions

The dynamics between two (or more) people can positively or negatively connect the individuals as they will support or oppose one another. Sometimes, people may join together, but create a negative atmosphere as they form pernicious alliances.

Alliances can add protection for aggressors to feel comfortable and confident to commit acts of violence without fear of repercussion or severe consequences. Informal alliances can be a mechanism for bullying as a group effort, and can increase the opportunities for mistreatment (Hutchinson et al. 2010; Blackstock et al. 2015). This is especially true if alliances are formed with superiors. Lower-level employees may unabashedly commit aggressive acts if they know they have the protection of senior employees (Hutchinson et al. 2009). In addition, these callous alliances contribute to the overall group atmosphere by demonstrating that aggressive behavior will be tolerated. Alliances perpetuate shared norms of tolerating serious misconduct, regardless of whether the violence is directed toward individuals or the organization (Hutchinson et al. 2009).

In addition, negative interactions or lack of positive relations between people are commonly seen as antecedents to violence. The absence of group identity, for example, can deprive employees of feelings of belonging or support. Feeling detached or alienated from the group, especially owing to ethnicity, education or nursing position, contributes to experiencing bullying (Simons and Mawn 2010). Further, aggression may result from differences in expectations or perceptions. Some expectations include quick care with no delay and receiving responsible and empathetic care, which serve as violence-predisposing factors (Najafi et al. 2018). In general, people use social rules regarding relationships to judge their interactions as appropriate or expected. These rules have expectations of equal workload, respect for privacy, cooperative behaviors for a shared environment, simple manners, eye contact and appropriate use of names, repayment of debt and favors and avoiding criticizing others. The rules provide a basis for expectations and, when broken, employees view the behavior as unfair, unpleasant aggression (Farrell 1997), regardless of whether it occurs to relationships between nurses and patients or between them and other healthcare staff. These rules can be violated in minor ways that may instigate backlash and verbal aggression, and lead to more drastic behavior, such as physical violence that neglects social standards and expectations.

Relationships with leaders, other employees and patients all have an impact on what people perceive as appropriate and inappropriate behaviors. Training workers to increase their communication skills, and encouraging beneficial rather than pernicious groups, would help minimize unpleasant interactions. Providing leaders with training for transformational and ethical leadership could also effectively manage negative interactions.

Organizational

Violence against nurses can easily be misunderstood as only a problem at the individual or interpersonal level, but organizational constraints and conditions foster the environment where aggression can be cultivated or inhibited. It is important to consider factors of aggression that are outside any individual's control but could be changed at the system level.

Job demands

The innately emotionally charged aspects of nursing can be demanding to individuals in the profession. The complexities of nursing, such as instability, variability and uncertainty, have been shown to positively predict intragroup relationship conflict; specifically, unpredictable changes in work and inconsistency among patients (Almost et al. 2010). In addition, perceived violence increases when conditions or the environment become more demanding, such as changes in patients' needs, increased workload, change in perceptions of nurse leadership, lowered nurse autonomy and worsened relations with doctors (Roche et al. 2010). Workload has repeatedly been connected with workplace aggression. Nurses believe that high workload and abundant patients or visitors increase the likelihood of experiencing violence (Brewer et al. 2013; Najafi et al. 2018). In the qualitative study by Najafi et al. (2018), all nurses believed nursing shortages and excessive workload obstructed them from providing effective care, which led to dissatisfaction among patients, relatives and physicians. Workload has been thought to partially cultivate an environment that will foster bullying (Islam et al. 2018). In addition, workload has been found to be a mediator between ethical leadership and bullying (Islam et al. 2018). Job demands in general have been repeatedly linked to workplace violence. Demands lead to vulnerability to aggression exposure and lower resistance to aggressive acts, and are related to higher frequencies of threat of assault by external sources such as patients or relatives (Demir and Rodwell 2012) and to higher frequencies of emotional abuse by external sources (Rodwell et al. 2013).

Organizational constraints

A shortage of nursing staff positively correlates with the threat of violence and physical violence (Roche et al. 2010). Multiple empirical studies have demonstrated that nursing samples identify inadequate staffing, and consequently increased workload and patient-to-nurse ratios, as contributors to aggression (AbuAlRub and Al-Asmar 2014; Bortoluzzi et al. 2014). In addition, the perception of needing more overtime to complete work per shift has been linked to perceptions of higher frequencies of the threat of violence (Roche et al. 2010). When nurses perceive that their organization provides enough staffing and resources, they report less incivility (Smith et al. 2018). Intensive interviews with nurses reveal that inadequate workforce, facilities and supplies to offer appropriate care are precursors to violence (Najafi et al. 2018). Poor working conditions aggravate bullying by creating space for it to develop (Islam et al. 2018). Specifically, poor working conditions such as working alone during certain shifts and working in male wards with an absence of male colleagues have been perceived by some nurses as enabling violence committed by patients and their relatives (Najafi et al. 2018). Similarly, circumstances such as frequent and noisy visits of family led to lower tolerance between nurses and patients, and subsequently higher levels of tension (Najafi et al. 2018).

Environment, climate and culture

Nurse work environment significantly relates to coworker incivility (Smith et al. 2018). Undisciplined working environments and poor organization between physicians and nurses have also been perceived as antecedents to violence (Najafi et al. 2018). The surrounding environment and overall culture dictate social norms. It is well known that the social norms in nursing include workplace violence; this notorious aspect of nursing can foster tolerance to the point that nurses believe it is a component of the job (Fisher et al. 1995). When an organizational climate tolerates bullying, reports of bullying seem to be trivialized or disbelieved, discouraging nurses from reporting their experiences and furthering the norms of tolerance (Deans 2004; Hutchinson et al. 2006). Hutchinson et al. (2010) found that higher rates of bullying occur in environments that encourage it through perks, promotion, or favorable treatment, rather than effective sanctions. In addition, they suggested that association with others who tolerate or engage in bullying socializes nurses to the norm of tolerance. Alliances sustain the norms of tolerance for violence regardless of individual or organizational consequences, and the environment can be heavily influenced by politics and political gain (Hutchinson et al. 2009). Organizational structure decision, allocation of equipment and operational means, and performance appraisal have been some political reasons contributing to bullying behaviors (Katrinli et al. 2010). Also, studies have supported the idea that climate perceptions relate to violence exposure, and that targeting violence prevention climate for interventions can be effective for reducing violence (Spector et al. 2015). This can be approached from the supervisor level; evidence by Yang and Caughlin (2017) demonstrated that aggression-preventive supervisor behaviors contributed to more positive violence prevention climate perceptions at the group level, which is subsequently linked to less violence exposure among nurses.

The physical environment has also been shown to be a precursor for workplace violence against nurses. In one study, the physical layout of a care ward contributed to aggression against nurses (Duxbury and Whittington 2005). In addition, multiple studies have looked at specific departments, determining that some have higher risk of violence than others. The emergency room, for example, was repeatedly reported as a high-risk department (Hesketh et al. 2003; Kamchuchat et al. 2008; Campbell et al. 2011; Speroni et al. 2014). Some other departments reported to have higher frequencies include psychiatric settings, medical-surgical settings, and nursing homes or long-term care facilities (Hesketh et al. 2003). Kamchuchat et al. (2008) classified the outpatient, emergency, operating, medical and surgical units as high risk and chose to classify the intensive care unit (ICU), pediatrics unit, postpartum unit, labor room, ophthalmological unit and private rooms as low risk, based on previous research. Furthermore, high-risk areas could be characterized by necessary care of acutely disturbed and violent individuals, unrestricted movement of the public in clinics and hospitals, and long waits. Kamchuchat et al. (2008) found that high-risk wards increased the risk of verbal abuse by 80 percent.

Organizational factors provide the basis for work environments and provide the opportunity for violence to be encouraged, ignored or inhibited. Nurses rely on resources to fulfill their responsibilities, but when organizational factors obstruct responsibilities or prevent them from being adequately completed, violence ensues. Targeting organizational factors will lead to healthier and more productive work environments that will subsequently curtail violence against nurses.

CONCLUSION

Summary

Factors at the individual, interpersonal and organizational levels can influence the likelihood of nurse violence. At the individual level, past research has demonstrated that nurses are more likely to be victims of violence if they are younger and/or less experienced (Pai and Lee 2011), belong to a historically marginalized group (Simons and Mawn 2010) or have a disposition characterized by negative affectivity (Rodwell et al. 2015) or anxiety (Pai and Lee 2011). In addition, the literature has largely focused on patient characteristics as individual-level precursors of violence perpetrators, finding that an altered state of mind – specifically substance abuse and mental illness (Crilly et al. 2004) – increases the likelihood of mistreating nursing staff.

At the interpersonal level, three particular factors have been identified as antecedents to nursing violence: leadership behaviors, miscommunication and social alliances. Passive leaders are likely to fail to respond appropriately to incidents of violence (Farrell 1997); worse yet, some leaders are often enablers and instigators of aggression themselves through misuse of authority (Hutchinson et al. 2009). Miscommunication has also been identified as a substantial contributor to aggression, especially from patients and visitors, because these individuals may feel that miscommunication with nursing staff is indicative of a lack of care and respect (Najafi et al. 2018). Finally, social alliances, especially with superiors, can contribute to social norms in which aggression is tolerated or even rewarded through institutional or financial gains (Hutchinson et al. 2009, 2010).

At the organizational level, aspects of the nursing work role and work environment may interact to cause violence. Nursing work is characterized by unpredictability (Almost et al. 2010), high workload (Roche et al. 2010; Najafi et al. 2018) and low autonomy (Roche et al. 2010), especially as a shortage of the nursing workforce plagues most nursing environments (Roche et al. 2010; Najafi et al. 2018). Aspects of the nursing environment also increase the likelihood of aggression, such as an undisciplined working environment (Najafi et al. 2018), social norms surrounding the tolerance and/or encouragement of aggression for political gain (Fisher et al. 1995), and aspects of the physical environment (Duxbury and Whittington 2005). Future interventions focused on violence prevention climate and aggression preventive supervisor behaviors leading to violence prevention climate may effectively reduce nurses' exposure to violence (Yang et al. 2014; Spector et al. 2015; Yang and Caughlin 2017).

Practical Implications for Organizations

The literature reviewed in this chapter has numerous practical implications for reducing and preventing violence in the field of nursing. Since violence is an inherently interactional process, and individual-level antecedents to violence are largely demographic and unchangeable in nature, it is more likely that successful interventions to reduce violence occur at the interpersonal or organizational level. Indeed, Hutchinson et al. (2010) and Smith et al. (2018) argue that changes should be implemented at the organizational level, instead of addressing interpersonal conflict or individual personality differences, to be most effective. Three recommendations emerge from the literature reviewed here: (1) to ensure adequate resources and staffing, (2) to adopt and enforce policies aimed at preventing and addressing violence, and (3) to provide training and continued support to nurses and nurse managers and supervisors to prevent, recognize and report violence.

Staffing
According to projections by the Bureau of Labor Statistics (Hogan and Roberts 2015), healthcare professions will see the fastest job growth in the United States through to 2024. These projections indicate that the nursing shortage at the time of writing is not likely to improve, creating an environment that will indirectly increase the likelihood of aggression. Therefore, it should be a priority of healthcare organizations to ensure adequate nurse staffing and resources (Smith et al. 2018), as these can reduce the high workload and time pressure that increase the likelihood of violence.

Policies and procedures
As social norms regarding violence are influenced by leader behaviors, it is important to provide clear policies and procedures for nurse managers and supervisors to address aggression among their staff (Hoffman and Chunta 2015). Moreover, because job demands are high for nurse managers and supervisors, it is important that clear policies are in place to ease the challenges that many nurse managers and supervisors face. In addition, required managerial actions should be clearly delineated within organizational policies to promote accountability among these leaders (Hoffman and Chunta 2015).

Training and organizational support
Provision of training and resources to nurses and nurse leaders can improve nursing staff's prevention, recognition and confrontation of violence (Yang and Caughlin 2017). Nurses in some studies have noted a need for training regarding management of violence and aggression (Pinar and Ucmak 2011). Some literature has demonstrated that training can prove effective in these goals. Preventing violence, especially from patients, can be accomplished through improved communication skills. Kamchuchat et al. (2008) demonstrated that training in communication skills helped nurses clarify patient needs and identify appropriate responses, reducing the risk of verbal abuse in their sample by 40 percent. In order to improve

recognition of violence at work, Nikstaitis and Simko (2014) offered a 60-minute educational program to registered nurses, and found that participants were better able to recognize and be aware of incivility in their workplaces after taking the program. An intervention focused on civility (courteous and considerate behavior toward other people; Andersson and Pearson 1999), respect and engagement at work using randomized control trials among healthcare workers conducted by Leiter et al. (2011) demonstrated successful improvement in coworker and supervisor civility and in management trust.

While training can be beneficial to preventing and recognizing violence, addressing these instances may require more consistent organizational support. Nurses have discussed fear of retaliation, lack of support and inaction at the organizational level as reasons for under-reporting aggressive incidents and creating an environment that is conducive to these incidents (Kvas and Seljak 2014). Hutchinson et al. (2010) argue that healthcare organizations should aim to support nurse managers in being alert to instances of violence, especially groups of individuals who may engage in structural forms of violence through restructuring, performance reviews or disciplinary procedures. Moreover, Smith et al. (2018) urge that hospital administrations must support nurse managers in monitoring, evaluating and addressing staff and resource adequacy, as a lack of appropriate staff and resources increases the likelihood of violence. In order to achieve these goals, Smith et al. (2018) recommend resources such as public group forums with other nurse managers to identify shared solutions to problems, and increasing communication with, and support for, nurse managers.

NOTE

1. A full list of the 64 articles found in our review is available from the corresponding author.

REFERENCES

* Indicates article was included in the review.

*AbuAlRub, R.F. and Al-Asmar, A.H. (2014), 'Psychological violence in the workplace among Jordanian hospital nurses', *Journal of Transcultural Nursing*, **25** (1), 6–14, doi:10.1177/1043659613493330.
*Almost, J., Doran, D.M., McGillis Hall, L. and Spence Laschinger, H.K. (2010), 'Antecedents and consequences of intra-group conflict among nurses',

Journal of Nursing Management, **18** (8), 981–92: doi:10.1111/j.1365-2834.2010. 01154.x.

Andersson, L.M. and Pearson, C.M. (1999), 'Tit for tat? The spiraling effect of incivility in the workplace', *Academy of Management Review*, **24** (3), 452–71, doi:10.5465/AMR.1999.2202131.

*Bilgin, H. (2009), 'An evaluation of nurses' interpersonal styles and their experiences of violence', *Issues in Mental Health Nursing*, **30** (4), 252–9, doi:10.1080 /01612840802710464.

*Blackstock, S., Harlos, K., Macleod, M.L.P. and Hardy, C.L. (2015), 'The impact of organisational factors on horizontal bullying and turnover intentions in the nursing workplace', *Journal of Nursing Management*, **23** (8), 1106–14, doi:10.1111/jonm.12260.

*Bortoluzzi, G., Caporale, L. and Palese, A. (2014), 'Does participative leadership reduce the onset of mobbing risk among nurse working teams?', *Journal of Nursing Management*, **22** (5), 643–52, doi:10.1111/jonm.12042.

Bowling, N.A. and Beehr, T.A. (2006), 'Workplace harassment from the victim's perspective: a theoretical model and meta-analysis', *Journal of Applied Psychology*, **91** (5), 998–1012.

*Brewer, C.S., Kovner, C.T., Obeidat, R.F. and Budin, W.C. (2013), 'Positive work environments of early-career registered nurses and the correlation with physician verbal abuse', *Nursing Outlook*, **61** (6), 408–16, doi:10.1016/j. outlook.2013.01.004.

*Budin, W.C., Brewer, C.S., Chao, Y.-Y. and Kovner, C. (2013), 'Verbal abuse from nurse colleagues and work environment of early career registered nurses', *Journal of Nursing Scholarship: An Official Publication of Sigma Theta Tau International Honor Society of Nursing*, **45** (3), 308–16, doi:10.1111/jnu.12033.

Cameron, L. (1998), 'Verbal abuse: a proactive approach', *Nursing Management*, **29** (8), 34–6.

*Campbell, J.C., Messing, J.T., Kub, J., Agnew, J., Fitzgerald, S., Fowler, B. et al. (2011), 'Workplace violence: prevalence and risk factors in the safe at work study', *Journal of Occupational and Environmental Medicine*, **53** (1), 82–9, doi:10.1097/JOM.0b013e3182028d55.

*Celik, S.S., Celik, Y., Ağirbaş, I. and Uğurluoğlu, O. (2007), 'Verbal and physical abuse against nurses in Turkey', *International Nursing Review*, **54** (4), 359–66, doi:10.1111/j.1466-7657.2007.00548.x.

*Chen, W.-C., Huang, C.-J., Hwang, J.-S. and Chen, C.-C. (2010), 'The relationship of health-related quality of life to workplace physical violence against nurses by psychiatric patients', *Quality of Life Research: An International Journal of Quality of Life Aspects of Treatment, Care and Rehabilitation*, **19** (8), 1155–61, doi:10.1007/s11136-010-9679-4.

*Chen, W.-C., Sun, Y.-H., Lan, T.-H. and Chiu, H.-J. (2009), 'Incidence and risk factors of workplace violence on nursing staffs caring for chronic psychiatric patients in Taiwan', *International Journal of Environmental Research and Public Health*, **6** (11), 2812–21, doi:10.3390/ijerph6112812.

Cook, J.K., Green, M. and Topp, R.V. (2001), 'Exploring the impact of physician verbal abuse on perioperative nurses', *AORN Journal*, **74** (3), 317–31, doi:10.1016/S0001-2092(06)61787-0.

*Cortina, L.M., Magley, V.J., Williams, J.H. and Langhout, R.D. (2001), 'Incivility in the workplace: incidence and impact', *Journal of Occupational Health Psychology*, **6** (1), 64–80.

*Crilly, J., Chaboyer, W. and Creedy, D. (2004), 'Violence towards emergency department nurses by patients', *Accident and Emergency Nursing*, **12** (2), 67–73, doi:10.1016/j.aaen.2003.11.003.

*Deans, C. (2004), 'Who cares for nurses? The lived experience of workplace aggression', *Collegian*, **11** (2), 32–6, doi:10.1016/S1322-7696(08)60453-9.

*Demir, D. and Rodwell, J. (2012), 'Psychosocial antecedents and consequences of workplace aggression for hospital nurses', *Journal of Nursing Scholarship: An Official Publication of Sigma Theta Tau International Honor Society of Nursing*, **44** (4), 376–84, doi:10.1111/j.1547-5069.2012.01472.x.

*Duxbury, J. and Whittington, R. (2005), 'Causes and management of patient aggression and violence: staff and patient perspectives', *Journal of Advanced Nursing*, **50** (5), 469–78, doi:10.1111/j.1365-2648.2005.03426.x.

Einarsen, S. (2000), 'Harassment and bullying at work: a review of the Scandinavian approach', *Aggression and Violent Behavior*, **5** (4), 379–401, doi:10.1016/S1359 -1789(98)00043-3.

Etienne, E. (2014), 'Exploring workplace bullying in nursing', *Workplace Health & Safety*, **62** (1), 6–11, doi:10.1177/216507991406200102.

*Farrell, G.A. (1997), 'Aggression in clinical settings: nurses' views', *Journal of Advanced Nursing*, **25** (3), 501–8, doi:10.1046/j.1365-2648.1997.1997 025501.x.

*Fisher, J., Bradshaw, J., Currie, B.A., Robins, P. and Smith, J. (1995), '"Context of silence": violence and the remote area nurse', research report, Central Queensland University, Rockhampton, accessed 22 May 2019 at https://pdfs. semanticscholar.org/9af3/2231b78172bdbc4b1dcb8e246aae9e36da6c.pdf?_ga=2 .66886252.391188819.1557181607-673549889.1557181607.

Hershcovis, M.S. and Barling, J. (2010), 'Comparing victim attributions and outcomes for workplace aggression and sexual harassment', *Journal of Applied Psychology*, **95** (5), 874–88.

*Hesketh, K.L., Duncan, S.M., Estabrooks, C.A., Reimer, M.A., Giovannetti, P., Hyndman, K. et al. (2003), 'Workplace violence in Alberta and British Columbia hospitals', *Health Policy (Amsterdam, Netherlands)*, **63** (3), 311–21.

Hoffman, R.L. and Chunta, K. (2015), 'Workplace incivility: promoting zero tolerance in nursing', *Journal of Radiology Nursing*, **34** (4), 222–7, doi:10.1016/j. jradnu.2015.09.004.

Hogan, A. and Roberts, B. (2015), 'Occupational employment projections to 2024: monthly labor review', US Bureau of Labor Statistics, accessed 26 July 2018 at https://www.bls.gov/opub/mlr/2015/article/occupational-employment-projec tions-to-2024.htm.

*Hutchinson, M., Vickers, M., Jackson, D. and Wilkes, L. (2006), 'Workplace bullying in nursing: towards a more critical organisational perspective', *Nursing Inquiry*, **13** (2), 118–26, doi:10.1111/j.1440-1800.2006.00314.x.

*Hutchinson, M., Vickers, M.H., Wilkes, L. and Jackson, D. (2009), '"The worse you behave, the more you seem, to be rewarded": bullying in nursing as organizational corruption', *Employee Responsibilities and Rights Journal*, **21** (3), 213–29, doi:10.1007/s10672-009-9100-z.

*Hutchinson, M., Wilkes, L., Jackson, D. and Vickers, M.H. (2010), 'Integrating individual, work group and organizational factors: testing a multidimensional model of bullying in the nursing workplace', *Journal of Nursing Management*, **18** (2), 173–81, doi:10.1111/j.1365-2834.2009.01035.x.

*Islam, T., Ahmed, I. and Ali, G. (2018), 'Effects of ethical leadership on bullying and voice behavior among nurses: mediating role of organizational identification, poor working condition and workload', *Leadership in Health Services*, **32** (1), 2–17, doi:0.1108/LHS-02-2017-0006.

*Kaiser, J.A. (2017), 'The relationship between leadership style and nurse-to-nurse incivility: turning the lens inward', *Journal of Nursing Management*, **25** (2), 110–18, doi:10.1111/jonm.12447.

*Kamchuchat, C., Chongsuvivatwong, V., Oncheunjit, S., Yip, T.W. and Sangthong, R. (2008), 'Workplace violence directed at nursing staff at a general hospital in southern Thailand', *Journal of Occupational Health*, **50** (2), 201–07.

*Katrinli, A., Atabay, G., Gunay, G. and Cangarli, B.G. (2010), '"Nurses" perceptions of individual and organizational political reasons for horizontal peer bullying', *Nursing Ethics*, **17** (5), 614–27, doi:10.1177/0969733010368748.

Keashly, L. (1997), 'Emotional abuse in the workplace: conceptual and empirical issues', *Journal of Emotional Abuse*, **1** (1), 85–117.

Kvas, A. and Seljak, J. (2014), 'Unreported workplace violence in nursing', *International Nursing Review*, **61** (3), 344–51, doi:10.1111/inr.12106.

Leiter, M.P., Laschinger, H.K.S., Day, A. and Oore, D.G. (2011), 'The impact of civility interventions on employee social behavior, distress, and attitudes', *Journal of Applied Psychology*, **96** (6), 1258–74.

*Najafi, F., Fallahi-Khoshknab, M., Ahmadi, F., Dalvandi, A. and Rahgozar, M. (2018), 'Antecedents and consequences of workplace violence against nurses: a qualitative study', *Journal of Clinical Nursing*, **27** (1–2), e116–e128, doi:10.1111/jocn.13884.

Nikstaitis, T. and Simko, L.C. (2014), 'Incivility among intensive care nurses: the effects of an educational intervention', *Dimensions of Critical Care Nursing*, **33** (5), 293–301.

*Pai, H.-C. and Lee, S. (2011), 'Risk factors for workplace violence in clinical registered nurses in Taiwan', *Journal of Clinical Nursing*, **20** (9–10), 1405–12, doi:10.1111/j.1365-2702.2010.03650.x.

*Park, L.S. and Martinez, L.R. (2018), 'Selective incivility and well-being in nursing', paper presented at the 2018 Annual APA Convention, San Francisco, CA, August.

*Pich, J.V., Kable, A. and Hazelton, M. (2017), 'Antecedents and precipitants of patient-related violence in the emergency department: results from the Australian VENT Study (violence in emergency nursing and triage)', *Australasian Emergency Nursing Journal: AENJ*, **20** (3), 107–13, doi:10.1016/j.aenj.2017.05.005.

Pinar, R. and Ucmak, F. (2011), 'Verbal and physical violence in emergency departments: a survey of nurses in Istanbul, Turkey', *Journal of Clinical Nursing*, **20** (3–4), 510–17, doi:10.1111/j.1365-2702.2010.03520.x.

*Roche, M., Diers, D., Duffield, C. and Catling-Paull, C. (2010), 'Violence toward nurses, the work environment, and patient outcomes', *Journal of Nursing Scholarship*, **42** (1), 13–22, doi:10.1111/j.1547-5069.2009.01321.x.

*Rodwell, J. and Demir, D. (2014), 'Addressing workplace violence among nurses who care for the elderly', *Journal of Nursing Administration*, **44** (3), 152–7, doi:10.1097/NNA.0000000000000043.

*Rodwell, J., Demir, D. and Gulyas, A. (2015), 'Individual and contextual antecedents of workplace aggression in aged care nurses and certified nursing assistants', *International Journal of Nursing Practice*, **21** (4), 367–75, https://doi.org/10.1111/ijn.12262.

*Rodwell, J., Demir, D. and Steane, P. (2013), 'Psychological and organizational impact of bullying over and above negative affectivity: a survey of two nursing contexts', *International Journal of Nursing Practice*, **19** (3), 241–48, doi:10.1111/ijn.12065.

Rowe, M.M. and Sherlock, H. (2005), 'Stress and verbal abuse in nursing: do burned out nurses eat their young?', *Journal of Nursing Management*, **13** (3), 242–8, doi:10.1111/j.1365-2834.2004.00533.x.

Simons, S. (2008), 'Workplace bullying experienced by Massachusetts registered nurses and the relationship to intention to leave the organization', *ANS. Advances in Nursing Science*, **31** (2), E48–59, doi:10.1097/01.ANS.0000319571.37373.d7.

*Simons, S.R. and Mawn, B. (2010), 'Bullying in the workplace – a qualitative study of newly licensed registered nurses', *AAOHN Journal: Official Journal of the American Association of Occupational Health Nurses*, **58** (7), 305–11, doi:10.3928/08910162-20100616-02.

*Smith, J.G., Morin, K.H. and Lake, E.T. (2018), 'Association of the nurse work environment with nurse incivility in hospitals', *Journal of Nursing Management*, **26** (2), 219–26, doi:10.1111/jonm.12537.

*Spector, P.E., Yang, L.Q. and Zhou, Z.E. (2015), 'A longitudinal investigation of the role of violence prevention climate in exposure to workplace physical violence and verbal abuse', *Work & Stress*, **29** (4), 325–40.

Spence Laschinger, H.K., Leiter, M., Day, A. and Gilin, D. (2009), 'Workplace empowerment, incivility, and burnout: impact on staff nurse recruitment and retention outcomes', *Journal of Nursing Management*, **17** (3), 302–11, doi:10.1111/j.1365-2834.2009.00999.x.

*Speroni, K.G., Fitch, T., Dawson, E., Dugan, L. and Atherton, M. (2014), 'Incidence and cost of nurse workplace violence perpetrated by hospital patients or patient visitors', *Journal of Emergency Nursing*, **40** (3), 218–28, doi:10.1016/j.jen.2013.05.014.

Task Force on the Prevention of Workplace Bullying (2001), *Report of the Task Force on the Prevention of Workplace Bullying: Dignity at Work: The Challenge of Workplace Bullying*, Dublin: Health and Safety Authority, accessed 20 July 2018 at http://www.hsa.ie/eng/Publications_and_Forms/Publications/Safety_and_Health_Management/Report_of_the_Task_Force_on_the_Prevention_of_Workplace_Bullying.html.

*Yang, L.Q. and Caughlin, D.E. (2017), 'Aggression-preventive supervisor behavior: implications for workplace climate and employee outcomes', *Journal of Occupational Health Psychology*, **22** (1), 1–18.

Yang, L.Q., Caughlin, D.E., Gazica, M.W., Truxillo, D.M. and Spector, P.E. (2014), 'Workplace mistreatment climate and potential employee and organizational outcomes: a meta-analytic review from the target's perspective', *Journal of Occupational Health Psychology*, **19** (3), 315–35.

PART III

High-Risk Occupations

9. Pesticide exposure and the health effects among Latino and other farmworkers

Joseph G. Grzywacz, John S. Luque and Alan Becker

INTRODUCTION

Pesticides are simultaneously a wonderful deliverance and perennial challenge for human health. Pesticides deliver public health benefits in many ways. Pesticides serve as control systems to manage disease vectors such as mosquitos and vermin and their implications for infectious disease such as malaria and bubonic plague. Indeed, it was contended in a 1948 Nobel Laureate award acceptance speech by Dr Paul Muller, a chief architect of dichlorodiphenyltrichloroethane (DDT), that the pesticide 'preserved the life and health of hundreds of thousands' from diseases like malaria, yellow fever and the like. More recently the World Health Organization estimated that DDT was responsible for saving 25 million lives, and there have been calls for reinstating it as an agent in the wake of recent outbreaks such as the Zika virus. Pesticides are also lauded because of their contributions to food security. That is, pesticides enable greater food production my minimizing damage to plants and their fruits from insects and fungus, and they minimize food loss by repelling animal and insect pests, thereby resulting in greater food availability to humans.

However, pesticides are also public health threats. The environmental health movement, birthed by the publication in 1962 of Rachel Carson's *Silent Spring*, was galvanized by widespread use of DDT and other organochlorine pesticides to control insects. The fundamental question pursued by environmental health researchers is, 'What are the unanticipated health-related consequences of widespread use of chemicals to control pests?' Early scientists and activists noted potential threats of acid rain, contaminated plant and animal food sources, chemical weapons and ecological imbalances introduced by pesticides. Recent debates about the use of neonicotinoids and its implications for bees and subsequent pollination of plants reflects the most recent installment of these concerns. Finally, in the realm of occupational health and safety, concerns about the potential health effects of occupational exposure to pesticides have persisted and several major bodies including the US National Institute for Occupational Safety and Health (NIOSH) and the World Health Organization (WHO) characterize pesticides as a major threat to the health and safety of farmers and farmworkers.

The overall goal of this chapter is to describe the state of the science in pesticide exposure among farmworkers and the potential health effects of that exposure. To achieve this goal, the chapter begins by providing a basic foundation about the role of agriculture and, by extension, pesticide in the world economy. This basic foundation also provides a primer on key issues needed to understand what pesticides are, what pesticide exposure is and how endemic pesticides are in most human environments. The chapter then moves into a summary of the literature on farmworker exposure to pesticides, including both a comparison of exposure among farmworkers relative to non-farmworkers and studies of within-occupation variability in pesticide exposure. Third, given the complexity of pesticides and human health, the chapter provides a broad overview of evidence linking pesticide exposure to health outcomes. This broad overview is necessarily offset by a clear case study that typifies the debate about the scientific underpinnings of the pesticide-health linkage, and important areas for future research in the field.

BASIC FOUNDATIONS

Agriculture is a hidden giant. Agriculture contributed $3.3 trillion dollars to the global economy in 2017, which is nearly twice the contribution made by agriculture in 2000. The financial contribution of agriculture to a country's gross domestic product (GDP) is inversely related to the strength of a nation's economy: in prosperous countries such as the US, the percentage of GDP attributed to agriculture is small, perhaps 1–2 percent, whereas the percentage of GDP attributed to agriculture in developing countries like those of sub-Saharan Africa is 40–60 percent (Roser 2019). Similarly, in the developed world, agriculture contributes a relatively small share to the total workforce, whereas in developing countries agriculture accounts for major portions of the workforce. Estimates from the International Labor Organization (ILO) for 2018 indicate agriculture accounts for 3.1 percent employment in high income countries, but fully 68.9 percent of employment in low income countries (ILO 2018).

The agricultural workforce is estimated to consist of over 860 million individuals globally. In developing economies where agriculture is frequently for subsistence, the workforce looks much like the general population, but with over-representation of women and children (Food and Agriculture Organization of the United Nations 2011). However, in more developed economies, the agricultural workforce is difficult to describe. For example, the United States Department of Agriculture (USDA 2018) estimates that over half (60.6 percent) of the hired agricultural workforce consists of US citizens of Hispanic ethnicity (52.5 percent). The USDA also reports that about 25 percent of the US agricultural workforce is female, and the greatest proportion of agricultural workers lack a high school diploma (43.5 percent). By contrast, estimates from the past seven years of National Agricultural Workers Survey (NAWS) data suggest that nearly three-fourths of farmworkers (73.8 percent) are foreign born, primarily from Mexico, four of five (80 percent) are Hispanic and over 60 percent have nine or fewer years of education (Grzywacz et al. 2018).

Any attempt to understand the health effects of pesticide exposure requires a basic understanding of several key ideas. First, pesticide is a broad and multifaceted concept. The US Environmental Protection Agency (EPA), the agency charged with overseeing regulations and laws related to pesticides, define them as '*any* substance or mixture of substances intended for preventing, destroying, repelling or mitigating *any* pest' (EPA n.d., emphases added). The broadness of this definition is remarkable both in terms of the target (that is, any pest) and the agent (that is, any substance or mixture of substances). Pests take a wide variety of forms including insects, plants (sometimes referred to as weeds), fungi, and rodents or other animals. Bleach, when used to disinfect (that is, kill bacteria) a kitchen counter is as much a pesticide as Agent Orange, the defoliant used as a chemical weapon in the Vietnam War. Indeed, virtually every household in the developed world uses one or more pesticides regularly, recognizing that all antibacterial agents (for example, hand sanitizer, household cleaners), insect repellants or foggers used during backyard picnics, and applications to keep lawns green and weed free are all pesticides.

In agriculture, which accounts for 90 percent of pesticide use in the US (EPA 2017), the complexity of pesticides increases exponentially. As Alavanja et al. (2004) report, most pesticides available on the market are compounds consisting of multiple active and other ingredients. The active ingredient typically refers to the substance or agent used to prevent, destroy, repel or mitigate the pest. The most common active ingredients for herbicides (pesticides targeting plants or weeds) are glyphosate and atrazine, whereas the most common active ingredient for insecticides (pesticides targeting insects) and fungicides (pesticides targeting fungi) are chlorpyrifos and chlorothalonil, respectively. The other ingredients are a broad range of substances to allow different modes of application (for example, spray versus dust versus granules) or make the pesticide more effective. The WHO (2018) estimates there are more than 1000 different pesticides used around the world, each comprised of a mixture of shared and unique active and other ingredients. The vast majority of other ingredients are unknown because the business creating the pesticide holds the ingredient list as proprietary information (Fenner-Crisp 2010).

Pesticide exposure simply means coming into contact with a pesticide. The Sentinel Event Notification System for Occupational Risks (SENSOR)-Pesticides, a surveillance system administered by NIOSH, classifies five distinct types of exposure (Calvert et al. 2004). Direct contact with pesticide can occur through a deliberate act (for example, drinking pesticide with the intent of harming yourself) or non-deliberate action such as a spill or leaking pesticide container. Exposure can occur through spray, either deliberately as when an individual applies an insect repellant to their skin, or inadvertently as when walking in an agricultural field when an airplane is engaged in aerial spraying overhead. Exposure to pesticide can occur by coming into contact with treated surfaces (for example, plants and soil), either immediately after application or while the residues of the pesticide remain active. Exposure can occur through contamination, such as air or water sources. Finally, exposure can occur through drift, wherein pesticide being applied to a target area is moved by wind or air vapor to another area.

Pesticides are everywhere. An estimated 50 million people in the US are believed to have pesticide in their water supply (Nielsen and Lee 1987), and the US Geological Service (2014) reported that five or more organophosphate pesticides were detected in the air, rain, snow and fog at sensor stations located in cities around the US. Pesticide presence in water, the water cycle and air illustrate the contamination route to exposure. A study of one rural state in the US conservatively estimated that one in five residences (22 percent) are exposed to pesticides because of proximity to fields (Ward et al. 2000), illustrating exposure via drift. Pesticides are, not surprisingly, in the food supply and illustrate an example of surface exposure to pesticides. The degree to which pesticide are present in the food supply is widely believed to have improved over the past 20 years; nevertheless, insecticides such as befenthrin have been detected in 30 percent of samples of kale and 18 percent of samples of canned tomatoes evaluated through the Pesticide Data Program (United States Department of Agriculture, Agricultural Marketing Service 2018). Similarly, over one in five samples of apple sauce had detected flubendlamide (an insecticide), and about one in seven samples of grapefruit (16.5 percent) and mango (14.1 percent) detected imidacloprid (14.1 percent). Consequently, although pesticide was not detected in 99 percent of samples across a wide variety of domestic and imported foods, and when they were detected residue levels were frequently below the EPA tolerance level, it is equally clear that pesticides are in food.

The ecological data just used to illustrate routes of pesticide exposure and the endemic presence in the environment are supported by observational estimates from the US National Health and Nutrition Examination Survey (NHANES). Organophosphates, a common class of pesticides used in US agriculture, breaks down into diaklyphosphate (DAP) metabolites observable in urine: these metabolites are part of the US Centers for Disease Control's (CDC) National Biomonitoring Program. Between 1999 and 2008, the average urine sample had quantifiable, non-zero levels of three of the six DAP metabolites, specifically diethylphosphate, diethylthiophosphate and dimethylthiophosphate (CDC 2018). Further, there was noteworthy variation in the geometric mean of DAP metabolites within the population. For example, in the most recent NHANES panel (2007–08), the average creatinine corrected levels of dimethylthiophosphate were 4.35 (3.78–5.01) and 2.02 (1.73–2.36) µg/g of creatinine for children aged 6–11 years and adults aged 20–59 years, respectively (CDC 2018). Greater exposure to pesticides among children is not unexpected given their greater proximity to the floor where pesticide residues accumulate, together with behavioral attributes such as hand-to-mouth behavior (Hyland and Laribi 2017).

PESTICIDE EXPOSURE AMONG FARMWORKERS: WHAT DO WE KNOW?

Farmworkers are widely believed to have greater exposure to pesticides. Calvert et al. (2004) used data from the SENSOR pesticide surveillance system and found the incidence of pesticide-related illness was 18 times higher for workers in the agricultural sector compared with all non-agricultural occupational sectors combined. Subsequent research using the same data source considered acute pesticide poisoning or any acute or sub-acute illness or injury resulting from pesticide exposure (Calvert et al. 2008). Over 3200 cases of acute pesticide poisoning were documented between 1998 and 2005; the vast majority (87 percent) were classified as 'low severity', one was 'fatal' and 20 (less than 0.1 percent) were 'high severity'. Total incidence of acute pesticide poisoning was consistently higher from 1998 through to 2005 among workers in the agricultural sector relative to workers in the non-agricultural sectors (Calvert et al. 2008).

Results obtained from surveillance systems such as the SENSOR need to be interpreted with caution. SENSOR cases likely under-count events (pesticide-related illness or acute pesticide poisoning) because they rely on reports from individuals who visited a healthcare provider and on the accuracy of diagnosis (Calvert et al. 2004; Mehler et al. 2006). Commonly reported symptoms of pesticide exposure, such as headaches, burning eyes, pain or soreness in muscles, joints and bones, or rashes and itchy skin (Strong et al. 2004), are generally associated with a variety of ailments. Farmworkers typically do not report exposure or visit healthcare providers over self-appraised minor health problems because of lost wages, fear of work dismissal or threats of deportation, limited English proficiency, and limited access to healthcare or public health authorities (Prado et al. 2017). Nevertheless, estimates obtained from surveillance systems such as SENSOR do provide evidence that employment in agriculture increases the risk of pesticide exposure.

A growing number of epidemiological studies also provide evidence that farmworkers have elevated exposure to pesticides. One study documented notably higher urinary metabolite concentrations for several pesticides among farmers who applied their own pesticides relative to non-farmers (Curwin et al. 2005). Evidence from a sample of workers in Yakima Valley, Washington in the US indicated substantially elevated concentrations of dimethylthiophosphate, a diaklyphosphate (DAP) metabolite, among workers performing apple thinning compared to farmworkers from whom urine samples were obtained outside of peak season (Fenske et al. 2005). A study in eastern Washington State of the US documented several differences in DAP urinary metabolites, suggesting that farmworkers had greater exposure to pesticides than non-farmworkers (Thompson et al. 2014).

There is also negative evidence suggesting that farmworkers are not at elevated risk for added occupational exposure to pesticide. Arcury et al. (2016, p. 1084), using data from a rigorous matched comparison design, reported '[W]ith few exceptions, the farmworkers and non-farmworkers had similar levels of detection and concentration for the DAP urinary metabolites'. Arcury and colleagues noted some instances where farmworkers had a greater frequency of DAP detections or higher concentrations of DAP metabolites than non-farmworkers, but these were exceptions. Plainly, there is some evidence of differential exposure among farmworkers, but the preponderance of evidence is that farmworkers may not face elevated exposure to pesticides compared with non-farmworkers.

A subsequent report focused on specific pesticide metabolites, as DAP metabolites are reflective of any organophosphate pesticide, yielded a similar pattern of results indicating similarity between farmworkers and non-farmworkers (Arcury et al. 2018).

INTRA-OCCUPATIONAL VARIABILITY IN PESTICIDE EXPOSURE AMONG FARMWORKERS

Putting differential exposure to pesticides by farmworkers relative to non-farmworkers aside, there is evidence of variation in pesticide exposure among farmworkers. Temporal variation in pesticide exposure seems to exist among farmworkers. Drawing on surveillance of pesticide poisoning cases in California across the 1990s, Reeves and Schafer (2003) reported a reduction from 665 cases per year in the early 1990s to 475 cases per year in the later 1990s. Calvert et al.'s (2008) analysis of national SENSOR data indicated meaningful year-to-year variation in incidence rates of acute pesticide poisoning, whereby case incidence declined from 1998 to 2001, spiked in 2002 but then returned to lower but still elevated (relative to the low in 2001) incidence levels from 2003 through to 2005. Within a single year, Thompson et al. (2014) documented variation in DAP urinary metabolites that coincided with times in the agricultural season where pesticides were typically sprayed. Consistent with this idea, others (Arcury et al. 2009) documented a notable linear trend wherein median urinary levels of dimethylthiophosphate, one of the DAP urinary metabolites, was lowest during the period when pesticides are not usually applied (that is, May–June) relative to later periods when pests (that is, insects and weeds) become more problematic and possibly require greater use of pesticide.

An interesting sub-text of Arcury et al. (2009) is that DAP metabolites were present in farmworkers' urine samples throughout the agricultural season, suggesting chronic exposure to pesticide across time. Further, they also noted that more than half the farmworkers were exposed to a variety of distinct classes of insecticide (organophosphate and pyrethroid) together with fungicides and herbicides throughout the year (Arcury et al. 2010), suggesting chronic exposure to a combination of pesticide. Nevertheless, studies of farmworkers' occupational exposure to pesticide typically do not have a comparison group of non-farmworkers. Given that pesticides are endemic in the environment and that drift is the primary route for exposure (Reeves and Schafer 2003), it is difficult to determine whether observations of pesticide metabolites in farmworkers' urine are reflective of high and frequent exposure to pesticides from work-related tasks, or whether it is risk of rural dwelling or proximity to agricultural operations (Coronado et al. 2011).

The re-entry interval is a time-based concept believed to affect farm-workers' occupational exposure to pesticide. The re-entry interval is the period of time between applying a pesticide and when workers are allowed to perform work tasks in the pesticide-treated area (EPA 2018). Baldi et al.'s (2014) study of farmworkers in the Bordeaux region of France reported higher levels of total exposure to pesticide among those performing agricultural tasks during the re-entry interval compared with farmworkers who performed harvest tasks. Others have suggested that farmworkers in the developing world, such as Africa, give little regard to the concept of re-entry intervals and may result in unnecessarily high exposure to pesticide (Ntow et al. 2009). Based on studies such as these, as well as laboratory-based toxicological studies, the US EPA's Worker Protection Standard requires that farmworkers be restricted from pesticide-treated areas until the re-entry interval stated on the pesticide label elapses.

Discrete job tasks create an increased risk of pesticide exposure. It seems that working in crops that bring the whole body into contact with plant foliage, such as grapes, oranges and cotton, increases the risk of exposure relative to crops with lesser foliage, such as strawberries or broccoli (Reeves and Schafer 2003). Tasks such as raising wires for grape vines and hand-picking grapes, both of which bring farmworkers' forearms into repeated contact with grape foliage, is associated with elevated pesticide exposure (Baldi et al. 2014). In their case-control study, Curwin et al. (2005) consistently showed elevated levels of five pesticide metabolites among farmers who applied pesticides themselves, relative to those who did not apply pesticides or had another person apply them. Calvert et al.'s (2008) study of acute pesticide poisoning demonstrated substantially elevated incidence rates for workers in agricultural processing or packing (362.6 per 100 000 full time workers) relative to workers in the field (74.8 per 100 000). Contrary to expectations, Coronado et al. (2004) found that farmworkers who reported recent mixing, loading or applying pesticides noted lower, although not statistically different, levels of pesticide in dust samples obtained from their home and vehicle than those who performed different tasks. Arcury et al. (2009, 2010) assessed job tasks, such as having mixed or loaded pesticides in the three days prior to the collection of farmworkers' urine samples, and the type of crop worked in but have not reported any variation in pesticide exposure by these attributes.

There is some evidence of demographic variability in pesticide exposure among farmworkers. The incidence of pesticide-related illness (Calvert et al. 2004) and acute pesticide poisoning (Calvert et al. 2008) is greatest among young adult (that is, ages 18–24 years) farmworkers and shows a linear decline across the rest of adulthood. For example, the incidence of acute pesticide poisoning among young adult farmworkers is 76.2 per 100 000 full-time workers (Calvert et al. 2008), whereas it is incrementally lower for workers aged 35–44 years (46.2 per 100 00 full-time workers) and 45–54 (32.9 per 100 000 full-time workers). Similarly, López-Gálvez et al. (2018) reported that urinary concentrations of one organophosphate pesticide, chlorpyrifos, was lower for older workers than for younger farmworkers in Mexico. Others (Kasner et al. 2012) have suggested that female farmworkers have twice the incidence rate of pesticide-related illness than males, although the absolute number of pesticide-related illness is substantially higher for males given their vast overrepresentation in the occupation. Baldi et al. (2014) also reported that women had greater exposure to pesticide than men in both the re-entry period and harvest periods, possibly reflecting gender segregation of work tasks. Arcury et al. (2010) reported no variation in detection or concentration of urinary pesticide metabolites, indicating potential pesticide exposure, by any farmworker demographic characteristic such as age, gender, educational attainment or worker type.

One study considered the role of farmworker-reported work characteristics in pesticide exposure (Grzywacz et al. 2010). Guided by the demand-control model (Karasek and Theorell, 1992), the investigators asked whether DAP urinary metabolites varied by demands experienced on the job (that is, psychological demand and physical exertion) and control over job tasks. Contrary to the main effects hypothesis, Grzywacz et al. (2010) found no evidence that detection of any of the DAP urinary metabolites was greater when psychological demand or physical exertion were elevated or control was low. However, a pattern of interactions emerged indicating that the association of physical exertion, a demand characteristic, was modified by psychosocial aspects. For example, when control was low the probability of detecting two of six DAP metabolites, diethyldithiophosphate and dimethylthiophosphate, increased as physical exertion increased. By contrast, when control was high, increases in physical exertion was associated with lower probability of detecting these DAP urinary metabolites.

Willful activities on the part of farmworkers can lessen exposure to pesticides. Clothing worn while working and personal hygiene behaviors have been linked to variation in pesticide exposure (Salvatore et al. 2008). Specifically, wearing long-sleeved shirts and trousers (to minimize contact with plant foliage), wearing clean clothes each work day (as opposed to re-wearing clothes without laundering them) and more frequent hand-washing with soap were all independently associated with lower urinary concentrations of dimethylalkylphosphate. Each of these activities, and others, are codified in the revised Worker Protection Standard, which requires annual pesticide training for all farmworkers (Brennan et al. 2015). Meta-analytic results suggest that layered clothing likely provides greater protection against exposure than single-layer clothing (Miguelino 2014). A Malaysian study noted variation in exposure among pesticide applicators based on type of equipment used, application technique and use of protective clothing (Baharuddin et al. 2011).

Farmworkers often inadvertently carry pesticide home and create para-occupational exposure opportunities for their family members. Farmworkers frequently have pesticide residues on their shoes or clothing, their bodies and objects brought into the fields (for example, lunch boxes or water bottles). Simple activities such as hugging children or loved ones upon arriving home from work create opportunity for pesticide exposure. Worker Protection Standard training has long incorporated behaviors for minimizing para-occupational exposure to pesticide though basic activities such as showering immediately after getting home from work, removing work clothes outside the dwelling and laundering work clothes separately from family clothing. Models using lay health worker outreach programs are effective methods to reduce potentially harmful take-home exposure pathways (Liebman et al. 2007; Trejo et al. 2013; Salvatore et al. 2015).

PESTICIDE EXPOSURE AND HEALTH OUTCOMES

Before launching into a review of the health-related implications of pesticide exposure, it is essential to put a series of disclaimers into place to caution readers. There is a wide assortment of factors that stand in the way of drawing firm conclusions about specific linkages between pesticide exposure and states of health. One factor already discussed in this chapter is the large number of pesticides available in the global marketplace, and the virtually unknown combination of active and other ingredients that are considered proprietary (Fenner-Crisp 2010). Next, the step between exposure (that is, coming into contact with pesticide) and the amount of pesticide that makes its way into the body (that is, the dose ingested) is exceedingly complex and often unknown (Arcury et al. 2006). Consequently, although a wide variety of animal studies may link pesticide exposure and dose to a health outcome, corresponding precise human studies are simply not possible. Indeed, among the most common limitations of occupational exposure studies and subsequent reviews and meta-analysis is the inherently crude assessment of pesticide exposure (for example, Van Maele-Fabry et al. 2010; Yan et al. 2016). Third, the health-related consequences of pesticide exposure are undoubtedly modified by a host of genetic, behavioral and environmental attributes. Behavioral attributes, such as those advocated by the Worker Protection Standard, likely minimize the health effects of exposure by reducing total dose ingested (Salvatore et al. 2008). Similarly for environmental conditions such as heat and humidity (Baldi et al. 2014) and pesticide drift from wind modify exposure levels (Suratman et al. 2015), and presumably dose and health effects. Fourth, with few minor exceptions, that vast majority of pesticide exposure involves chronic exposure to small amounts of pesticide, rather than acute exposure to highly toxic concentrations. Finally, the combination of chronic low-dose exposure to pesticides over time, combined with the reality that most disease states have a complex etiology, make linking pesticide exposure to health exceedingly difficult.

Human health is a complex phenomenon that is matched by the dizzying complexity of pesticides and pesticide exposure. Therefore, it is not possible to identify and elaborate all the biological and physiological processes affected by pesticide exposure. Moreover, the number of indirect pathways by which pesticide may affect human health through modification or destruction to plant or animal ecosystems is likely unfathomable. Nevertheless, it is important to note that several major classes of pesticides work by inhibiting nervous system function: pesticides make it difficult for one cell to 'talk' to another. The inability to talk among cells often manifests as symptoms and illnesses in the respiratory, gastrointestinal, cardiovascular, integumentary and musculoskeletal systems and other areas in the body (Reigart and Roberts 1999).

Case studies and farmworkers' descriptions of how they feel after exposure to pesticides highlight the common manifestations of cells inability to talk. In a 1989 case study in Balm, Florida, farmworkers returned to cauliflower fields before the recommended re-entry interval for a pesticide whose active ingredient was mevinphos. Over 100 workers became ill with headaches, dizziness and muscle tremors (Hyland and Laribi 2017). An incident in 2014 in Hillsborough County, Florida, was investigated where 66 people exposed to dimethyl disulfide, a soil fumigant that is considered a replacement for methyl bromide, complained of low to moderate severity pesticide-related illness, with symptoms including eye pain, throat irritation, dizziness, headache and fatigue (Mulay et al. 2016). Results from case studies comport directly with studies of common illness experiences reported by farmworkers after pesticide exposure. For example, in a study of Lebanese agricultural workers, those exposed to pesticides reported more upper-respiratory tract, gastrointestinal and immunoallergic complaints (Salameh and Abi Saleh

2004). Similarly, frequent symptoms reported by Latino farmworkers in the US include headaches, burning eyes, pain or soreness in muscles, joints and bones, and rash or itchy skin (Strong et al. 2004). Nigatu et al. (2016) found that pesticide exposure increased the risk of chronic cough and shortness of breath, whereas others have linked exposure to pyrethroid insecticide to respiratory symptoms (Quansah et al. 2016). An exposure–response relationship for respiratory effects has been observed for commonly used herbicides, insecticides and a rodenticide (Hoppin et al. 2017).

The nervous system is often targeted by pesticides, particularly insecticides, consequently a substantial amount of research has focused on neurological disease states. Low-dose cumulative exposure to pesticides have been implicated in Parkinson's disease in the US (Kamel et al. 2007; Tanner et al. 2011; Van Maele-Fabry et al. 2012), as well as neuropathy and Parkinsonism in a British birth cohort (Povey et al. 2014). Risk for all-cause dementia is 38 percent higher among those exposed to pesticide, and the risk of Alzheimer's disease is more than 50 percent greater for those exposed to organophosphate pesticide (Hayden et al. 2010; Yan et al. 2016). A meta-analysis of a possible link between pesticide exposure and risk of amyotrophic lateral sclerosis (ALS) found elevated rates of ALS among farmers relative to non-farmers, and among those who self-reported exposure to pesticide (Kang et al. 2014). Importantly, Kang's et al.'s results found no difference in ALS among those living in rural areas.

Several studies focus on pre-clinical indicators of neurologic disease. Rohlman et al. (2007) noted that years in agriculture, as a proxy for cumulative pesticide exposure, was associated with poorer neurobehavioral performance such as reaction time and latency in matching symbols and digits among women. Even completing the neurobehavioral assessment battery was more difficult for women with longer agricultural work histories (Rohlman et al. 2007). Functional connectivity in the brain of farmworkers is linked with cholinesterase activity (Bahrami et al. 2017), an enzyme frequently targeted by organophosphate pesticide. Greater physical symptoms suggestive of possible exposure to organophosphate pesticide has been associated with greater impulsivity, aggression and increased odds of suicide attempt (Lyu et al. 2018). Similarly, pesticide exposure has been associated with a two-fold increase risk of depression among Korean and Brazilian agricultural workers (Kim et al. 2013; Campos et al. 2016), and with elevated suicidality (Faria et al. 2014). Park et al. (2012) noted an association of self-reported number of times pesticides were applied across a farmer's lifetime with nerve conduction velocity such that greater pesticide applications was linked with slower nerve conduction velocity. Rates of idiopathic rapid eye movement sleep behavior disorder were nearly twice as high for individuals reporting occupational exposure to pesticide (Postuma et al. 2012), and others reported high rates of sleepiness among Latino farmworkers (Grzywacz et al. 2011).

There has been substantial interest in the potential role of pesticide exposure in various forms of cancer. Long-term low-dose exposure to pesticides has been associated with increased risk for hematopoietic cancers (leukemia and multiple myeloma) and non-Hodgkin's lymphoma (Merhi et al. 2007), and myeloid leukemia (Van Maele-Fabry et al. 2007). Similarly, maternal exposure to pesticide is associated with increased risk of leukemia among children of the exposed (Wigle et al. 2009), and parental occupational exposure to pesticides increases the risk for brain tumors in exposed individuals' children reaching into young adulthood (Van Maele-Fabry et al. 2013). The combination of pesticide exposure with family history synergistically increases the risk of prostate cancer (Lewis-Mikhael et al. 2016). Collectively, these results suggest pesticide exposure may increase cancers of the blood and soft tissues.

Active and other ingredients in pesticides also frequently have endocrine disruption potential, leading scientists to study the fertility-related implications of occupational exposure to pesticides. Increased exposure to organophosphate pesticide is associated with more chromosomal abnormalities in human sperm; individuals at the highest concentrations of urinary DAP metabolites (relative to those with the lowest concentrations) had greater risk of disomy or having sperm with two copies of the same chromosome (Figueroa et al. 2015). Exposure to pyrethroid and organophosphate pesticide is associated with several indicators of poor sperm quality including low sperm concentration, decreased motility and abnormal morphology (Martenies and Perry 2013). Mancozeb, a widely used fungicide, is believed to impair both male and female fertility (Runkle et al. 2017), and occupational exposure to pesticide interferes with ovarian function, particularly increased follicular depletion and earlier age of menopause (Vabre et al. 2017). Finally, prospective evidence from the Shanghai Birth Cohort study indicates that women with the highest concentrations of one urinary DAP metabolite (that is, diethylthiophosphate) and pyrethroid metabolite had the greatest odds of being classified 'infertile' (Hu et al. 2018), and others reported that greater exposure to pesticide may lengthen time to achieve pregnancy (Snijder et al. 2012).

A review of the putative health effects of pesticide exposure cannot ignore the obvious issue, glyphosate. Glyphosate, which is the active ingredient in Roundup®, provides a perfect case study of how difficult it is to link pesticide exposure to human health outcomes. Media outlets such as CNN and popular sources of information such as YouTube brought widespread attention to the possibility that glyphosate is linked with a host of negative health outcomes ranging from cancer to birth defects. The scientific foundations for the alleged linkages are debated. On the one hand, evidence from large epidemiologic studies such as the Agricultural Health Study suggest no link between glyphosate and either cancer (Andreotti et al. 2018) or poor birth outcomes (de Araujo et al. 2016). On the other hand, prospective evidence from a birth cohort study in the US Midwest suggest glyphosate exposure during pregnancy may result in shortened fetal gestation (Parvez et al. 2018), and results from laboratory studies suggest that glyphosate may increase risk of hormone-dependent breast cancer but not hormone-independent breast cancer (Thongprakaisang et al. 2013).

The debate over the health effects of glyphosate is vigorous. The International Agency for Research on Cancer (IARC 2015, p. 398) classified glyphosate as 'probably carcinogenic to humans', based on 'sufficient evidence of carcinogenicity in animals, limited evidence of carcinogenicity in humans, and strong mechanistic data'. This conclusion has been accused of being based on a 'flawed and incomplete summary of the experimental evidence' enabled by an insufficient vetting of potential conflicts of interest by panel members (Tarone 2018, p. 82). The European Food Safety Authority (Portier et al. 2016, p. 743) maintains the IARC's classification is 'not warranted' and that 'glyphosate is devoid of genotoxic potential'. There are undoubtedly many issues contributing to the vigorous debate over the potential health effects of glyphosate, but one contributing factor is that Roundup®, for which glyphosate is the primary active ingredient, is the most commonly used herbicide in the world (Benbrook 2016).

IMPORTANT NEXT STEPS AND CONCLUSIONS

The literature linking farmworkers' exposure to pesticides, and the health effects of pesticide exposure, is perhaps best labeled as a hot mess. Data from surveillance studies that rely on strong evidence of an exposure together with a physical complaint consistent with pesticide toxicology suggest clear differential exposure to pesticides among farmworkers relative to non-farmworkers (Calvert et al. 2008). However, epidemiological studies of farmworkers often demonstrate that farmworkers have evidence of being exposed to pesticides, but those levels of exposure are frequently not different from other rural-dwelling non-farmworkers (Arcury et al. 2016, 2018). A large number of meta-analytic studies have linked pesticide exposure with a variety of neurological and endocrinological outcomes, including fertility and cancer; yet, debate persists about the putative dangers of commonly used pesticides such as glyphosate (Portier et al. 2016). The current state of the literature (that is, a hot mess) highlights a need for additional research to bring clearer understanding to the overarching issue of pesticide exposure among farmworkers.

Alternative forms of biomonitoring for pesticide exposure are needed. The vast majority of this biomonitoring occurs through the collection of urine samples to test for urinary metabolites, and occasionally through serum samples to monitor cholinesterase inhibition. Several practical realities make each of these strategies problematic. The greatest practical impediment is simply the ability to reliably obtain urine or serum samples from large cohorts of individuals. This impediment is further compounded by the fact that most commonly used pesticides do not bio-accumulate, but are metabolized within 48–72 hours (Arcury et al. 2006). This means that capturing pesticide exposure in a biological sample requires attentiveness to both when exposure occurs, and within a narrow window before the parent compound is metabolized. Strategies that allow for continuous sampling of cells or fluids and that are scalable for large cohort studies are needed. With additional validation work, the use of sweat (Rosenberg et al. 1985; Kapka-Skrzypczak et al. 2015) or saliva (Weber Set al. 2017) may enable more complete assessment of pesticide exposure.

Additional studies of pesticide exposure among farmworkers are needed. Importantly, future studies must learn from previous studies. A fundamental shift in the data collection protocol between Arcury et al.'s cohort (2009, 2010) and cohort comparison (Arcury et al. 2016, 2018) studies is informative. In the cohort study involving only farmworkers, serial first-morning void urine samples were collected from participants across a series of days, and each sample was retained and analyzed for the presence of pesticide metabolites. By contrast, in the cohort comparison study, serial first-morning void urine samples were again collected, but they were pooled prior to laboratory analyses for pesticides. The use of pooled samples may have masked temporal variation in metabolite concentrations between farmworkers and non-farmworkers. Similarly, collecting samples from the same workers during both high and low pesticide-use periods, together with assessments of specific tasks (for example, Baldi et al. 2014) performed during those periods, would clarify potentially ambiguous findings (Fenske et al. 2005). Equally important though, given the endemic nature of pesticide in human environments, is the need for more complete and refined assessment procedures for pesticide exposure.

A wider variety of scientists need to engage in pesticide exposure and health research. Human pesticide exposure and health research has largely been undertaken by environmental and occupational epidemiologists with support from toxicologists and analytic chemists. Although these are undoubtedly essential disciplines, researchers from other disciplines are also needed. The potential of informatics, big data and machine learning are beginning to be applied to pesticide exposure and health research (Wang et al. 2015; Tomiazzi et al. 2018). The US EPA has ongoing monitoring programs of air and water quality (United States Geological Service 2014), for which the presence and concentration of pesticide is continuously assessed. Data such as these, together with the growing availability of health information exchange data should be exploited to discern whether variability in water and air concentrations of pesticide are linked to acute indicators of possible pesticide exposure (for example, upper-respiratory or gastrointestinal complaints) as well as longer-term patterns of disease. A wider array of research designs is also needed. Whereas between-person comparisons are the norm of epidemiological inquiry, between-person comparisons of exposed versus unexposed can mask important within-person variation among both the exposed and unexposed individuals. Analytic techniques that allow disaggregating between-person and within-person (Curran and Bauer 2011; Wang and Maxwell 2015) effects of pesticide exposure and health outcomes would probably yield new insights.

Despite the hot mess of the literature, it seems clear that farmworkers around the world are exposed to pesticides. Whether or not absolute level of exposure differs between farmworkers and non-farmworkers remains open to interpretation of the available data. Similarly, there is a substantial body of research linking occupational exposure to pesticide with a variety of negative health outcomes among farmworkers and others involved with pesticide, including neurological conditions such as Parkinson's disease, ALS and Alzheimer's disease, as well as a variety of cancers. As with the interpretation of the pesticide exposure data, the definitive health effects of specific pesticides are clouded by the ambiguous ingredient list of active and inactive compounds in most pesticide formulations, and by the number and quality of specific studies of the diverse array of pesticides available on the market. More research is clearly needed, and the observed individual health effects of chronic exposure to pesticide must be placed alongside the parallel reality that pesticides play a tremendous public health value through their control of animal and insect disease vectors and their contributions to worldwide food security.

REFERENCES

Alavanja, M.C.R., Hoppin, J.A. and Kamel, F. (2004), 'Health effects of chronic pesticide exposure: cancer and neurotoxicity', *Annual Review of Public Health*, **25** (April), 155–97, doi:10.1146/annurev.publhealth.25.101802.123020.

Andreotti, G., Koutros, S., Hofmann, J.N., Sandler, D.P., Lubin, J.H., Lynch, C.F. et al. (2018), 'Glyphosate use and cancer incidence in the agricultural health study', *Journal of the National Cancer Institute*, **110** (5), 509–16, doi:10.1093/jnci/djx233.

Arcury, T.A., Chen, H., Laurienti, P.J., Howard, T.D., Barr, D.B., Mora, D.C. et al. (2018), 'Farmworker and nonfarmworker Latino immigrant men in North Carolina have high levels of specific pesticide urinary metabolites', *Archives of Environmental & Occupational Health*, **73** (4), 219–27, doi:10.1080/19338244.2017.1342588.

Arcury, T.A., Grzywacz, J.G., Chen, H., Vallejos, Q.M., Galvan, L., Whalley, L.E. et al. (2009), 'Variation across the agricultural season in organophosphorus pesticide urinary metabolite levels for Latino farmworkers in eastern North Carolina: project design and descriptive results', *American Journal of Industrial Medicine*, **52** (7), 539–50, doi:10.1002/ajim.20703.

Arcury, T.A., Grzywacz, J.G., Talton, J.W., Chen, H., Vallejos, Q.M., Galvan, L. et al. (2010), 'Repeated pesticide exposure among North Carolina migrant and seasonal farmworkers', *American Journal of Industrial Medicine*, **53** (8), 802–13, doi:10.1002/ajim.20856.

Arcury, T.A., Laurienti, P.J., Chen, H., Howard, T.D., Barr, D.B., Mora, D.C. et al. (2016), 'Organophosphate Pesticide urinary metabolites among Latino immigrants: North Carolina farmworkers and non-farmworkers compared', *Journal of Occupational and Environmental Medicine*, **58** (11), 1079–86, doi:10.1097/JOM.0000000000000875.

Arcury, T.A., Quandt, S.A., Barr, D.B., Hoppin, J.A., McCauley, L., Grzywacz, J.G. et al. (2006), 'Farmworker exposure to pesticides: methodologic issues for the collection of comparable data', *Environmental Health Perspectives*, **114** (6), 923–28.

Baharuddin, M.R.B., Sahid, I.B., Noor, M.A.B.M., Sulaiman, N. and Othman, F. (2011), 'Pesticide risk assessment: a study on inhalation and dermal exposure to 2,4-D and paraquat among Malaysian paddy farmers', *Journal of Environmental Science and Health. Part. B, Pesticides, Food Contaminants, and Agricultural Wastes*, **46** (7), 600–607, doi:10.1080/03601234.2011.589309.

Bahrami, M., Laurienti, P.J., Quandt, S.A., Talton, J., Pope, C.N., Summers, P. et al. (2017), 'The impacts of pesticide and nicotine exposures on functional brain networks in Latino immigrant workers', *Neurotoxicology*, **62** (September), 138–50, doi:10.1016/j.neuro.2017.06.001.

Baldi, I., Lebailly, P., Bouvier, G., Rondeau, V., Kientz-Bouchart, V., Canal-Raffin, M. et al. (2014), 'Levels and determinants of pesticide exposure in re-entry workers in vineyards: results of the PESTEXPO study', *Environmental Research*, **132** (July), 360–69, doi:10.1016/j.envres.2014.04.035.

Benbrook, C.M. (2016), 'Trends in glyphosate herbicide use in the United States and globally', *Environmental Sciences Europe*, **28** (1), 3, doi:10.1186/s12302-016-0070-0.

Brennan, K., Economos, J. and Salerno, M.M. (2015), 'Farmworkers make their voices heard in the call for stronger protections from pesticides', *New Solutions: A Journal of Environmental and Occupational Health Policy: NS*, **25** (3), 362–76, doi:10.1177/1048291115604428.

Calvert, G.M., Karnik, J., Mehler, L., Beckman, J., Morrissey, B., Sievert, J. et al. (2008), 'Acute pesticide poisoning among agricultural workers in the United States, 1998–2005', *American Journal of Industrial Medicine*, **51** (12), 883–98, doi:0.1002/ajim.20623.

Calvert, G.M., Plate, D.K., Das, R., Rosales, R., Shafey, O., Thomsen, C. et al. (2004), 'Acute occupational pesticide-related illness in the US, 1998–1999: surveillance findings from the SENSOR-pesticides program', *American Journal of Industrial Medicine*, **45** (1), 14–23, doi:10.1002/ajim.10309.

Campos, Ÿ., Dos Santos Pinto da Silva, V., Sarpa Campos de Mello, M. and Barros Otero, U. (2016), 'Exposure to pesticides and mental disorders in a rural population of Southern Brazil', *Neurotoxicology*, **56** (September), 7–16, doi:10.1016/j.neuro.2016.06.002.

Carson, R. (1962), *Silent Spring*, Boston, MA: Houghton Mifflin.

Centers for Disease Control (CDC) (2018), 'Fourth national report on human exposure to environmental chemicals: updated tables, March 2018, volume one', accessed 29 December 2018 at https://www.cdc.gov/exposurereport/pdf/Fourth Report_UpdatedTables_Volume1_Mar2018.pdf.

Coronado, G.D., Holte, S., Vigoren, E., Griffith, W.C., Barr, D.B., Faustman, E. et al. (2011), 'Organophosphate pesticide exposure and residential proximity to nearby fields: evidence for the drift pathway', *Journal of Occupational and Environmental Medicine*, **53** (8), 884–91, doi:10.1097/JOM.0b013e318222f03a.

Coronado, G.D., Thompson, B., Strong, L., Griffith, W.C. and Islas, I. (2004), 'Agricultural task and exposure to organophosphate pesticides among farmworkers', *Environmental Health Perspectives*, **112** (2), 142–7, doi:10.1289/ehp.6412.

Curran, P.J. and Bauer, D.J. (2011), 'The disaggregation of within-person and between-person effects in longitudinal models of change', *Annual Review of Psychology*, **62** (January), 583–619, doi:10.1146/annurev.psych.093008.100356.

Curwin, B.D., Hein, M.J., Sanderson, W.T., Barr, D.B., Heederik, D., Reynolds, S.J. et al. (2005), 'Urinary and hand wipe pesticide levels among farmers and non-farmers in Iowa', *Journal of Exposure Analysis and Environmental Epidemiology*, **15** (6), 500–508, doi:10.1038/sj.jea.7500428.

De Araujo, J.S.A., Delgado, I.F. and Paumgartten, F.J.R. (2016), 'Glyphosate and adverse pregnancy outcomes, a systematic review of observational studies', *BMC Public Health*, **16** (June), 472, doi:10.1186/s12889-016-3153-3.

Environmental Protection Agency (EPA) (2017), 'Pesticide industry sales and usage: 2008–2012 market estimates', accessed 29 December 2018 at https://www.epa.gov/sites/production/files/2017-01/documents/pesticides-industry-sales-usage-2016_0.pdf.

Environmental Protection Agency (EPA) (2018), 'Restrictions to protect workers after pesticide applications', accessed 29 December 2018 at https://www.epa.gov/pesticide-worker-safety/restrictions-protect-workers-after-pesticide-applications.

Environmental Protection Agency (EPA) (n.d.), 'What is a pesticide?', accessed 22 May 2019 at https://www.epa.gov/minimum-risk-pesticides/what-pesticide.

Faria, N.M.X., Fassa, A.G. and Meucci, R.D. (2014), 'Association between pesticide exposure and suicide rates in Brazil', *Neurotoxicology*, **45** (December), 355–62, doi:10.1016/j.neuro.2014.05.003.

Fenner-Crisp, P.A. (2010), 'Risk assessment and risk management: the regulatory process', in R. Krieger (ed.), *Hayes' Handbook of Pesticide Toxicology* 3rd edn, London and San Diego, CA: Elsevier, pp. 1371–80.

Fenske, R.A., Lu, C., Curl, C.L., Shirai, J.H. and Kissel, J.C. (2005), 'Biologic monitoring to characterize organophosphorus pesticide exposure among children and workers: an analysis of recent studies in Washington State', *Environmental Health Perspectives*, **113** (11), 1651–7, doi:10.1289/ehp.8022.

Figueroa, Z.I., Young, H.A., Meeker, J.D., Martenies, S.E., Barr, D.B., Gray, G. et al. (2015), 'Dialkyl phosphate urinary metabolites and chromosomal abnormalities in human sperm', *Environmental Research*, **143** (part A), 256–65, doi:10.1016/j.envres.2015.10.021.

Food and Agriculture Organization of the United Nations (2011), 'The state of food and agriculture: women in agriculture – closing the gender gap for development', accessed 29 December 2018 at http://www.fao.org/3/a-i2050e.pdf.

Grzywacz, J.G., Chatterjee, A.B., Quandt, S.A., Talton, J.W., Chen, H., Weir, M. et al. (2011), 'Depressive symptoms and sleepiness among Latino farmworkers in eastern North Carolina', *Journal of Agromedicine*, **16** (4), 251–60, doi:10.1080/1 059924X.2011.605722.

Grzywacz, J.G., Gabbard, S., Fung, W., Salvatore, A.L., Georges, A. and Carroll, D. (2018), 'Social determinants of farmworker health: evidence from the NAWS', roundtable discussion at the Annual Meeting of the American Public Health Association, San Diego, CA, November.

Grzywacz, J.G., Quandt, S.A., Vallejos, Q.M., Whalley, L.E., Chen, H., Isom, S. et al. (2010), 'Job demands and pesticide exposure among immigrant Latino farmworkers', *Journal of Occupational Health Psychology*, **15** (3), 252–66, doi:10.1037/a0019303.

Hayden, K.M., Norton, M.C., Darcey, D., Ostbye, T., Zandi, P.P., Breitner, J.C.S. et al. (2010), 'Occupational exposure to pesticides increases the risk of incident AD: the Cache County study', *Neurology*, **74** (19), 1524–30, doi:10.1212/ WNL.0b013e3181dd4423.

Hoppin, J.A., Umbach, D.M., Long, S., London, S.J., Henneberger, P.K., Blair, A. et al. (2017), 'Pesticides are associated with allergic and non-allergic wheeze among male farmers', *Environmental Health Perspectives*, **125** (4), 535–43, doi:10.1289/EHP315.

Hu, Y., Ji, L., Zhang, Y., Shi, R., Han, W., Tse, L.A. et al. (2018), 'Organophosphate and pyrethroid pesticide exposures measured before conception and associations with time to pregnancy in Chinese couples enrolled in the Shanghai birth cohort', *Environmental Health Perspectives*, **126** (7), 077001, doi:10.1289/EHP2987.

Hyland, C. and Laribi, O. (2017), 'Review of take-home pesticide exposure pathway in children living in agricultural areas', *Environmental Research*, 156 (July), 559–70, doi:10.1016/j.envres.2017.04.017.

International Agency for Research on Cancer (IARC) (2015), *Some Organophosphate Insecticides and Herbicides: Tetrachlorvinphos, Parathion, Malathion, Diazinon and Glyphosate*, Lyon: IARC, accessed 16 May 2019 at https://monographs.iarc. fr/wp-content/uploads/2018/07/mono112.pdf.

International Labor Organization (ILO) (2018), 'Employment by sector: ILO modelled estimates', May, accessed 29 December 2018 at https://www.ilo.org/ilostat/ faces/oracle/webcenter/portalapp/pagehierarchy/Page3.jspx?locale=en&MBI_ID =33&_adf.ctrl-state=kib4xktuj_51&_afrLoop=687270306442349&_afrWindow Mode=0&_afrWindowId=null#!%40%40%3F_afrWindowId%3Dnull%26loca le%3Den%26_afrLoop%3D687270306442349%26MBI_ID%3D33%26_afrWin dowMode%3D0%26_adf.ctrl-state%3DrjI2e4hgy_4.

Kamel, F., Tanner, C., Umbach, D., Hoppin, J., Alavanja, M., Blair, A. et al. (2007), 'Pesticide exposure and self-reported Parkinson's disease in the agricultural health study', *American Journal of Epidemiology*, **165** (4), 364–74, doi:10.1093/ aje/kwk024.

Kang, H., Cha, E.S., Choi, G.J. and Lee, W.J. (2014), 'Amyotrophic lateral sclerosis and agricultural environments: a systematic review', *Journal of Korean Medical Science*, **29** (12), 1610–17, doi:10.3346/jkms.2014.29.12.1610.

Kapka-Skrzypczak, L., Sawicki, K., Czajka, M., Turski, W.A. and Kruszewski, M. (2015), 'Cholinesterase activity in blood and pesticide presence in sweat as biomarkers of children`s environmental exposure to crop protection chemicals', *Annals of Agricultural and Environmental Medicine: AAEM*, **22** (3), 478–82, doi:10.5604/12321966.1167718.

Karasek, R. and Theorell, T. (1992), *Healthy Work: Stress, Productivity, and the Reconstruction of Working Life*, revd edn, New York: Basic Books.

Kasner, E.J., Keralis, J.M., Mehler, L., Beckman, J., Bonnar-Prado, J., Lee, S.-J. et al. (2012), 'Gender differences in acute pesticide-related illnesses and injuries among farmworkers in the United States, 1998–2007', *American Journal of Industrial Medicine*, **55** (7), 571–83, doi:10.1002/ajim.22052.

Kim, J., Ko, Y. and Lee, W.J. (2013), 'Depressive symptoms and severity of acute occupational pesticide poisoning among male farmers', *Occupational and Environmental Medicine*, **70** (5), 303–9, doi:10.1136/oemed-2012-101005.

Lewis-Mikhael, A.-M., Bueno-Cavanillas, A., Ofir Giron, T., Olmedo-Requena, R., Delgado-Rodríguez, M. and Jiménez-Moleón, J.J. (2016), 'Occupational exposure to pesticides and prostate cancer: a systematic review and meta-analysis', *Occupational and Environmental Medicine*, **73** (2), 134–44, doi:10.1136/oemed-2014-102692.

Liebman, A.K., Juarez, P.M., Leyva, C. and Corona, A. (2007), 'A pilot program using promotoras de salud to educate farmworker families about the risk from pesticide exposure', *Journal of Agromedicine*, **12** (2), 33–43, doi:10.1300/J096v12n02_04.

López-Gálvez, N., Wagoner, R., Beamer, P., de Zapien, J. and Rosales, C. (2018), 'Migrant farmworkers' exposure to pesticides in Sonora, Mexico', *International Journal of Environmental Research and Public Health*, **15** (12), doi:10.3390/ijerph15122651.

Lyu, C.P., Pei, J.R., Beseler, L.C., Li, Y.L., Li, J.H., Ren, M. et al. (2018), 'Case control study of impulsivity, aggression, pesticide exposure and suicide attempts using pesticides among farmers', *Biomedical and Environmental Sciences: BES*, **31** (3), 242–46, doi:10.3967/bes2018.031.

Martenies, S.E. and Perry, M.J. (2013), 'Environmental and occupational pesticide exposure and human sperm parameters: a systematic review', *Toxicology*, **307** (May), 66–73, doi:10.1016/j.tox.2013.02.005.

Mehler, L.N., Schenker, M.B., Romano, P.S. and Samuels, S.J. (2006), 'California surveillance for pesticide-related illness and injury: coverage, bias, and limitations', *Journal of Agromedicine*, **11** (2), 67–79, doi:10.1300/J096v11n02_10.

Merhi, M., Raynal, H., Cahuzac, E., Vinson, F., Cravedi, J.P. and Gamet-Payrastre, L. (2007), 'Occupational exposure to pesticides and risk of hematopoietic cancers: meta-analysis of case-control studies', *Cancer Causes & Control: CCC*, **18** (10), 1209–26, doi:10.1007/s10552-007-9061-1.

Miguelino, E.S. (2014), 'A meta-analytic review of the effectiveness of single-layer clothing in preventing exposure from pesticide handling', *Journal of Agromedicine*, **19** (4), 373–83, doi:10.1080/1059924X.2014.946636.

Mulay, P.R., Cavicchia, P., Watkins, S.M., Tovar-Aguilar, A., Wiese, M. and Calvert, G.M. (2016), 'Acute illness associated with exposure to a new soil fumigant containing dimethyl disulfide – Hillsborough County, Florida, 2014', *Journal of Agromedicine*, **21** (4), 373–9, doi:10.1080/1059924X.2016.1211574.

Nielsen, E.G. and Lee, L.K. (1987), 'The magnitude and costs of groundwater contamination from agricultural chemicals: a national perspective', *Economic Research Service*, US Department of Agriculture, accessed 29 December 2018 at https://naldc.nal.usda.gov/download/CAT88907300/PDF.

Nigatu, A.W., Bråtveit, M. and Moen, B.E. (2016), 'Self-reported acute pesticide intoxications in Ethiopia', *BMC Public Health*, **16** (July), 575, accessed 7 May 2019 at https://bmcpublichealth.biomedcentral.com/articles/10.1186/s12889-016-3196-5.

Ntow, W.J., Tagoe, L.M., Drechsel, P., Kelderman, P., Nyarko, E. and Gijzen, H.J. (2009), 'Occupational exposure to pesticides: blood cholinesterase activity in a farming community in Ghana', *Archives of Environmental Contamination and Toxicology*, **56** (3), 623–30, doi.org/10.1007/s00244-007-9077-2.

Park, S.K., Kong, K.A., Cha, E.S., Lee, Y.J., Lee, G.T. and Lee, W.J. (2012), 'Occupational exposure to pesticides and nerve conduction studies among Korean farmers', *Archives of Environmental & Occupational Health*, **67** (2), 78–83, doi:10.1080/19338244.2011.573022.

Parvez, S., Gerona, R.R., Proctor, C., Friesen, M., Ashby, J.L., Reiter, J.L. et al. (2018), 'Glyphosate exposure in pregnancy and shortened gestational length: a prospective Indiana birth cohort study', *Environmental Health: A Global Access Science Source*, **17** (1), 23, doi:10.1186/s12940-018-0367-0.

Portier, C.J., Armstrong, B.K., Baguley, B.C., Baur, X., Belyaev, I., Bellé, R. et al. (2016), 'Differences in the carcinogenic evaluation of glyphosate between the International Agency for Research on Cancer (IARC) and the European Food Safety Authority (EFSA)', *Journal of Epidemiology and Community Health*, **70** (8), 741–5, doi:10.1136/jech-2015-207005.

Postuma, R.B., Montplaisir, J.Y., Pelletier, A., Dauvilliers, Y., Oertel, W., Iranzo, A. et al. (2012), 'Environmental risk factors for REM sleep behavior disorder: a multicenter case-control study', *Neurology*, **79** (5), 428–34,doi:10.1212/WNL.0b013e31825dd383.

Povey, A.C., McNamee, R., Alhamwi, H., Stocks, S.J., Watkins, G., Burns, A. et al. (2014), 'Pesticide exposure and screen-positive neuropsychiatric disease in British sheep farmers', *Environmental Research*, **135** (November), 262–70, doi:10.1016/j.envres.2014.09.008.

Prado, J.B., Mulay, P.R., Kasner, E.J., Bojes, H.K. and Calvert, G.M. (2017), 'Acute pesticide-related illness among farmworkers: barriers to reporting to public health authorities', *Journal of Agromedicine*, **22** (4), 395–405, doi:10.1080/10599924X.2017.1353936.

Quansah, R., Bend, J.R., Abdul-Rahaman, A., Armah, F.A., Luginaah, I., Essumang, D.K. et al. (2016), 'Associations between pesticide use and respiratory symptoms: a cross-sectional study in Southern Ghana', *Environmental Research*, **150** (October), 245–54, doi:10.1016/j.envres.2016.06.013.

Reeves, M. and Schafer, K.S. (2003), 'Greater risks, fewer rights: U.S. farmworkers and pesticides', *International Journal of Occupational and Environmental Health*, **9** (1), 30–39, doi:10.1179/107735203800328858.

Reigart, J.R. and Roberts, J.R. (1999), *Recognition and Management of Pesticide Poisoning*, 5th edn, Baltimore, MD: Environmental Protection Agency.

Rohlman, D.S., Lasarev, M., Anger, W.K., Scherer, J., Stupfel, J. and McCauley, L. (2007), 'Neurobehavioral performance of adult and adolescent agricultural workers', *Neurotoxicology*, **28** (2), 374–80, doi:10.1016/j.neuro.2006.10.006.

Rosenberg, N.M., Queen, R.M. and Stamper, J.H. (1985), 'Sweat-patch test for monitoring pesticide absorption by airblast applicators', *Bulletin of Environmental Contamination and Toxicology*, **35** (1), 68–72.

Roser, M. (2019), 'Employment in agriculture', Our World in Data online graph, accessed 22 May 2019 at https://ourworldindata.org/grapher/gdp-vs-agriculture-gdp.

Runkle, J., Flocks, J., Economos, J. and Dunlop, A.L. (2017), A systematic review of Mancozeb as a reproductive and developmental hazard', *Environment International*, **99** (February), 29–42, doi:10.1016/j.envint.2016.11.006.

Salameh, P.R. and Abi Saleh, B. (2004), 'Symptoms and acute pesticide intoxication among agricultural workers in Lebanon', *Le Journal Medical Libanais. The Lebanese Medical Journal*, **52** (2), 64–70.

Salvatore, A.L., Bradman, A., Castorina, R., Camacho, J., López, J., Barr, D.B. et al. (2008), 'Occupational behaviors and farmworkers' pesticide exposure: findings from a study in Monterey County, California', *American Journal of Industrial Medicine*, **51** (10), 782–94, doi:10.1002/ajim.20622.

Salvatore, A.L., Castorina, R., Camacho, J., Morga, N., López, J., Nishioka, M. et al. (2015), 'Home-based community health worker intervention to reduce pesticide exposures to farmworkers' children: a randomized-controlled trial', *Journal of Exposure Science & Environmental Epidemiology*, **25** (6), 608–15, doi:10.1038/jes.2015.39.

Snijder, C.A., te Velde, E., Roeleveld, N. and Burdorf, A. (2012), 'Occupational exposure to chemical substances and time to pregnancy: a systematic review', *Human Reproduction Update*, **18** (3), 284–300, doi:10.1093/humupd/dms005.

Strong, L.L., Thompson, B., Coronado, G.D., Griffith, W.C., Vigoren, E.M. and Islas, I. (2004), 'Health symptoms and exposure to organophosphate pesticides in farmworkers', *American Journal of Industrial Medicine*, **46** (6), 599–606, doi:10.1002/ajim.20095.

Suratman, S., Edwards, J.W. and Babina, K. (2015), 'Organophosphate pesticides exposure among farmworkers: pathways and risk of adverse health effects', *Reviews on Environmental Health*, **30** (1), 65–79, doi:10.1515/reveh-2014-0072.

Tanner, C.M., Kamel, F., Ross, G.W., Hoppin, J.A., Goldman, S.M., Korell, M. et al. (2011), 'Rotenone, paraquat, and Parkinson's disease', *Environmental Health Perspectives*, **119** (6), 866–72, doi:10.1289/ehp.1002839.

Tarone, R.E. (2018), 'On the International Agency for Research on Cancer classification of glyphosate as a probable human carcinogen', *European Journal of Cancer Prevention: The Official Journal of the European Cancer Prevention Organisation (ECP)*, **27** (1), 82–7, doi:10.1097/CEJ.0000000000000289.

Thompson, B., Griffith, W.C., Barr, D.B., Coronado, G.D., Vigoren, E.M. and Faustman, E.M. (2014), 'Variability in the take-home pathway: farmworkers and non-farmworkers and their children', *Journal of Exposure Science & Environmental Epidemiology*, **24** (5), 522–31, doi:10.1038/jes.2014.12.

Thongprakaisang, S., Thiantanawat, A., Rangkadilok, N., Suriyo, T. and Satayavivad, J. (2013), 'Glyphosate induces human breast cancer cells growth via estrogen receptors', *Food and Chemical Toxicology: An International Journal Published for the British Industrial Biological Research Association*, **59** (September), 129–36, doi:10.1016/j.fct.2013.05.057.

Tomiazzi, J.S., Judai, M.A., Nai, G.A., Pereira, D.R., Antunes, P.A. and Favareto, A.P.A. (2018), 'Evaluation of genotoxic effects in Brazilian agricultural workers exposed to pesticides and cigarette smoke using machine-learning algorithms',

Environmental Science and Pollution Research International, **25** (2), 1259–69, doi:10.1007/s11356-017-0496-y.

Trejo, G., Arcury, T.A., Grzywacz, J.G., Tapia, J. and Quandt, S.A. (2013), 'Barriers and facilitators for promotoras' success in delivering pesticide safety education to Latino farmworker families: La Familia Sana', *Journal of Agromedicine*, **18** (2), 75–86, doi:10.1080/1059924X.2013.766143.

United States Department of Agriculture (USDA) (2018), 'Demographic characteristics of hired farmworkers and all wage and salary workers, 2016', accessed 29 December 2018 at https://www.ers.usda.gov/topics/farm-economy/farm-labor/#demographic.

United States Department of Agriculture, Agricultural Marketing Service (2018), 'Pesticide data program: annual summary, calendar year 2017', accessed 29 December 2018 at https://www.ams.usda.gov/sites/default/files/media/2017PDP AnnualSummary.pdf.

United States Geological Service (2014), 'Pesticides found in the atmosphere', accessed 29 December 2018 at https://water.usgs.gov/nawqa/pnsp/pubs/fs152-95/atmos_4.html.

Vabre, P., Gatimel, N., Moreau, J., Gayrard, V., Picard-Hagen, N., Parinaud, J. et al. (2017), 'Environmental pollutants, a possible etiology for premature ovarian insufficiency: a narrative review of animal and human data', *Environmental Health: A Global Access Science Source*, **16** (1), 37, doi:10.1186/s12940-017-0242-4.

Van Maele-Fabry, G., Duhayon, S. and Lison, D. (2007), 'A systematic review of myeloid leukemias and occupational pesticide exposure', *Cancer Causes & Control: CCC*, **18** (5), 457–78, doi:10.1007/s10552-007-0122-2.

Van Maele-Fabry, G., Hoet, P. and Lison, D. (2013), 'Parental occupational exposure to pesticides as risk factor for brain tumors in children and young adults: a systematic review and meta-analysis', *Environment International*, **56** (June), 19–31, doi:10.1016/j.envint.2013.02.011.

Van Maele-Fabry, G., Hoet, P., Vilain, F. and Lison, D. (2012), 'Occupational exposure to pesticides and Parkinson's disease: a systematic review and meta-analysis of cohort studies', *Environment International*, **46** (October), 30–43, doi:10.1016/j.envint.2012.05.004.

Van Maele-Fabry, G., Lantin, A.-C., Hoet, P. and Lison, D. (2010), 'Childhood leukaemia and parental occupational exposure to pesticides: a systematic review and meta-analysis', *Cancer Causes & Control: CCC*, **21** (6), 787–809, doi:10.1007/s10552-010-9516-7.

Wang, L. and Maxwell, S.E. (2015), 'On disaggregating between-person and within-person effects with longitudinal data using multilevel models', *Psychological Methods*, **20** (1), 63–83, doi:10.1037/met0000030.

Wang, X., Zhang, M., Ma, J., Zhang, Y., Hong, G., Sun, F. et al. (2015), 'Metabolic changes in paraquat poisoned patients and support vector machine model of discrimination', *Biological & Pharmaceutical Bulletin*, **38** (3), 470–75, doi:10.1248/bpb.b14-00781.

Ward, M.H., Nuckols, J.R., Weigel, S.J., Maxwell, S.K., Cantor, K.P. and Miller, R.S. (2000), 'Identifying populations potentially exposed to agricultural pesticides using remote sensing and a Geographic Information System', *Environmental Health Perspectives*, **108** (1), 5–12, doi:10.1289/ehp.001085.

Weber, T.J., Smith, J.N., Carver, Z.A. and Timchalk, C. (2017), 'Non-invasive saliva human biomonitoring: development of an in vitro platform', *Journal of Exposure Science & Environmental Epidemiology*, **27** (1), 72–7, doi:10.1038/ jes.2015.74.

Wigle, D.T., Turner, M.C. and Krewski, D. (2009), 'A systematic review and meta-analysis of childhood leukemia and parental occupational pesticide exposure', *Environmental Health Perspectives*, **117** (10), 1505–13, accessed 22 May 2019 at https://doi.org/10.1289/ehp.0900582.

World Health Organization (WHO) (2018), 'Pesticide residues in food, accessed 29 December 2018 at https://www.who.int/en/news-room/fact-sheets/detail/pesticide -residues-in-food.

Yan, D., Zhang, Y., Liu, L. and Yan, H. (2016), 'Pesticide exposure and risk of Alzheimer's disease: a systematic review and meta-analysis', *Scientific Reports*, **6** (September), 32222, doi:10.1038/srep32222.

10. Occupational health and safety in the mining sector

Carmel Bofinger and David Cliff

INTRODUCTION

The mining sector covers a range of operational types and commodities. The scale and complexity of operations and equipment in the mining sector varies from operations owned by multinational companies with large and automated equipment to small mines owned by the single operator with basic equipment. There are open cut and underground operations and the range of materials and resources extracted is diverse – coal, metals, uranium, quarry material and gemstones. The sector covers exploration, development, extraction and closure. The location of the sites varies from built-up areas to some of the most remote areas of the world and if we include oil and gas, there are also off-shore operations.

Some hazards faced in the mining sector are common to other industrialised sectors, for example, confined spaces and working at heights, and some are specific to mining, for example, falls of ground.

This chapter focuses on the health and safety issues that are most relevant to the mineral industry operations and the community health and safety issues that are directly affected by the mining sector.

HEALTH AND SAFETY LEGISLATION IN MINING

The rate of health and safety incidents in mining varies considerably by country, and these are rarely comparable owing, in part, to different health and safety legislation (Nowrouzi-Kia et al. 2018). The legislation tends to be of two dominant regulatory approaches, either compliance based or risk based.

Compliance-Based Legislation

An example of the compliance-based approach is the US where the development of mine safety and health regulations over the past half century has created a complex regulatory system that is largely predicated on a compliance-based structure. These regulations represent the mandatory health and safety guidelines and standards, the dos and don'ts, governing how all mining facilities under the Mine Safety and Health Administration (MSHA) jurisdiction must operate (Poplin et al. 2008). Despite these efforts, mining reported a fatality rate per 100 000 full-time workers of 10.1 in 2016 as shown in Figure 10.1 (US BLS 2018).

Risk-Based Legislation

Mining health and safety legislation in Australia is risk-based. There is no national mining health and safety legislation in Australia and each state has established its own legislative framework. All legislation encourages the development of management systems and processes and is based on the principles of:

- duty of care;
- risk management; and
- workforce representation.

The primary responsibility for the provision of a safe place of work reside with the operator of the mine site. Government inspectors act as both enforcers of regulations and mentors who encourage good health and safety performance. Enforcement protocols are generally risk based, with action being defined by the level and immediacy of the risk.

Management systems provide a framework and structure for the development, implementation and review of the plans and processes needed to manage health and safety in the workplace. The Australian mining industry reported a fatality rate per 100 000 full-time workers of 2.7 in 2016 as shown in Figure 10.1 (SafeWork Australia 2018).

The implementation of legislation based on risk management in the mining sector has coincided with a fall in accidents and injuries. From when legislation was first introduced in 2000–2001 to 2015–16, the frequency rate (serious claims per million hours worked) for the Australian mining industry dropped by 62% from 10.7 to 4.1 (SafeWork Australia 2018).

RISK MANAGEMENT PRINCIPLES AND PRACTICE

Risk management processes are used across an entire organisation and within its many areas and levels, in addition to specific functions, projects

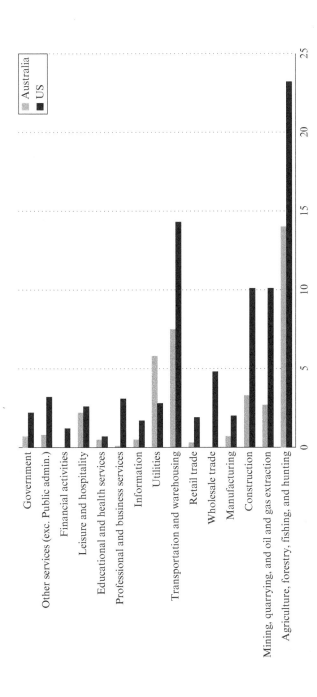

Sources: SafeWork Australia (2018); US BLS (2018).

Figure 10.1 Comparison of Australian and United States fatal work injury rate per 100000 full-time equivalent workers, 2016

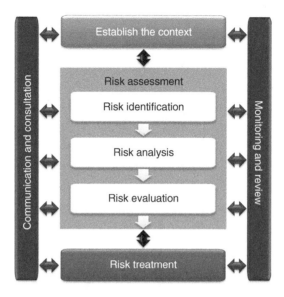

Figure 10.2   The risk management process (ISO 31000:2018)

and activities to identify the risks that need to be managed to ensure the sustainability of the organisation (ISO 2018). Moreover, the risk management process should not only be an integral part of management but also 'embedded in the culture and practices, and tailored according to the processes of the organisation' (ISO 2018, p. 9). The activities that constitute the risk management process are shown in Figure 10.2.

The mining sector has been progressively adopting risk management at a corporate, site, major mine process and day-to-day practice. The approaches vary in detail and formality but mining companies consider risks in some way when making health and safety related decisions. The degree to which the process is systematic, formal and accurate varies depending on country, company and site.

Current best practice is a fully integrated risk management system. Integrated risk management involves the formal placement of risk management activities throughout the key management and engineering processes of the site, including day-to-day work planning and control. This is shown in Figure 10.3.

Achieving an effective integrated risk management system requires evolution of the sector's culture and systems over time. It is necessary for the systems to evolve through a series of stages and, importantly, the culture or people factors must evolve in parallel. The expectations of the risk management system should outline the principles and methods customised

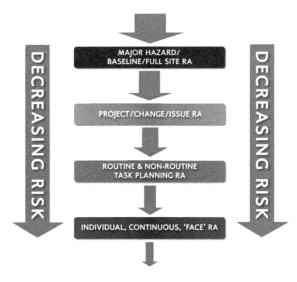

Note: RA = risk assessment.

Figure 10.3 Integrated risk management framework

for site requirements and be designed to suit the relevant decisions that are made by the site engineering, management, supervisory and individual workforce members.

A fully integrated risk management system will have a carefully designed set of expectations for risk management for process safety, personal safety and health. However, one of the continuing conflicts for the effective management of safety in the mining sector is between process safety and personal safety.

PROCESS SAFETY AND PERSONAL SAFETY

Process Safety

Process safety considers the major and catastrophic hazards inherent to the industry, for example, explosions, heavy mobile equipment, fires, strata control and falls of ground. These occur infrequently but the potential consequences are very high. In large-scale mining operations, mining disasters with multiple fatalities are infrequent; however, the sector continues to have fatalities from known hazards, often single fatalities, for example, crush, vehicles and pedestrians.

These major and catastrophic hazards are generally identified in legislation. They are the issues that have the potential to result in multiple fatalities and even where risk-based legislation is in place, they are considered to be so important that they are specifically identified in legislation as requiring management – usually through the development of specific hazard management plans which includes both systems and hardware to manage the hazard.

In more recent years this has included the use of bow-tie analysis with a focus on the identification and monitoring of controls, particularly the controls identified as critical. The International Council on Mining and Metals (ICMM) has produced guidance materials on the management of major hazards and the associated control processes (ICMM 2015). This process has been widely embraced internationally in the minerals and resource industries for companies and regulators. This has driven a tighter definition of what constitutes a control and is working to strengthen the management of the major hazards.

There can be differences in the legislative, corporate and site definitions of major hazards. Legislative and corporate definitions tend to refer to potential multiple fatalities (sometimes five or more). The sites tend to use potential single fatalities as the cut-off for the implementation of critical control processes. This gives a greater site focus on the single fatalities and, hopefully, we will see less of those as the management of these hazards is more effective through similar processes to those used for major hazards.

Personal Safety

Personal safety involves the hazards that occur frequently with less severe consequence – injuries rather than fatalities – for example, slips, trips and manual tasks. These can be managed by the systems approach but there are also limitations associated with individual decision-making that need consideration (Bofinger et al. 2015). These limitations are not specific to the mining sector and are addressed in similar ways through training and awareness as well as the hardware and systems controls.

In addition to training and awareness for individuals, mining sites have responsibility for ensuring that working hours and conditions of the work environment do not compromise safety. Fatigue owing to work, environment and roster arrangements has been identified as a safety issue. A fatigued worker is recognised as a safety risk for themselves and fellow workers. This is a complex issue to manage, with both work and non-work issues affecting fatigue, and the effective identification of fatigue within the workplace remains problematic.

HEALTH ISSUES

As with workplace health issues in many high-risk sectors, there are problems associated with the identification and management of miner's health, including long latency, multiple work and non-work causes, lack of follow-up after employment, and so on. The historical health issues associated with mining, for example, pneumoconiosis and mesothelioma, have generated a lasting image of the mining industry as hazardous to health and it has become increasingly unacceptable to the community for perceived hazardous industries to continue unless the hazards are demonstrably managed. As the extract in Box 10.1 from the 'Queensland mines and quarries safety performance and health report 2016–2017' demonstrates, the importance of health issues is increasingly being recognised (DNRM 2017).

However, the reported occupational health issues mostly reflect occupational exposures to hazards that occurred in the past, which may no longer exist or are now well recognised and potentially minimised. The use

---

BOX 10.1   EXTRACT FROM THE 'QUEENSLAND MINES AND
QUARRIES SAFETY PERFORMANCE AND
HEALTH REPORT 2016–2017'

- There has been a major reduction in measured respirable dust levels in underground coal mines in the first half of 2017 compared to the same period in 2016, following regulatory changes that commenced in January 2017. These changes require coal mines to report respirable dust data to the Chief Inspector of Mines, and to report single sample exceedances of the occupational exposure limit.
- Monitoring respirable crystalline silica in mineral mines and quarries is now required following the gazettal of QGL02 Guideline for Management of Respirable Crystalline Silica in Queensland Mineral Mines and Quarries, which prescribes mandatory monitoring and reporting to the Queensland Mines Inspectorate.
- Hearing loss is an area of concern, with 10 of the 19 permanent incapacities in surface coal mines in 2016–17 attributed to noise induced hearing loss. Mine sites should review their audiometry surveillance and the effectiveness of their noise abatement and management controls.
- Worker well-being is recognised as an increasingly important priority for the mining industry. Mental health, obesity, drug and alcohol use and worker fatigue can all adversely affect safety and health outcomes in the mining and quarrying industry, with potential for serious and tragic consequences.
- The Queensland Mines Inspectorate will survey the industry on drug and alcohol abuse in 2017–18 and report the findings in next year's report.

Source: DNRM (2017, p. 47).

of exposure monitoring for the potential health hazards in the minerals industry can be an effective management technique where there are well-established cause and effect and dose and response relationships. Exposure monitoring can be used as a lead indicator of health issues, since waiting for reporting of disease is too late.

Work-Related Health Issues

The minerals industry has a long history of recognised workplace health issues. The most obvious are the respiratory issues associated with dust exposures – coal workers pneumoconiosis (black lung) and silicosis.

Recent identification of cases of pneumoconiosis in the Australian coal industry after an absence of 30 years has shown that even these well-recognised health issues have not been well managed in a mining industry that is usually considered to have good practice risk management processes (Coal Workers' Pneumoconiosis Select Committee 2017). A number of contributing factors to the lack of control of dust and diagnosis have been identified including:

- complacency – it was not considered to be a risk with modern mining;
- familiarity with the hazard of respirable dust;
- loss of focus on the controls and monitoring of the control effectiveness;
- collection of data but lack of analysis;
- loss of skill-set in diagnosis; and
- focus on short term safety issues.

In South Africa, the high rates of silicosis are of major concern in the gold-mining industry, and this is exacerbated by the association of silicosis and tuberculosis. South African gold miners have one of the highest incidence rates of pulmonary tuberculosis in the world (Glynn et al, 2008). As miners continue to age and work for longer periods, the burden of silicosis and other diseases will continue to rise. This has far-reaching implications for health services that need to be prepared for increasing morbidity and mortality in current and ex-miners (Nelson 2013).

Another well-recognised health issue is hearing loss. Considerable work has been done to try to control noise exposures in what is a noisy environment with cutting, breaking and crushing as part of the mining process. However, noise levels remain high and noise-induced hearing loss is one of the most common occupational health issues in mining (NIOSH 2010; Edwards 2011; Coal Services 2013).

There has also been an increasing focus on mental health issues in the minerals industry. It has been recognised that the remote fly in/fly out or camp-based operations have the potential to contribute to mental health issues (Bowers et al. 2018).

Lifestyle Health Issues

As with all industries, the extra complication of lifestyle on health issues also affects the identification and management of health in mining. Many companies have introduced healthy lifestyle diet and exercise programmes and incentives, with varying degrees of success (Goater et al. 2013; Mining Safety 2018).

Communicable Diseases

Communicable diseases are a major problem for some mining operations. In some countries, tuberculosis (TB) and HIV have been identified as significant risks in the mining sector. For example, South Africa's National Strategic Plan for human immunodeficiency virus (HIV) (South African National AIDS Council 2017), TB and sexually transmitted infections (STIs) 2017–22 identified mine workers as a vulnerable population for HIV, TB and STIs, and mine workers are included in the customised and targeted interventions outlined in the plan.

COMMUNITY SAFETY AND HEALTH ISSUES

The effect of the mining sector on the health and safety of associated communities also needs consideration and management. These effects can range from multiple fatalities to mental health impacts. There is a need to balance personal and community responsibility with company responsibility for managing the health and safety of the community.

Major Safety Hazards for Communities

There have been a number of tragic consequences for communities when major hazards have not been managed in the minerals sector, for example, 19 people were killed owing to the Samarco dam failure in Brazil in 2015 and more than 200 killed due to the Brumadinho dam failure in 2019. The following are considered to be the most significant issues for community safety related to mining operations:

- road safety;
- mine site access;
- tailings dam safety;
- blasting safety; and
- transport of explosives and other hazardous materials (Leading Practice Sustainable Development Program for the Mining Industry 2016)

Road safety
The road safety challenges can be greatest when the mining sector first enters, or grows rapidly, in a region. Local residents may not be used to heavy vehicle traffic, increased light vehicle traffic that corresponds to shift work, an increase in population generally, changing conditions of roads (improvements and degradation) and shifts in patterns of traffic (for example, more vehicles on what were less travelled roads).

Mining companies' road safety interventions may need to extend beyond their fleets of company vehicles and their workers' commute to and from work, and include consideration of the driving, walking and riding practices of community members in a locality.

Mine site access
Controlling access to mine sites, whether active or abandoned, is another area where the safety of the community must be considered. This is potentially a serious problem where informal small-scale mining is undertaken or where other site trespass could result in accidents leading to injuries and even fatalities.

Tailings dams
There are two dominant safety issues associated with tailings dams:

- the physical safety of the dam; and
- safe containment of any toxic substance.

Any tailings dam must ensure physical, radioactive and chemical safety for both the environment and the community, during mine operation and after closure, and considering long-term stability, extreme events and slow deterioration.

Blasting safety
Blasting at mine sites, particularly when close to site boundaries, can have impacts on the surrounding community, infrastructure and environment owing to vibration through the air (overpressure) and earth (ground

vibration), the generation of dust, fume, noise, odour and flyrock. Flyrock is the undesirable throw of debris from a blast and can cause severe injury and property damage.

Transport of explosives and other hazardous materials

Mining and processing operations transport, store and use a range of hazardous materials including fuels, process reagents, lubricants, solvents and explosives. Where operations are located close to communities, or transport of materials has to pass through a community, there can be safety issues for that community if not properly controlled. For example, in Peru on 2 June 2000, an accident occurred when a truck of a carrier contracted by Minera Yanacocha spilled 11.1 litres of mercury while transporting elemental mercury to Lima via the Cajamarca-Pacasmayo road (Yanacocha 2013).

These hazards need to be managed using similar processes to those used at site, for example, the development and implementation of management systems including the identification and management of controls.

Health Issues for Communities

Mining operations can also impact on the health of communities – both physically and mentally (Leading Practice Sustainable Development Program for the Mining Industry 2016).

The physical effects can be caused by a range of workplace hazards reaching nearby communities including:

- Airborne contaminants – airborne particulates and gaseous emissions have the potential to cause personal health problems in the community and dust and odour can cause annoyance and complaints.
- Noise – noise is one of the most significant issues for communities located near mining projects particularly owing to 24 hour, seven-day operations. Mining activities such as blasting, drilling and digging, coal loading, the operation of excavators, trucks, conveyor belts and other machinery all contribute to elevated levels of environmental noise.
- Lighting – excessive or obtrusive artificial light from mining operations, or other steps in the logistics chain such as transport, can affect nearby communities through impacting on the quality of sleep. This can be in the form of fixed lighting around infrastructure, mobile lighting plants or mobile plant and equipment lights.
- Waterborne contamination – water contaminated with high concentrations of metals, sulphide minerals, dissolved solids or salts

---

BOX 10.2   COMMUNITY CONSEQUENCES OF HAZELWOOD FIRE

The Hazelwood mine fire in 2014 has been linked to a spike in doctor visits by Latrobe Valley residents, as well as a jump in rates of prescription medicine being dispensed. The fire burned inside the mine for 45 days and shrouded surrounding towns in smoke and ash.

The Monash University-led health study, which is being funded by the Victorian Government, found there were an extra 5,137 visits to GPs in the Latrobe Valley in a month when the coal mine fire was alight in 2014. This included 405 cardio-vascular visits, 174 respiratory visits and 286 mental health consultations. The increase in GP visits was attributed to the fine particles emitted by the coal fire four years ago. The Hazelwood health study was established to monitor any long-term health effect of that smoke event.

The study estimated there was an extra 1,429 mental health-related medications being dispensed during the period examined, along with an additional 2,501 cardiovascular prescription medications and 574 respiratory medications.

Source: Field (2018).

---

can negatively affect surface water quality and groundwater quality. Impacts on human health can occur where the quality of water supplies used for irrigation, drinking and/or industrial applications is affected.

These exposures can be caused by routine mining operations. Communities can face additional health risks when things go wrong at a mine site. Consider, for example, the health effects of the Hazelwood fire in Victoria, Australia. Box 10.2 is from a news report in April 2018.

The boom and bust cycle of mining can impact on the physical and mental health of communities. Overarching community health issues prominent during both boom and bust periods include burdens to health and social services, family stress, violence towards women and addiction issues (Shandro et al. 2011). There are either not enough resources and services (housing and health) during the boom cycle to address the demand, or services and resources are taken away during the bust cycle (shops close and house values drop).

SMALL-SCALE MINING

Health and safety issues in the artisanal and small-scale mining (ASM) sector differ from the large scale operations in a number of ways. Artisanal and small-scale mining activities cover an array of minerals from gemstones to gravel and often operate outside the legal framework. With the exception of mercury use in the gold sector, understanding of the safety and health risks faced by the workers in this sector is often limited. The majority of ASM accidents, fatalities and illnesses are not likely to be reported or recorded, and therefore are not included in national and international statistics (Hermanus 2007; Miserendino et al. 2013).

Smith et al. (2016) identified a comprehensive range of causes and outcomes of health and safety issues facing ASM miners, as shown in Table 10.1.

Many of these issues are also prevalent for large-scale mining; however, the understanding of the causal relationships and the resources to address them may not be present in SSM. There is also often not the regulatory understanding and support for this sector owing to limited resources of the local inspectorate.

There has been a significant focus on the use of mercury in SSM with alternate processes for separating the gold being introduced. The other health and safety exposures that SSM face have received less attention, despite poor health and safety on an SSM site significantly impacting on the development and management of environmental and community issues. For example, if a worker is injured on site and can no longer work, the wife or child may have to step into the role to keep the mine going. Also, effective management of chemicals, including mercury, will assist in preventing environmental contamination.

INDICATORS OF HEALTH AND SAFETY IN MINING

There are a number of sources of evidence regarding safety and health in the mining sector with safety being more detailed than health. Over time, this information has provided an indication of the safety and health performance, and indicators have been used for many years in mining and other high-risk industries (Grabowski et al. 2007). Traditionally, performance was generally based on lagging indicators, such as fatality rates, incident statistics and health problems. Ironically, the better the health and safety performance, the fewer incidents that occur and the harder it is to use them to further assist in improving occupational health and safety (OHS) performance.

Leading indicators, however, can be created to produce forward-looking analysis in order to implement effective proactive control measures. The ICMM has released a report providing an overview on leading indicators and how they can be applied for occupational health and safety in the

*Table 10.1	Health and safety issues facing ASM miners*

Diseases and injuries	Hazards and exposures	Chemicals
Ergonomic stresses	Landslides	Mercury (used in gold processing)
Musculoskeletal disorders and diseases (arthritis)	Decompression sickness (from diving)	Cyanide (used in gold processing)
Respiratory diseases	Airborne pollutants (equipment exhaust, dust)	Arsenic (naturally occurring)
Noise and hearing loss	Heat and cold stress	Chromium (naturally occurring)
Parasitic infections (malaria)	Poor air quality or ventilation	Radon (naturally occurring)
Blood borne infectious diseases (HIV/ AIDS, hepatitis B or C, Ebola)	Blasting or explosives	Aluminum (naturally occurring)
Cancer (occupational)	Rock falls	Copper (naturally occurring)
Neurotoxicity	Flooding	Manganese (naturally occurring)
Airborne infectious diseases (TB)	Stumbling, slipping and falling	Nickel (mined and naturally occurring)
Dengue fever	Unstable underground structures	Zinc (used in gold processing)
Diseases of blood and skin	Obsolete, inappropriate or damaged equipment	Lead (emitted during processing and naturally occurring)
Traumatic injury	Poor visibility and light	Cadmium (naturally occurring)
Water, soil and food contamination-related diseases (cholera, typhoid)	Poorly built tunnels	Cobalt (mined and naturally occurring)
Enteric (intestinal) infections	Lack of exits	Selenium (naturally occurring)
Lifestyle factors (smoking-related diseases, inadequate nutrition, alcohol and drugs, STDs, HIV/AIDS)	Gender-based violence and abuse	Uranium (mined)

Table 10.1   (continued)

Diseases and injuries	Hazards and exposures	Chemicals
Genital corrosions and miscarriages (from prolonged standing in water)	Dense living arrangements	Methane (naturally occurring)
Skeletal fractures	Remoteness of work	
Cardiovascular diseases	Poor sanitation	
Mental impairments; psychological effects	Water, soil and food contamination Improper use of chemicals Social conflicts	

mining industry (ICMM 2012). Leading indicators contribute to the ability of an organisation to develop appropriate proactive strategies to prevent harm through recognising early signals and taking action. They are particularly beneficial for health-related issues as they do not rely on measuring the development of disease many years after exposure.

Unfortunately, this information is not reliably available for small-scale and artisanal operations.

FUTURE FOR HEALTH AND SAFETY IN MINING

For the mining industry, the vision of safe and healthy production is well recognised and supported at the corporate level for large-scale mining operations. It is the practical implementation of that vision that remains a challenge. There can still be conflict between safety and production at the site level. This may not be overt and often not consciously done, but it does remain. This is a great challenge and must be supported by consistent and visual site and front-line leadership which puts safety first.

In addition to leadership, the following points need recognition:

- One culture does not fit all. The mining industry is international and diverse. Good culture needs to be developed for different operations.
- It is not possible to simply wave a magic wand and change culture – it takes time and commitment and the promotion of ownership of safety and health at all levels within an organisation and operation.

This is often included in mining safety and health legislation, but sometimes safety and health is still perceived as the responsibility of the safety and health department instead of being the responsibility of all workers.

- Effective safety and health management requires trained people, fit-for-purpose equipment and safe work practices. Mines need the support structure and resources to create and maintain systems to promote and sustain a positive safety and health culture.
- The economic value of safe and healthy production has to be demonstrated. This is an aspect that has not been done well for the mining industry. Safety is most noticeable by an absence of incidents and accidents, but for organisations that do not have a level of maturity regarding safety and health, that tends to lead to down-sizing of the safety department as they are perceived as not being needed anymore. The value of not having incidents and accidents and the consistent effort needed to maintain that state needs to be better quantified.

The minerals industry needs standards, codes of practice, work instructions and other documents and resources for training and to support workers. However, human behaviour cannot be codified. Workers need to think and be alert to the work environment. Given the well-recognised ways in which human error can occur, equipment and process design needs to ensure that there are adequate barriers so that if human errors occur, controls are in place to prevent or minimise the potential for injury.

Through the industry working together to improve processes and exchange information and working with the surrounding communities, safety and health in the mining industry will continue to improve.

REFERENCES

Bofinger, C., Hayes, J., Bearman, C. and Viner, D. (2015), 'OHS risk and decision-making', in Safety Institute of Australia, *The Core Body of Knowledge for Generalist OHS Professionals*, Tullamarine, VIC: Safety Institute of Australia.

Bowers, J., Lo, J., Miller, P., Mawren, D. and Jones, B. (2018), 'Psychological distress in remote mining and construction workers in Australia', *Medical Journal of Australia*, **208** (9), 391–7.

Coal Services (2013), *Managing Noise in the Coal Industry to Protect Hearing*, Argenton, NSW: Coal Services.

Coal Workers' Pneumoconiosis Select Committee (2017), 'Black lung white lies: inquiry into the re-identification of coal workers pneumoconiosis in Queensland', Report No. 2, 55th Parliament, May, Coal Workers' Pneumoconiosis Select Committee, Brisbane.

Department of Natural Resources and Mines (DNRM) (2017), 'Queensland mines and quarries safety performance and health report 2016–2017', Queensland Government, Brisbane.

Edwards, A.L., Dekker, J.J, Franz, R.M., van Dyk, T. and Banyini, A. (2011), 'Profiles of noise exposure levels in South African mining', *Journal of the Southern African Institute of Mining and Metallurgy*, **111**, 315–22.

Field, E. (2018), 'Hazelwood mine fire linked to spike in doctor visits and prescription medication use', Australian Broadcasting Commission Gippsland, accessed 28 October 2018 at http://www.abc.net.au/news/2018-04-10/hazelwood-mine-fire-linked-to-increased-gp-visits-health-study/9637116.

Grabowski, M., Ayyalasomayajula, P., Merrick, J.R.W., Harrald, J.R. and Roberts, K. (2007), 'Leading indicators of safety in virtual organisations', *Safety Science*, **45** (10), 1013–43.

Glynn, J.R., Murray, J., Bester, A., Nelson, G., Shearer, S. and Sonnenberg, P. (2008), 'Effects of duration of HIV infection and secondary tuberculosis transmission on tuberculosis incidence in the South African gold mines', *AIDS*, **22** (14), 1859–67.

Goater, S., Goater, I., Trivell, H., Knowles, M., Leveritt, M. and Lynas, D. (2013), 'Health promotion in FIFO and resident mine workforces: a case for a Wellness-Watch program', paper presented at the Twelfth National Rural Health Conference, Adelaide, 7–10 April.

Hermanus, M.A. (2007), 'Occupational health and safety in mining – status, new developments and concerns', *Journal of the Southern African Institute of Mining and Metallurgy*, **107** (8), 531–8.

International Council on Mining and Metals (ICMM) (2012), 'Overview of leading indicators for occupational health and safety in mining', International Council on Mining and Metals, London.

International Council on Mining and Metals (ICMM) (2015), 'Health and safety critical control management', International Council on Mining and Metals, London.

International Organization for Standardization (ISO) (2018), 'Risk management – guidelines', ISO 31000:2018, October, International Organization for Standardization, Geneva.

Leading Practice Sustainable Development Program for the Mining Industry (2016), *Community Health and Safety Handbook*, Canberra: Australian Government.

Mining Safety (2018), 'Learn about your health: nutrition', accessed September 2018 at http://www.miningsafety.co.za/dynamiccontent/17/Learn-About-Your-Health -Nutrition.

Miserendino, R.A., Bergquist, B.A., Adler, S.E., Davée Guimarães, J.R., Lees, P.S., Niquen, W., et al. (2013), 'Challenges to measuring, monitoring, and addressing the cumulative impacts of artisanal and small-scale gold mining in Ecuador', *Resources Policy*, **38** (4), 713–22.

National Institute of Occupational Safety and Health (NIOSH) (2010), 'Mining topic: hearing loss prevention overview', accessed September 2018 at https://www.cdc.gov/niosh/mining/topics/hearinglosspreventionoverview.html.

Nelson, G. (2013), 'Occupational respiratory disease in the South African mining industry', *Global Health Action*, **6** (1), 19520, doi:10.3402/gha.v6i0.19520.

Nowrouzi-Kia, B., Gohar, B., Casole, J., Chidu, .C, Dumaon, J., McDougall, A., et al. (2018), 'A systematic review of lost-time injuries in the global mining industry', *Work*, **60** (1), 49–61.

Poplin, G.S., Miller, H.B., Ranger-Moore, J., Bofinger, C.M., Kurzius-Spencer,

M., Harris, R.B., et al. (2008), 'International evaluation of injury rates in coal mining: a comparison of risk and compliance-based regulatory approaches', *Safety Science*, **46** (8), 7922–37.

SafeWork Australia (2018), 'Number and incidence rate of work-related traumatic injury fatalities by industry 2012–2016', 19 April, accessed September 2018 at https://www.safeworkaustralia.gov.au/doc/number-and-incidence-rate-work-rela ted-traumatic-injury-fatalities-industry-2012-2016.

Shandro, J., Veiga, M., Shoveller, J., Scoble, M. and Koehoorn, M. (2011), 'Perspectives on community health issues and the mining boom–bust cycle', *Resources Policy*, **36** (2), 178–86.

Smith, N.M., Ali, S., Bofinger, C. and Collins, N. (2016), 'Human health and safety in artisanal and small-scale mining: an integrated approach to risk mitigation', *Journal of Cleaner Production*, **129** (August) 43–52.

South African National AIDS Council (2017), 'Let our actions count: South Africa's National Strategic Plan for HIV, TB and STIs 2017–22', April, South African National AIDS Council, Hatfield, Pretoria.

US Bureau of Labor Statistics (US BLS) (2018), 'Number and rate of fatal work injuries, by industry sector', US Department of Labor, accessed September 2018 at https://www.bls.gov/charts/census-of-fatal-occupational-injuries/number-and-rate-of-fatal-work-injuries-by-industry.htm.

Yanacocha (2013), 'Fact sheet: Choropampa', January, accessed 30 October 2018 at https://s1.q4cdn.com/259923520/files/doc_downloads/south_america/yanecocha/Choropampa-Fact-Sheet-01-25-2013.pdf.

11. Occupational health and safety in the construction sector*

Helen Lingard

OCCUPATIONAL HEALTH AND SAFETY IN CONSTRUCTION

Construction work is inherently dangerous, combining a constantly changing work environment with work at height, continuous movement of people, materials and vehicles, the use of power-driven machinery, plant and equipment and exposures to a wide range of hazardous substances.

Occupational health and safety (OH&S) risks experienced by construction workers are many and varied and construction is often singled out by legislators and policy-makers as an industry deserving special treatment in terms of the prevention of work-related deaths, injuries and illnesses.

The types of incidents and mechanisms of injury impacting construction workers are similar the world over, whether these relate to low frequency-high impact or high frequency-low impact injury outcomes. Falling from height, being struck by falling or mobile objects, being trapped or crushed by collapsing structures or coming into contact with electricity are the most common incident types associated with fatal or serious injury, while the occurrence of slips, trips and falls (on the same level) and musculoskeletal problems linked to manual handling are frequently cited in non-fatal injury reports.

Traditionally, a great deal of focus has been placed on understanding and addressing the causes of immediate-effect safety incidents. Historically, occupational health risks have received less attention than safety risks. This is the case despite deaths arising from work-related illness being estimated to far outweigh deaths arising from acute effect injuries. For example, on average, over 250 Australian workers die from an injury sustained at work each year, yet it is estimated that over 2000 workers die from a work-related illness each year (Safe Work Australia 2012). This is consistent with UK estimates that construction workers are at least 100 times more likely to die from a disease caused or made worse by their work as they are from a work-related injury (IOSH 2015).

Key occupational health risks applicable to construction work include:

- asbestos;
- dusts including silica and lead;
- hazardous chemicals;
- sunlight, and ultraviolet radiation exposure;
- diesel engine exhaust emissions;
- loud noise;
- the use of vibrating tools;
- excessive manual handling of loads; and
- stress and fatigue (IOSH 2015).

International studies show that rates of occupational illness are high among construction workers. For example, Stocks et al. (2011) analysed instances of medically reported work-related ill health among construction workers in the UK and found elevated rates of contact dermatitis, all types of skin neoplasma, non-malignant pleural mesothelioma, lung cancer, pneumoconiosis and musculoskeletal disorders.

Sherratt (2015) observes that legislation in the UK (and in Australia) bundles work-related health and safety together, which she sees as being unhelpful because, unlike safety, which occurs at the site at a fixed point in time, health occurs over extended periods and is not specific to a single worksite. Occupational health and safety legislation requires the protection of workers from occupational health risk exposures. Yet, as Sherratt observes, employers sometimes confuse this responsibility with the implementation of health promotion programmes targeting lifestyle risk factors, such as alcohol consumption, smoking cessation, diet and physical activity. Although programmes focused on improving workers' health-related behaviours may produce benefits, their implementation does not remove the employers' duty to reduce exposure to occupational health risk factors as far as is reasonably practicable.

Thus, risk reduction should focus on providing the highest level of protection that it is reasonable to implement in a particular situation, and be guided by the hierarchy of control. The hierarchy of control is applied to OH&S risks and arranges control measures in descending order of effectiveness. The top three levels are preferred as they address aspects of the physical workplace, while the bottom two levels rely on workers' behaviour for their effectiveness and are generally considered to be less reliable. Figure 11.1 shows an example hierarchy of control for managing the risk of diesel exhaust emissions in the construction industry.

When applied to occupational health, a failure to apply the hierarchy of control properly in decision-making is likely to perpetuate the selection

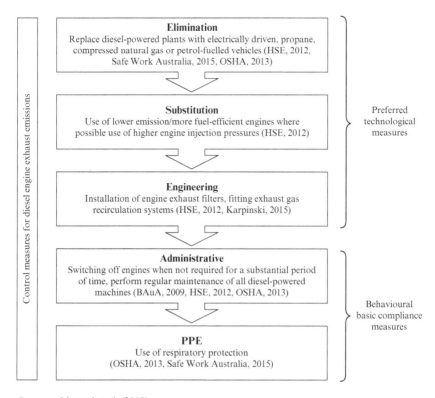

Source: Lingard et al. (2018).

Figure 11.1   Example risk control hierarchy for diesel exhaust emissions

and implementation of behavioural controls for occupational health risks, such as worker training and a reliance on the use of personal protective equipment. Although these controls are important, they should be seen as a supplement (rather than alternative) to the implementation of more effective technological controls that actively reduce exposures to hazardous conditions in the workplace. Behavioural forms of control for OH&S risks are often limited in their effectiveness, particularly in the construction context because:

- they protect workers engaged in a particular activity but may not protect those working in the surrounding environment, or
- they are inconsistently applied by workers who may be engaged to work at a site for a limited period of time (Sherratt 2015).

However, barriers to the selection of more effective forms of control for occupational health risks in the construction context have been identified, including labour market forces, productivity pressures, and a normalisation of risk, pain, injury and illness among the industry's workforce (Stergiou-Kita et al. 2015).

Research also highlights the importance of considering OH&S before construction work commences, for the selection and implementation of more effective technological controls for OH&S risks (see case study 1).

Case Study 1: Timing of Decision-Making and Effective Risk Control Selection and Implementation

An international benchmarking study of design decision-making and its impact on construction workers' OH&S tested the proposition that early consideration of risk control measures in project decision-making would produce better outcomes, in terms of the selection of technological, rather than behavioural forms of control. The study:

- measured the effectiveness of risk controls in case study construction projects;
- identified whether risk control measures were identified and developed before the commencement of construction; and
- examined whether risk control decisions made early in the project life-cycle, that is, before the commencement of construction, were more likely to produce effective (upper level) controls for identified risks.

Data were collected from a total of 23 construction projects, ten of which were in Australia and New Zealand, and 13 were in the USA. The research design involved replication and cross-validation across two diverse and different samples (that is, the US and Australasian project samples). The relationship between the timing of project decisions and the effectiveness of risk controls was evaluated independently in the Australian and the American data.

In-depth interviews were conducted with stakeholders involved in the planning, design and construction of selected 'features of work' particular to each project. Interviews explored the timing and sequence of key decisions and the quality of risk control solutions that were ultimately implemented during the construction stage of the project. The findings were based on 288 interviews (185 conducted in Australia and 103 conducted in the USA).

A statistically significant, positive relationship was found between the selection of risk control measures in the early stages of a construction project (before construction work commenced) and the quality of risk controls implemented in the construction stage of the project. Thus, when risks were identified and control decisions taken before the commencement of construction, it was more likely that risks would be controlled at source, through the implementation of technological control measures.

When decisions about risk control were left until the construction stage, it was more likely that measures implemented to control OH&S risks would be lower levels of control that rely for their effectiveness on workers' behaviour, for example administrative controls or personal protective equipment (Lingard et al. 2015b).

PROJECT-BASED WORK

Occupational health and safety impacts arising from the processes, equipment and materials of construction have been well documented elsewhere (see, for example, Lingard and Rowlinson 2005). However, hazards can also arise as a result of the organisation and management of construction projects and, in some cases, the absence of effective management and integration of OH&S into project decision-making.

Unlike many other industrial sectors, construction is characterised by complicated and unstable supply relationships. Construction projects are commissioned by clients and delivered through temporary coalitions of organisations, working interdependently to achieve project outcomes. Project timelines are often tight and financial penalties are applied for late delivery. Contractor selection decisions made by construction clients are often based upon a number of criteria (including OH&S), but price is usually a heavily weighted factor. The construction industry's reliance on competitive tendering means that project delivery teams constantly change as different constellations of actors are formed and then disbanded. This instability limits opportunities for learning from experience (Briscoe and Dainty 2005). Further, Gadde and Dubois (2010) observe that relationships tend to be contractual and arm's-length and the development of trust, a shared culture and a mutual orientation to issues, including OH&S, can be hard to achieve within the duration of a single project.

Construction projects are organisationally fragmented, both vertically and horizontally (Fellows and Liu 2012). Horizontal fragmentation describes a reliance on many actors (individuals, organisations and business units) to carry out different functions at the same stage of a construction project. For example, the engagement of specialist subcontractors has been identified as a characteristic of construction projects that creates coordination challenges for OH&S. Vertical fragmentation describes a situation in which different stages of a construction project involve contributions from different functional actors; for example, the quality and nature of interaction between a commissioning client, members of the design team and contractors engaged to provide construction services can impact OH&S.

Vertical and horizontal fragmentation can make it difficult to develop cultural alignment and a common purpose with regard to OH&S in construction project organisations. Indeed, Lingard et al. (2015c) report that constructors, design professionals (architects and engineers) and OH&S specialists varied significantly in their understanding of, and willingness to accept, OH&S risk. It is perhaps not surprising that the complex supply arrangements typical of construction project delivery are reported to have a detrimental impact on OH&S (Swuste et al. 2012). For example, Priemus and Ale (2011) describe how, the design and construction of a mixed-use development of commercial, residential and recreational facilities in the Netherlands was divided into three parts, each requiring a separate building permit. Responsibility for delivery was split between two developers, two building agencies (without a senior structural engineer), three architects' firms (with no consistent overall and final responsibility), one main contractor and around 50 subcontractors. Time pressures in the project were intense, the coherence of decision-making and effectiveness of communication were compromised and project monitoring control systems failed, resulting in a significant structural safety incident (Priemus and Ale 2010).

Alignment and integration of OH&S into project decision-making is also influenced by clients' selection of a project delivery approach, which shapes the contractual relationships, roles and responsibilities of the parties. In some cases, design and construction services are bundled into a single contract, with one organisation engaged to deliver both of these services. In other situations, design work is undertaken by a specialist consultant before a construction contractor is selected. Eriksson and Westerberg (2011) argue that choices made by clients at all stages in the buying process affect the extent to which the project governance structure operates on the basis of competition or collaboration. An emphasis on cooperation, rather than competition, in project procurement has been linked to improved OH&S (Lingard and Rowlinson 2005; El Asmar et al. 2016). Traditional competitive approaches emphasise design specification by the client, competitive tendering, high emphasis on price in contractor selection and output-based pay. In contrast, cooperative procurement emphasises shared (that is, client and contractor) responsibility for design, selection of a contractor through direct negotiation, and emphasis on soft parameters (not just price) to inform contractor selection decisions and the use of performance incentives (Eriksson and Westerberg 2011).

CALLS FOR GREATER CLIENT INVOLVEMENT IN PROJECT OH&S

In recent years there has been an increased focus on the role of construction clients in driving OH&S in the projects they procure. In her review of deaths in the UK construction industry, Rita Donaghy argued that 'public procurement is important because of its size and its potential for insisting on driving up standards including health and safety' (Donaghy 2009, p. 12). The *Australian Work Health and Safety Strategy 2012–2022* (Safe Work Australia 2012) also recognises the potential role of public sector clients in driving OH&S performance improvements. This strategy specifically calls for commercial relationships to be used to improve OH&S, and for Australian governments to use their investment and purchasing power to improve OH&S (Safe Work Australia 2012).

As the initiators of projects and purchasers of the construction industry's product, clients are in an influential position to drive the cultural change needed to improve OH&S in the construction industry. Clients make key decisions concerning project budgets, timelines, objectives and performance criteria. Clients also select the project delivery method and contracting strategy, and choose design and construction team members. All these decisions can potentially impact OH&S in a construction project (Gibb et al. 2014). Further, when clients have considerable construction volume, they are in a stronger position to initiate major improvements in the construction supply chain than when they engage in single one-off projects (Vrijhoef and Koskela, 2000).

In the USA, Huang and Hinze empirically evaluated the impact of a range of client-led OH&S initiatives on construction project safety performance. Their work revealed that high levels of OH&S performance were significantly linked to the involvement of the client in pre-project planning, financially supporting constructors' safety programmes, and participating in project OH&S activities (Huang and Hinze 2006a, 2006b). Clients' OH&S leadership behaviour is important for developing strong and positive safety climates in construction projects (Zhang et al. 2018). Examples of construction industry clients taking a more proactive stance to improve OH&S in construction projects are becoming more prominent, particularly in delivering large public infrastructure construction projects (Eban 2016).

However, the mechanisms through which clients influence project OH&S are not well understood, and few attempts have been made to evaluate client OH&S activities in construction projects. A notable exception is the work undertaken by Spangenberg et al. (2002) who conducted an analysis of the impact of a client-led OH&S program implemented during the construction of the Øresund rail link between Denmark and Sweden. They found that a 25 per cent reduction in the number of injuries resulting from safety incidents was produced through a multifaceted programme which included a large-scale information campaign, a twice yearly monetary award, and specific themed campaigns aimed at improving OH&S-related behaviour. However, Spangenberg et al. (2002) also note that the campaign's impact was limited because it focused too heavily on trying to change attitudes to OH&S (through providing information) rather than changing OH&S practices, and contractors were only involved in the project for relatively short periods of time, limiting their exposure to the campaign.

The important OH&S role to be played by government clients has also been recognised in Australia, where a model client framework was developed (on behalf of the Office of the Federal Safety Commissioner) to help government agencies drive safety in projects through procurement and project management practices. The framework recommended that public sector construction clients undertake some key management actions and entrench these actions into their safety culture (Lingard et al. 2009; Votano and Sunindijo 2014).

In recent years, public sector clients in the Australian construction industry have sought to commercially incentivise OH&S performance by:

- establishing OH&S as a project key result area;
- establishing minimum levels of performance for OH&S;
- using lead and lag metrics to measure contractors' OH&S performance in project delivery; and
- establishing complex mechanisms to financially reward exceptional OH&S performance and/or penalise substandard OH&S performance.

These measures have been embedded in the commercial frameworks used to deliver a number of large infrastructure construction projects in Australia (Lingard et al. 2016).

DESIGNING FOR OH&S

The recognition that decisions made during the design stage of construction projects can significantly impact construction workers' OH&S has led to the implementation of statutory OH&S responsibilities for persons who design buildings or structures, in many industrialised countries, including Australia (Behm 2005; Toole 2007; Driscoll et al. 2008). However, design work in construction is a complex process, involving multiple parties, iterative decision-making and many exchanges of information (Lingard et al. 2012). Further, product complexity and high levels of specialisation have segregated design and construction processes that were historically directed from inception to completion by a 'master builder' (Yates and Battersby 2003).

Commentators have identified significant implementation issues with safety in design in the construction industry. For example, Atkinson and Westall (2010) note that many widely cited safety in design solutions, such as designing anchorage points for fall arrest devices in structures and providing guard rails do not eliminate an inherently dangerous activity, that is, working at height. They suggest that these measures produce a modest reduction in OH&S risk experienced by workers but fail to eliminate hazards or optimise the reduction of risk. Researchers also comment that design professionals in the construction industry (architects and engineers) possess limited knowledge of construction processes (Yates and Battersby 2003). Even in the UK, where designers' responsibilities for construction workers' OH&S were established many years ago under the Construction Design and Management Regulations, Brace et al. (2009, p. 12) report that 'many designers still think that safety is "nothing to do with me," although there are a small cohort who want to engage and are having difficulty doing this because they do not fully understand what good practice looks like'.

The decoupling of design and construction work (implicated in traditional forms of procurement) has been identified as a barrier to the effective management of OH&S in design work. For example, a recent review of OH&S in the UK construction industry identifies separation and poor communication between the design and construction functions as a causal factor in construction fatalities (Donaghy 2009).

Atkinson and Westall (2010) also note that, even in more integrated projects utilising a design and construction delivery approach, the design of the product to be constructed is often outsourced to a specialist team of professional designers, and positive OH&S outcomes are not guaranteed.

Lingard et al. (2014) examined the quality of OH&S risk control outcomes associated with decisions made during the design stage of construction projects and found evidence that more effective technological forms of risk control (for example, elimination, substitution and engineering controls) were more likely to be implemented when people with construction-process knowledge were involved in decision-making. When construction-process knowledge was not available to decision-makers, there was a heavier reliance on behavioural forms of risk control (for example, administrative measures and the use of personal protective equipment). This is attributed to the detailed knowledge of construction methods, materials and technologies which often resides with specialist subcontractors who may not be engaged when design decisions (with the potential to impact OH&S) are made (see, for example, Franz et al. 2013). Given the importance of informing design decision-making with in-depth knowledge of construction processes, Lingard et al. (2018) evaluated the use of infographics to convey this information and report that design professionals who utilised the infographics were better able to understand less visible OH&S issues inherent in their designs, such as the risk of work-related musculoskeletal disorder.

PROCESS AS WELL AS PRODUCT DESIGN

The following case study also highlights the importance of considering OH&S in the design of construction processes and the opportunities afforded by technologies that can eliminate (or significantly reduce) hazards associated with conventional, manual construction work processes.

Case Study 2: Process Design for Cropping Concrete Piles

Breaking down the tops of concrete piles to expose steel reinforcement bars has traditionally been carried out using hand-held pneumatic breakers (see Figure 11.2).

Source: RMIT University (2018).

Figure 11.2   Using a hand-held pneumatic breaker to break down concrete pile heads

This method of pile-breaking involves several serious occupational health hazards, including exposure to hand-arm vibration (HAV), dust, noise and the risk of work-related musculoskeletal disorder (Gibb et al. 2007).

A recent study in the Australian construction industry (RMIT University 2018) revealed that systems of work designed to reduce the exposure to these hazards while breaking down piles were not effective. In particular, an integrative passive risk control method was intended to be adopted. This method involved the installation of a layer of non-bonding material (foam) at the cut-off point during construction of the concrete piles. This material is installed before the concrete is poured.

If performed effectively, the incorporation of the non-bonding material would significantly reduce the duration of the mechanical breaking activity. One participant reported that failure to install the material correctly increased the duration of the task from one hour to approximately four to five hours per pile.

The study revealed that the pile construction work and pile-breaking work were undertaken by two different subcontracted work crews. In many instances, the pile construction crew did not install the non-bonding material correctly, resulting in a substantial increase in health risk exposure for the workers who would subsequently break the pile heads.

This indicates how a lack of attention to the coordination of work crews and poor work process design produced an unnecessary and unanticipated exposure to occupational health risks for subcontracted construction workers. In this case the risk control measure that was specified was rendered ineffective by on-site work practices.

However, if considered at the design stage of a project, technological risk controls are available, including several commercially available hydraulic pile-breaking technologies (EFFC 2015). In addition, integrated active systems have been developed. These systems incorporate an active pile-breaking system within the pile. This system is activated when the concrete is cured. An example of an integrated active pile-breaking method is the Recepieux system, which: uses a system of breakers installed at the desired cut-off position before pouring concrete (Figure 11.3a); introduces an expanding agent into the pile (at least 72 hours after pouring) through carefully positioned ducts which deliver the chemicals to the breakers (Figure 11.3b), enabling the pile top to be cut off within the ground, the breakers working like a jack (Figure 11.3c), and finally enabling the pile top to be mechanically lifted off without the need for jackhammering (Figure 11.3d).

The adoption of integrated active systems of pile-breaking require intervention at the early design and planning stages of construction work (Gibb

Source: Recepieux. Images reproduced with permission.

Figure 11.3   The Recepieux pile-breaking method

et al. 2007). A failure to adequately consider, at the design stage, the health risks associated with mechanical pile-breaking using a pneumatic breaker is likely to be a barrier to the adoption of these methods. Furthermore, a failure to fully appreciate the potential production efficiency gains associated with these methods is also likely to act as a barrier to their specification.

MANAGING SUBCONTRACTORS' OH&S

Another feature of construction work that has the potential to impact OH&S is the heavy reliance on subcontracted labour. Subcontracting involves 'the subletting of the execution of a section(s) of a project to a contractor(s) who in most cases is a specialist in that section(s) of work' (Manu et al. 2013, p. 1018). Most site-based construction work is performed by subcontractors (Dubois and Gadde 2000; Hartemann and Caerteling 2010).

Subcontracting can create challenges for the management of OH&S (Arditi and Chotibhongs 2005). For example, in Australia, Loosemore and Andonakis (2007, p.580) argue that subcontractors 'lack the resources, culture and skills' to manage OH&S risks effectively. In a study of the residential building sector of the construction industry, Wadick (2010) found that the dangers of subcontracting were made worse by poor communication between different trade-based subcontractors and ineffective consultation between management and workers in relation to OH&S. Further, subcontractors often work on a payment-by-results basis which, when combined with competitive-tendering processes and the emphasis on price in selecting subcontractors, can encourage subcontractors to cut corners with regard to OH&S compliance (Mayhew et al. 1997).

McDermott and Hayes (2018) examined the role of civil construction contractors engaged in an infrastructure construction project in which work was performed in close proximity to buried infrastructure (that is, high-pressure gas pipelines). The use of multiple levels of subcontracting, selection of subcontractors based on price and payment by square metres of work completed led subcontracted workers to routinely break OH&S-related rules. Detailed procedures to identify the location of buried infrastructure (for example, using dial-before-you-dig and non-destructive pot-holing methods) were in place. However, these could not be followed if subcontractors were to meet tight production targets and remain profitable when working on this project (McDermott and Hayes 2018).

While many studies have identified subcontracting practices as having the potential to negatively impact OH&S, relatively few studies have examined the ways that principal contractors manage the OH&S of subcontractors. Manu et al. (2013) investigated the way that UK-based principal contractors manage subcontractor OH&S. Some of the subcontractor management practices used by the UK-based principal contractors were driven by regulatory requirements, including: undertaking competence assessment for subcontractors; ensuring the preparation of risk assessments by subcontractors; conducting safety training and induction for subcontractor workers; and consultation with the subcontractor workers/ representatives on site safety matters. However, Manu et al. (2013) also found other measures are used to reduce the safety risks inherent in subcontracting, including:

1. restricting layers/tiers of subcontracting;
2. engaging a regular chain of subcontractors;
3. introducing a reward system for subcontractors; and
4. the establishment of specific site-supervisory arrangements.

THE CRITICAL ROLE OF SUPERVISORS

Site-based supervisors plan, organise and facilitate daily work activities in construction projects and have a strong influence on the way work is performed, including in relation to workplace safety (Hardison et al. 2014). Supervisors operate primarily at the principal contractor–subcontractor interface and their role is critical in shaping work practices and safety. For example, in a study of the London Olympic construction programme, Finneran et al. (2012) identified supervisors' capabilities and competencies as a critical factor driving project safety performance. In Denmark, Jeschke et al. (2017) found that training construction supervisors in communication behaviours changed the way they interacted with workers and produced higher levels of safety-related cooperation and performance among subcontractor workers.

The influence that supervisors have on site-based safety practices is also believed to be significantly greater than that of senior managers owing mainly to the frequency of supervisors' interactions with workers (Therkelsen and Fiebich 2004). While workers rarely interact with senior managers, they receive daily instructions and performance feedback from their supervisors (Zohar 2000). Fugas et al. (2011) also describe how supervisor–worker interactions produce social norms that determine the safety-related behaviours that are expected of work-group members.

Although subcontractors are engaged to perform physical construction work, in most cases the principal contractor provides a foreperson to oversee subcontractors' operations at a worksite (Sherratt 2016). This foreperson's supervisory role is in addition to the role of the supervisors who are directly employed by individual subcontractors to manage the work of their own work groups. Previous research suggests the interaction between these two levels of supervision has the potential to shape safety practices in the workplace. For example, in a study conducted in the Australian construction industry, Lingard et al. (2010) report the safety climate of a subcontracted work group was most strongly influenced by the emphasis placed on safety by the subcontractor supervisor. However, subcontractor supervisors' safety-related practices were also significantly shaped by the extent to which they perceive the principal contractor's foreperson to be concerned about safety (Lingard et al. 2010). The engagement of a foreperson whose role is to oversee and coordinate the work of subcontractors has also been identified as a strategy adopted by principal contractors to reduce the negative impact of subcontracting on safety performance in the UK construction industry (Manu et al. 2013).

**Case Study 3: Managing OH&S as the Principal Contractor–
Subcontractor Interface (Lingard et al. 2017)**

Researchers spent over 100 hours observing interactions between the general foreperson, subcontractor supervisors, workers and OH&S professionals at four large commercial construction sites in Melbourne, Australia. Safety climate surveys were conducted among these work groups and social network data was collected to analyse the communication patterns between supervisors and workers. Survey data was collected from 39 work groups across four worksites between May 2017 and October 2017. A total of 154 workers and 39 supervisors participated in the data collection. The density of communication within subcontracted work groups was significantly and positively associated with the group's safety climate (Lingard et al. 2019). Density of communication captures the extent and frequency of occupational health and safety (OH&S)-related communication that occurs between the supervisor and all other members of each work group (that is, worker-to-supervisor and supervisor-to-worker communication), as well as the frequency of OH&S-related communication that occurs between the workers who make up each work group and all other worker members of the group (that is, coworker-to-coworker communication). Thus, the greater the extent and frequency of OH&S-related communication between all members of a work group (including the supervisor), the more positive the safety climate will be.

Site-based observation of interactions also revealed that general forepersons and subcontracted supervisors play a critical role in shaping OH&S behaviour within work groups and that the development of respectful and working relationships between forepersons and subcontractor supervisors is critical to ensuring that OH&S procedures are correctly followed. In particular, higher levels of subcontractor cooperation, compliance and participation in relation to OH&S are achieved when the foreperson:

- is respectful of subcontractors' technical and often highly specialised knowledge;
- is willing to listen and engage in active problem-solving in relation to OH&S with subcontractor supervisors; and
- effectively communicates the reasons behind the principal contractor's OH&S policies and requirements (Oswald and Lingard, 2019).

ENGAGING THE WORKFORCE

The management of OH&S is based on underlying theories about people, organisations and behaviour (Nielsen 2000). In many instances, workers are seen as having limited competence and a high propensity for error. If this is assumed, then it follows that workers' behaviours need to be constrained through the establishment of detailed rules and formal procedures developed by technical specialists, professionals and managers (Hale et al. 2003). However, this paternalistic approach is increasingly challenged as it devalues workers' knowledge and the important insights gained from situated work practices. A gap between work as imagined (and documented in formal procedures) and the practical reality of the workplace can undermine the usefulness and acceptance of safety rules by those expected to follow them (Knudson 2009; Lofquist et al. 2017).

For example, Vidal-Gomel (2017) describes how French concrete delivery workers rely on experience and judgement in adapting rules to suit the needs of the different construction site environments they visit, balancing delivery demands and situational constraints against rule compliance.

Construction workers offer a valuable source of OH&S knowledge that, if accessed, can be used to improve processes and performance. However, in a study of the UK construction industry, Sherratt et al. (2013) reveals that, despite statements of collective responsibility for OH&S, the way in which OH&S is spoken and written about on construction sites reflects an emphasis on top-down enforcement rather than engagement with the workforce. In Australia, Ayers et al. (2013) similarly describe how, despite a statutory requirement to consult workers about OH&S matters, most construction organisations do not involve workers in making strategic decisions about the way that health and safety are managed in the workplace. For example, it is rare for workers to be asked to participate in the design of systems of work or the selection of equipment.

This is a missed opportunity because workers possess a wealth of knowledge about OH&S hazards associated with the work tasks they perform and about ways to work safely. Much of this knowledge is tacit, that is, it is difficult to transfer to another person by means of writing it down or verbalising it (Polanyi 1958). This type of knowledge can be described as know-how, rather than know-what. In many cases, the people who possess tacit knowledge may not be aware of the knowledge or know how valuable it could be to others. Construction workers may not appreciate the extent and value of their knowledge and, in some cases, may not possess the skills to easily communicate this knowledge to others.

Recent developments in OH&S have recognised the benefits associated with engaging workers in OH&S improvement processes. Integral to a worker engagement approach is the understanding that providing workers with input into decisions that affect them will produce better OH&S outcomes. For example, participatory (worker-led) training approaches have produce demonstrably better results in terms of knowledge acquisition and injury prevention than traditional OH&S training approaches (Burke et al. 2006).

Case Study 4: Using Participatory Video to Improve Work Processes

Participatory video (PV) is a 'group-based activity that develops participants' abilities by involving them in using video equipment creatively to record themselves and the world around them and to produce their own videos' (Shaw and Robertson 1997, p. 1). It is used across a range of contexts and has been particularly successful in projects that seek to support otherwise disempowered groups in generating new forms of understanding and of representing themselves. It is also seen as an interventional method because it can produce change (Mitchell et al. 2012, pp. 1–2).

A team of construction workers whose work involved installing insulating materials in the residential construction sector were engaged in a PV intervention. During the process of making videos about their work, the insulation installation workers shared their frustrations concerning the impracticality of following the company standard operating procedure (SOP) for accessing ceiling spaces to install insulation. Many ceiling manholes that the workers need to enter are at a height of between 2.4 and 2.7 metres from the floor. The SOP required workers to place a straight ladder at an angle of 75 degrees, extending 900 millimetres beyond the step off point. In the video, the workers showed how this is practically impossible to achieve owing to conduits, cables, beams and other obstructions. The small size of the manholes did not allow adequate entry for the ladder, the worker and the pack of insulation to be installed. Having viewed the footage, the company OH&S manager observed how information contained in the SOP may be theoretically correct but the video materials highlighted the gap between documented work methods and situated work practices. The manager halted the filming and engaged workers in a project to find an industrial-rated solution for accessing ceiling manholes safely at the required height.

While making a similar film about the use of mobile tower scaffolds, the workers identified a point in the scaffold erection process at which workers were working adjacent to an unprotected edge. The workers redesigned the erection sequence for the scaffold, using a system of temporary platforms and horizontal members, to eliminate the hazard. The new erection sequence was video-recorded and has become the standard procedure for erecting scaffolds of this type within the company. The original and modified methods of erection are depicted in Figure 11.4.

The insulation installation workers were interviewed about their experience of the PV intervention. They described the benefits associated with

Source: Lingard et al. (2015).

Figure 11.4   Scaffold erection sequence (a) before and (b) after the modification

visual methods to understand, review, critique and revise work processes. Importantly, the PV process prompted the identification of OH&S issues inherent in the company's work processes as workers and managers were able to identify hazards and engage in creative problem-solving through the production and review of the video materials. This led them to collaboratively identify ways to eliminate identified. The insulation installation workers also spoke positively about having a genuine input, through the PV process, into the redesign and improvement of the company's work procedures (Lingard et al. 2015a).

CONCLUSIONS

The construction industry is a hazardous industry in which to work. Many hazards relate to the physical work processes, equipment and materials used in construction work. However, there is an increasing focus on hazards arising from the ways that construction projects are designed and delivered, and that construction work is organised and managed. Client organisations are increasingly recognising a need to become more active and to use their purchasing power to shape improvements in OH&S performance. The consideration of construction OH&S in the pre-construction and design stages of projects can also improve the quality of risk controls implemented during the construction stage. At the local worksite, effective frontline supervision is critically important to ensuring the physical work undertaken by subcontractors is well managed and OH&S requirements are met. Participatory forms of management are also helpful to identify work process improvements, drawing on workers' practical knowledge and worksite experience.

NOTE

*   The research described in this chapter was supported by an Australian Research Council Future Fellowship (FT0990337). Some of the research described was also supported by Australian Research Council Linkage Grants (LP0668012; LP120100587) and Discovery Grant (DP0881321). Case study 2 was jointly funded by WorkSafe Victoria and the Major Transport Infrastructure Program, Department of Economic Development, Jobs, Transport and Resources, Victorian Government. Case study 4 was supported by the Victorian Department of Business and Innovation and CodeSafe Solutions. Case study 1 was supported by Cooperative Agreement Number U60 OH009761, under which RMIT is a subcontractor to Virginia Tech, from the US Centers for Disease Control and Prevention (CDC), National Institute for Occupational Safety and Health (NIOSH). Its contents are solely the responsibility of the authors and do not necessarily represent the official views of the CDC NIOSH.

REFERENCES

Arditi, D. and Chotibhongs, R. (2005), 'Issues in subcontracting practice', *Journal of Construction Engineering and Management*, **131** (8), 866–76.

Atkinson, A.R. and Westall, R. (2010), 'The relationship between integrated design and construction and safety on construction projects', *Construction Management and Economics*, **28** (9), 1007–17.

Ayers, G., Culvenor, J., Sillitoe, J. and Else, D. (2013), 'Meaningful and effective consultation and the construction industry of Victoria', *Construction Management and Economics*, **31** (6), 542–67.

Behm, M. (2005), 'Linking construction fatalities to the design for construction safety concept', *Safety Science*, **43** (8), 589–611.

Brace, C., Gibb, A.G.F., Pendlebury, M. and Bust, P.D. (2009), 'Health and safety in the construction industry: underlying causes in construction fatal accidents – external research', Secretary of State for Work and Pensions, inquiry into the underlying causes of construction fatal accidents, Loughborough University, July, Health and Safety Executive report.

Briscoe, G. and Dainty, A. (2005), 'Construction supply chain integration: an elusive goal?', *Supply Chain Management: An International Journal*, **10** (4), 319–26.

Bundesanstalt für Arbeitsschutz- und Arbeitsmedizin (BauA) (2009), 'Technische Regeln für Gefahrstoffe. Abgase von Dieselmotoren. TRGS 554' ('Technical regulations for hazardous substances. Diesel engine exhausts. TRGS 554'), Bundesanstalt für Arbeitsschutz- und Arbeitsmedizin (Federal Institute for Occupational Safety and Health), Dortmund.

Burke, M.J., Sarpy, S.A., Smith-Crowe, K. Chan-Serafin, S., Salvador, R.O. and Islam. G. (2006), 'Relative effectiveness of worker safety and health training methods', *American Journal of Public Health*, **96**, (2), 315–24.

Donaghy, R., (2009), *Rita Donaghy's Report to the Secretary of State for Work and Pensions. One Death Is Too Many: Inquiry into the Underlying Causes of Construction Fatal Accidents*, Cmd 7657, Norwich: The Stationery Office.

Driscoll, T.R., Harrison, J.E., Bradley, C. and Newson, R.S. (2008), 'The role of design issues in work-related fatal injury in Australia', *Journal of Safety Research*, **39** (2), 209–14.

Dubois, A. and Gadde, L.E. (2000), 'Supply strategy and network effects – purchasing behaviour in the construction industry', *European Journal of Purchasing & Supply Management*, **6** (3–4), 207–15.

Eban, G., (2016), 'Major construction projects at airports: client leadership of health and safety', *Journal of Airport Management*, **10** (2), 131–7.

El Asmar, M., Hanna, A.S. and Loh, W.Y. (2016), 'Evaluating integrated project delivery using the project quarterback rating', *Journal of Construction Engineering and Management*, **142** (1), 04015046.

Eriksson, P.E. and Westerberg, M. (2011), 'Effects of cooperative procurement procedures on construction project performance: a conceptual framework', *International Journal of Project Management*, **29** (2), 197–208.

European Federation of Foundation Contractors (EFFC) (2015), 'Breaking down of concrete piles', accessed 2 March 2018 at https://www.effc.org/content/uploads/2015/12/Breaking_Down_of_Piles_May2015.pdf.

Fellows, R., and Liu, A.M. (2012), 'Managing organizational interfaces in engineering construction projects: addressing fragmentation and boundary issues across multiple interfaces', *Construction Management and Economics*, **30** (8), 653–71.

Finneran, A., Hartley, R., Gibb, A., Cheyne, A. and P. Bust (2012), 'Learning to adapt health and safety initiatives from mega projects: an Olympic case study', *Policy Practice Health Safety*, **10** (2), 81–102.

Franz, B.W., Leicht, R.M. and Riley, D.R. (2013), 'Project impacts of specialty mechanical contractor design involvement in the health care industry: comparative case study', *Journal of Construction Engineering and Management*, **139** (9), 1091–7.

Fugas, C.S., Meliá, J.L. and Silva, S.A. (2011), 'The "is" and the "ought": how do perceived social norms influence safety behaviors at work?', *Journal of Occupational Health Psychology*, **16** (1), 67–79.

Gadde, L.E. and Dubois, A. (2010), 'Partnering in the construction industry – problems and opportunities', *Journal of Purchasing and Supply Management*, **16** (4), 254–263.

Gibb, A., Haslam, R., Pavitt, T. and Horne, K. (2007), 'Designing for health – reducing occupational health risks in bored pile operations', *Construction Information Quarterly, Special Issue: Health and Safety*, **9** (3), 113–23.

Gibb, A., Lingard, H., Behm, M. and Cooke, T. (2014), 'Construction accident causality: learning from different countries and differing consequences', *Construction Management and Economics*, **32** (5), 446–59.

Hale, A.R., Heijer, T. and Koornneef, F. (2003), 'Management of safety rules: the case of railways', *Safety Science Monitor*, **7** (1), 1–11.

Hardison, D., Behm, M., Hallowell, M.R. and Fonooni, H. (2014), 'Identifying construction supervisor competencies for effective site safety', *Safety Science*, **65** (June), 45–53.

Hartmann, A. and Caerteling, J. (2010), 'Subcontractor procurement in construction: the interplay of price and trust', *Supply Chain Management: An International Journal*, **15** (5), 354–62.

Health and Safety Executive (2012), 'Control of diesel engine exhaust emissions in the workplace', accessed 7 February 2018 at http://www.hse.gov.uk/pubns/priced/hsg187.pdf.

Huang, X. and Hinze, J. (2006a), 'Owner's role in construction safety', *Journal of Construction Engineering and Management*, **132** (2), 164–73.

Huang, X. and Hinze, J. (2006b), 'Owner's role in construction safety: guidance model', *Journal of Construction Engineering and Management*, **132** (2), 174–181.

Institute of Occupational Safety and Health (2015), 'Occupational health risk management in construction', Institute of Occupational Safety and Health, Wigston, Leics.

Jeschke, K.C., Kines, P., Rasmussen, L., Andersen, L.P.S., Dyreborg, J., Ajslev, J., et al. (2017), 'Process evaluation of a Toolbox-training program for construction foremen in Denmark', *Safety Science*, **94** (April), 152–60.

Karpinski, E.-A. (2015), *Control Measures for Diesel Engine Exhaust Emissions in the Workplace*, HM Government of Canada, accessed 7 February 2018 at canada.ca/publicentre-ESDC.

Knudsen, F. (2009), 'Paperwork at the service of safety? Workers' reluctance against written procedures exemplified by the concept of "seamanship"', *Safety Science*, **47** (2), 295–303.

Lingard, H. and Rowlinson, S. (2005), *Occupational Health and Safety in Construction Project Management*, Abingdon: Routledge.

Lingard, H., Pirzadeh, P. and Oswald, D. (2019), 'Talking safety: health and safety communication and safety climate in subcontracted construction workgroups', *Journal of Construction Engineering and Management*, **145** (5), 1–11.

Lingard, H., Blismas, N., Cooke, T. and Cooper, H. (2009), 'The model client framework: resources to help Australian Government agencies to promote safe construction', *International Journal of Managing Projects in Business*, **2** (1), 131–40.

Lingard, H., Blismas, N., Harley, J., Stranieri, A., Zhang, R.P. and Pirzadeh, P. (2018), 'Making the invisible visible: stimulating work health and safety-relevant thinking through the use of infographics in construction design', *Engineering, Construction and Architectural Management*, **25** (1), 39–61.

Lingard, H., Cooke, T. and Blismas, N. (2010), 'Safety climate in conditions of construction subcontracting: a multi-level analysis', *Construction Management and Economics*, **28** (8), 813–25.

Lingard, H., Edirisinghe, R. and Harley, J. (2015), 'Using digital technology to share health and safety knowledge', Centre for Construction Work Health and Safety Research, RMIT University, Melbourne.

Lingard, H., Le, T., Oswald, D., Pirzadeh, P. and Harley, J. (2016), 'The use of commercial frameworks to drive exceptional health and safety performance in the construction industry', Centre for Construction Work Health and Safety Research, RMIT University, Melbourne.

Lingard, H., Pink, S., Harley, J. and Edirisinghe, R. (2015a), 'Looking and learning: using participatory video to improve health and safety in the construction industry', *Construction Management and Economics*, **33** (9), 741–52.

Lingard, H., Pirzadeh, P., Blismas, N., Saunders, L., Kleiner, B. and Wakefield, R. (2015b), 'The relationship between pre-construction decision-making and the quality of risk control: Testing the time-safety influence curve', *Engineering, Construction and Architectural Management*, **22** (1), 108–24.

Lingard, H., Pirzadeh, P., Blismas, N., Wakefield, R. and Kleiner, B. (2014), 'Exploring the link between early constructor involvement in project decision-making and the efficacy of health and safety risk control', *Construction Management and Economics*, **32** (9), 918–31.

Lingard, H., Zhang, R., Blismas, N., Wakefield, R. and Kleiner, B. (2015c), 'Are we on the same page? Exploring construction professionals' mental models of occupational health and safety', *Construction Management and Economics*, **33** (1), 73–84.

Lofquist, E.A., Dyson, P.K. and Trønnes, S.N. (2017), 'Mind the gap: a qualitative approach to assessing why different sub-cultures within high-risk industries interpret safety rule gaps in different ways', *Safety Science*, **92** (February), 241–56.

Loosemore, M. and Andonakis, N. (2007), 'Barriers to implementing OHS reforms – the experiences of small subcontractors in the Australian construction industry', *International Journal of Project Management*, **25** (6), 579–88.

Manu, P., Ankrah, N., Proverbs, D. and Suresh, S. (2013), 'Mitigating the health and safety influence of subcontracting in construction: the approach of main contractors', *International Journal of Project Management*, **31** (7), 1017–26.

Mayhew, C., Quinlan, M. and Ferris, R. (1997), 'The effects of subcontracting/ outsourcing on occupational health and safety: survey evidence from four Australian industries', *Safety Science*, **25** (1–3), 163–78.

McDermott, V. and Hayes, J. (2018), 'Risk shifting and disorganization in multi-tier contracting chains: the implications for public safety', *Safety Science*, **106** (July), 263–72.

Mitchell, C., Milne, E. and de Lange, N. (2012), 'Participation + video: an introduction to the Handbook on Participatory Video', in E.-J. Milne, C. Mitchell and N. deLange (eds), *The Handbook of Participatory Video*, Lanham, MD: AltaMira Press.

Nielsen, K. (2000), 'Organizational theories implicit in various approaches to OHS management', in K. Frick, P L Jensen, M. Quinlan and T. Wilthagen (eds), *Systematic Occupational Health and Safety Management: Perspectives on an International Development*, Amsterdam: Pergamon, pp. 99–124.

Occupational Health and Safety Administration (OHSA) (2013), 'Hazard alert – diesel exhaust/diesel particulate matter', accessed 7 February 2018 at https:// www.osha.gov/dts/hazardalerts/diesel_exhaust_hazard_alert.html.

Oswald, D. and Lingard, H. (2019), 'Development of a frontline H&S leadership maturity model in the construction industry', *Safety Science*, **118**, 674–86, accessed at https://doi.org/10.1016/j.ssci.2019.06.005.

Polanyi, M. (1958), *Personal Knowledge – Towards a Post-Critical Philosophy*, London: Routledge and Kegan Paul.

Priemus, H. and Ale, B. (2010), 'Construction safety: an analysis of systems failure: the case of the multifunctional Bos & Lommerplein estate, Amsterdam', *Safety Science*, **48** (2), 111–22.

RMIT University (2018), 'Musculoskeletal risk reduction – jackhammering and shotcreting', Construction Work Health and Safety @ RMIT, RMIT University, Melbourne, accessed 13 May 2019 at rmit.edu.au/musculoskeletalrisk reductionresearch.

Safe Work Australia (2012), *Australian Work Health and Safety Strategy 2012–2022*, Canberra: Safe Work Australia, p. 1, accessed 27 April 2018 at www. safeworkaustralia.gov.au/system/files/documents/1804/australian-work-health-safety-strategy-2012-2022v2_1.pdf.

Safe Work Australia (2015), 'Guide to managing risks of exposure to diesel exhaust in the workplace', accessed 7 February 2018 at https://www.safeworkaustralia.

gov.au/system/files/documents/1702/guidance-managing-risks-exposure-diesel-ex haust-in-the-workplace.pdf.

Shaw, J. and Robertson, C., (1997), *Participatory Video: A Practical Guide to Using Video Creatively in Group Development Work*, London: Routledge.

Sherratt, F. (2015), 'Legitimizing public health control on sites? A critical discourse analysis of the Responsibility Deal Construction Pledge', *Construction Management and Economics*, **33** (5–6), 444–52.

Sherratt, F. (2016), *Unpacking Construction Site Safety*, Chichester: Wiley Blackwell.

Sherratt, F., Farrell, P. and Noble, R. (2013), 'UK construction site safety: Discourses of enforcement and engagement', *Construction Management and Economics*, **31** (6), 623–35.

Spangenberg, S., Mikkelsen, K.L., Kines, P., Dyreborg, J. and Baarts, C. (2002), 'The construction of the Oresund Link between Denmark and Sweden: the effect of a multi-faceted safety campaign', *Safety Science*, **40** (5), 457–65.

Stergiou-Kita, M., Mansfield, E., Bezo, R., Colantonio, A., Garritano, E., Lafrance, M., et al. (2015), 'Danger zone: men, masculinity and occupational health and safety in high risk occupations', *Safety Science*, 80 (December), 213–20.

Stocks, S.J., Turner, S., McNamee, R., Carder, M., Hussey, L. and Aguis, R.M. (2011), 'Occupation and work-related ill-health in UK construction workers', *Occupational Medicine*, **61** (6), 407–15.

Swuste, P., Frijters, A. and Guldenmund, F. (2012), 'Is it possible to influence safety in the building sector? A literature review extending from 1980 until the present', *Safety Science*, **50** (5), 1333–43.

Therkelsen, D.J. and Fiebich, C.L. (2004), 'The supervisor: the linchpin of employee relations', *Journal of Communication Management*, **8** (2), 120–29.

Toole, T.M. (2007), 'Design engineers' responses to safety situations', *ASCE Journal of Professional Issues in Engineering Education and Practice*, **133** (2), 126–31.

Vidal-Gomel, C. (2017), 'Training to safety rules use. Some reflections on a case study', *Safety Science*, **93** (September), 134–42.

Votano, S. and Sunindijo, R.Y. (2014), 'Client safety roles in small and medium construction projects in Australia', *Journal of Construction Engineering and Management*, **140** (9), 04014045.

Vrijhoef, R. and Koskela, L. (2000), 'The four roles of supply chain management in construction', *European Journal of Purchasing & Supply Management*, **6** (3–4), 169–78.

Wadick, P. (2010), 'Safety culture among subcontractors in the domestic housing construction industry', *Structural Survey*, **28** (2), 108–20.

Yates, J.K. and Battersby, L.C. (2003), 'Master builder project delivery system and designer construction knowledge', *Journal of Construction Engineering and Management*, **129** (6), 635–44.

Zhang, R.P., Pirzadeh, P., Lingard, H. and Nevin, S. (2018), 'Safety climate as a relative concept: exploring variability and change in a dynamic construction project environment', *Engineering, Construction and Architectural Management*, **25** (3), 298–316.

Zohar, D. (2000), 'A group-level model of safety climate: testing the effect of group climate on microaccidents in manufacturing jobs', *Journal of Applied Psychology*, **85** (4), 587–96.

12. The case for the psychosocial safety climate to be recognized in mining disaster investigations

Tony Pooley, Silvia Pignata and Maureen F. Dollard

INTRODUCTION

This chapter examines the findings of inquiries into internationally recognized hazardous industry disasters around the world. Many of the disaster investigations have identified individual findings that clearly relate to job stress, including harsh work environments, long working hours and high production targets, but do not recognize these as cumulative components of a wider organizational condition. The chapter puts forward a body of evidence to argue that an organization's psychosocial safety climate (PSC) (a snapshot of psychosocial safety culture) should be viewed as the overall job-stress leading indicator when gathering data and undertaking analyses in major incident investigations. Doing so will avoid the fragmentation of job stress into its individual stressors such as production versus safety, shift work, unclear communication, physical stress and harassment. The whole of job stress is different to, and probably greater than, the sum of its parts.

The objective of this chapter is complex, in that the psychosocial safety culture is the pre-eminent cause of a cocktail of separate factors that are commonly treated as stand-alone issues. It is hypothesized that, if the level of safety disaster investigators' support for organizational PSC profiling was similar to police investigators' support for criminal profiling (Snook et al. 2007), the findings and recommendations of investigation reports would be substantially improved and result in a decline in the number of safety disasters. The chapter concludes that all hazardous industries would benefit from this approach, but that the mining industry has special considerations that would make the cost–benefit case even higher than that for other high-risk industries, including aviation, oil and gas, chemicals and nuclear power.

PSYCHOSOCIAL SAFETY CLIMATE

Job stress is a significant issue which impacts negatively on individuals, industry and society (Hall et al. 2010). Job-related stress is 'the adverse reaction people have to excessive pressure or other types of demand placed on them' (HSE 2007, p. 7). As part of an organization's approach to corporate social responsibility, it is important to address job stress and other psychosocial risk factors that contribute to human capital and organizational costs. Job stress is one of the biggest challenges in workplaces globally (Skakon et al. 2010), is increasing (Mucci et al. 2016) and impacts adversely on worker health and organizational performance (Leiter et al. 2014). As organizations have competing operational demands (for example, safety versus productivity) it can be argued that the best indicator of an organization's priorities is their performance against their articulated policies, procedures and practices. The PSC refers to shared perceptions regarding an organization's policies, practices and procedures to protect worker psychological health and safety (Dollard and Bakker 2010). PSC predicts a range of work-related stressors and, ultimately, job stress. While this chapter discusses the benefits of utilizing the PSC concept to help avoid high-profile health, safety and environmental disasters in high-risk industries, ample data exists on the less dramatic cost of ignoring PSC which include poor health and workplace injuries, lost productivity and rehabilitation costs, and work–family conflict.

Low levels of PSC cost Australian employers an estimated $6 billion annually (Becher and Dollard 2016). Large costs to the economy associated with poor mental health indicate an urgent need to address job stress. For example, a national survey by the Australian Psychological Society (APS 2015) found rising levels of stress over a five-year period reported across the country, with 35 per cent of Australians reporting a significant level of distress, and increasing anxiety symptoms. In the study, 72 per cent of Australians reported that stress impacts their physical health and 64 per cent reported an impact on their mental health. In a recent SafeWork Australia report, Becher and Dollard (2016, p. 5) found that 'workers with psychological distress took four times as many sick days per month and had a 154 per cent higher performance loss at work' with an average cost of $6309 per year more than those not experiencing psychological distress. As depressed workers in Australia cost employers an average of $2791 per annum for mild depression to $23 143 per annum for severe depression (Becher and Dollard 2016), there is a need for new insights for industries, particularly high-risk sectors, so that best practices can be implemented to reduce these high levels and ensure that employees work at their full potential. Understanding the sources of psychological ill health can mitigate potential sources of accident risk (HSE 2006), and being able to recognize the effects of stress can indicate the degree to which workers will place themselves in error-inducing conditions (Sexton et al. 2000).

SUM OF THE PARTS

Prior to compiling this chapter, the authors realized that a great deal of its credibility was based on the premise that the combined stress from a cocktail of workplace stressors was far more influential in the unfolding of disaster events than the vast majority of employers, safety professionals and disaster investigators realize. Published investigation reports repeatedly identify one or more such stressors (as is shown later in this chapter) but the inability to aggregate all of the stressors experienced by the most exposed workers into a single PSC-type value continually reduces the chance that recommended actions from an investigation report will effectively prevent future occurrences.

The sum of the parts statement is to an extent intuitive, but difficult to prove. We were unable to identify substantial research into changes in an individual's performance under multiple stressors. One reason for this may be that there are limits to how many stressors might be applied to humans for research without exceeding ethical boundaries. For this reason, we looked to what is probably the most stressful socially acceptable workplace imaginable as a potential source of credible research on multiple stressors in the workplace, that is, the defence forces during combat or peacekeeping operations.

In-depth research into US military deployments from 1993 to 1996, by Bartone et al. (1998), found five psychological stress dimensions of soldier stress. Moreover, Castro and Adler (1999) argue that workload is a sixth factor. The factors are:

1. Isolation – remote location away from family, cultural differences.
2. Ambiguity – multiple conflicting objectives, role confusion, different language or accents.
3. Powerlessness – restricted to camp, unable to integrate with locals, dress codes.
4. Boredom – often high activity periods are followed by long repetitive and/or dull periods.
5. Danger – varies with industry. (While military work is more stressful than other work, the participants may be more conditioned and trained to avoid traumatization.)
6. Workload – long work cycles, possibility of being called out at any time.

It is important to note that in hazardous industries, particularly in fly-in/fly-out (FIFO) mining situations, all of these factors can be present, although the danger component is clearly more acute for a soldier than for an oil worker or miner. Importantly, for the sum of the parts argument, Bartone (2006) notes that while the six dimension labels are distinct, in practice they overlap and interact. This does not state that the whole is more than the sum of the parts, but it does mean that they cannot be aggregated easily, and certainly not by the typical skill set of a disaster investigator (see next section).

It is recognized that fatigue can impact the body and the brain, and lead to errors by the workforce, but it is important to note that they require different solutions, as muscles simply need time away from the activity but the brain requires sound sleep (SafeWork Australia 2013). Research going back to the 1950s shows that a hot working environment directly impacts human performance (Frazer 1955; Bell and Provins 1962; Pepler 1963) and, on completion of tests on the effect of dehydration from ambient heat, it was found that as little as a 2 per cent reduction in dehydration caused a highly significant deterioration in mental performance (Gopinathan et al. 1988).

A different way of demonstrating the case for considering PSC as a complex combination of stressors, comes from looking at the three stressor types in the form of a Venn diagram (Figure 12.1), where distinct examples of overlap can be identified as follows:

- Physical stressors directly impact the components that make up the body, including bones, muscles and organs.

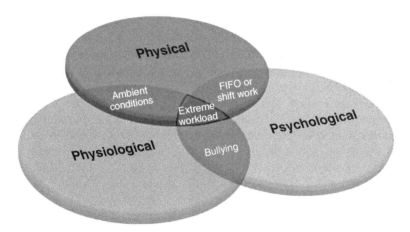

Figure 12.1   PSC stressor types

- Physiological stressors impact a healthy state, for example, blood circulation, body temperature, food processing and oxygen production.
- Psychological stressors affect our emotions which in turn may affect behaviour and decision-making.

ASSESSING AN ORGANIZATION'S PSC

The PSC of an organization reflects how senior management value worker psychological health. Management values influence workplace policies and procedures, and set the tone for the organization. This has implications, for example, for how jobs are designed, and the standards for workplace behaviours which may affect a worker's psychology (for example, bullying behaviour). In a high PSC context (psychologically healthy workplaces) jobs are designed where demands are manageable, and resources are adequate. However, in a low PSC context, work is more stressful because demands are high and/or resources are low (Dollard and Bakker 2010). The PSC can be measured and monitored on what is called the PSC-12 scale which involves four domains that provide insights about where to intervene: management commitment, management priority, organizational communication and organizational participation. Each domain is assessed by seeking answers to three questions about how effectively each domain is being addressed by management (Hall et al. 2010).

Psychosocial risk levels, working conditions and worker health can be measured, and the results used to reduce extreme safety events. Strategies in industries with high-demand work include: increasing PSC scores, increasing job resources (such as supervisor support), lowering job demands (reducing hard and fast-paced work), establishing PSC as a key performance indicator, encouraging workers' involvement in developing systems and safe work conditions, developing communication processes about psychosocial risks and promoting psychological health.

Individual employee behaviours such as job crafting (proactively improving the work environment as an individual employee) has the potential to provide practical skills and knowledge to workers in various employment settings to improve their own outcomes. This may enhance person–demand fit, and reduce job stress in a high-PSC organization. For example, in a significant study of over 15000 employees from more than 1200 workplaces, Croon et al. (2015) found that job-enrichment (for example, targeted training) human resource practices (which presumably coincide with higher levels of PSC) indirectly enhanced organizational productivity through employee job satisfaction. Hence, we suggest that good production performance and safety enhancement may not necessarily be opposing forces.

ROOT CAUSES OF SAFETY DISASTERS

Three broad categories of stressors that can limit human performance are (1) physical, such as extreme temperature and humidity, noise and vibration, (2) physiological, such as fatigue, poor physical condition and hunger, and (3) psychological stressors such as work–life imbalance, poor relationships with superiors and or colleagues, high demands (workload), low control and effort–reward imbalance (CASA 2009). With regard to safety, stressed workers may be distracted or suffer from a diminished ability to regulate their thoughts and actions (HSE 2006).

PSC is a separate construct but is related to safety climate (Zohar 1980), sometimes referred to as physical safety climate (Idris et al. 2012). PSC relates to worker psychological health and well-being, whereas safety climate concerns safety from accidents and injuries. Research has examined the effect of PSC on work injuries in healthcare (Zadow et al. 2017). They found that low physical safety climate, low PSC and high levels of emotional exhaustion all separately predicted registered injury rates in the second (Time 2) wave of a longitudinal study controlling for the level of registered injury rates in the first (Time 1) wave of the study. For self-reported injuries, when PSC, physical safety climate and emotional exhaustion were simultaneously in a multilevel model, emotional exhaustion was the strongest predictor of survey-reported total injuries and under-reporting. Since PSC was the only predictor of emotional exhaustion, they concluded that low PSC was the origin of higher rates of self-reported work injuries and injury under-reporting (physical and psychological) operating through emotional exhaustion. Zadow et al. (2017) highlighted that to understand injury events a dual physical–psychosocial safety explanation was required.

Two safety management leaders of the late twentieth century lend credence to the importance of enhancing human performance. Dr Tony Barrell was instrumental in the formation of UK draft major hazards legislation following the Flixborough disaster in 1974, which formed the basis of the European Communities 1982 Seveso Directive. Barrell was the first Head of the UK Health and Safety Executive's (HSE's) Major Hazards Assessment Unit, and was involved in several disaster enquiries including the North Sea flooding of Canvey Island, the Kings Cross underground fire, and the Piper Alpha platform fire/explosion (IChemE 2017). In 1996, Barrell was asked to comment on the Piper Alpha experience in a BBC documentary and re-enactment of the disaster. He offered the following wisdom long before the concept of a psychosocial safety culture was recognized by major hazard industry companies (Sanders 2015, p. 88): 'There is actually an awful sameness about these incidents. They're nearly always characterized by lack of forethought, lack of analysis, and nearly always the problem comes down to poor management. It is not due to one particular person not following a procedure or doing something wrong.'

The second example of uncommon clarity on this complex subject came from James Reason who published multiple books and papers on human error and organizational processes from 1977 to date, mostly in his role as Emeritus Professor at the University of Manchester. His work has seen him recognized as a Fellow of many psychological societies including the British Academy, and he received the Distinguished Foreign Colleague Award from the US Human Factors and Ergonomics Society (Texas Medical Institute of Technology – Safetyleaders 2018). Reason (2000, p. 768) stated that: 'We cannot change the human condition, but we can change the conditions under which humans work.'

Both insights recognize that organizations that have experienced disasters have always missed opportunities at the organizational and systems level to assist the movement of its people to a better position on the *'part of the problem'* vs. *'part of the solution'* spectrum. PSC measures an organization's commitment to protecting workers' psychological health and well-being, including the psychological risk assessment of management commitment, priority of safety, communication and involvement in psychological health and safety (Dollard et al. 2012). Since PSC predicts psychosocial risks, it is referred to as a cause of the causes of job stress (Becher and Dollard 2016) and is therefore a 'root cause' and a primary focus of a good safety investigation (HSE 2004, p. 5).

In 2006, the HSE commissioned a broad review of 14 disasters or near disasters that had worldwide importance including: the Flixborough chemical explosion in England in 1974; the Three Mile Island nuclear radiation release in Pennsylvania in 1979; the Bhopal toxic gas leak in India in 1984; the Space Shuttle *Challenger* explosion in the US in 1986; the Chernobyl reactor radiation release in Ukraine, also in 1986; the sinking of the *Herald of Free Enterprise* in Zeebrugge Port, Belgium in 1987; the Kings Cross underground fire in London, again in 1987; the Piper Alpha offshore explosion in the UK North Sea in 1988; the Exxon Valdez oil release in Prince William Sound, Alaska in 1989; and the Space Shuttle *Columbia* disintegration in the US in 2003. The review looked at the formal investigation findings and reviews by noted academics to conclude the primary and the underlying causes of the incidents (Bell and Healey 2006).

The summary of 15 key factors causing the major incidents included nine engineering issues and six behaviour issues (Bell and Healey 2006). Four of the six behaviour issues were clear PSC contributors: poor management practices, for example, inadequate supervision; pressure to meet production targets; organizational communication issues, for example, between shifts, between personnel and management; and excessive working hours resulting in mental fatigue.

The remaining two behaviour issues (complacency and violations or non-compliance behaviour) beg the question 'Why?' to be asked, because PSC aspects are highly likely to be involved. Of the nine system or engineering issues there are also instances where the 'Why?' question may identify more PSC contributors. For example, the review makes the point that 60 per cent of all incidents identified from maintenance (one of the nine engineering factors) were identified as human-related factors (Bell and Healey 2006), but there is no evidence that any attempt was made in any investigation or subsequent review to seek to understand how much of this non-compliant behaviour is caused by highly stressed team members, or how much stress is created in the remaining compliant members.

In Australia in 1998, the Longford gas plant explosion resulted in a Royal Commission (Victoria Longford Royal Commission 1999). The report found three workplace stressors to be among the main findings, that is, the company:

- failed to train its site people in this particularly hazardous process;
- failed to effectively monitor adherence to basic operating practices; and
- reduced supervision at the plant, resulting in too much reliance on operators' experience, and did so specifically to control operating costs.

While the above findings would indicate to a neutral observer a possible organizational culture issue, all of the recommendations of the report went to the correction of technical or oversight shortcomings. Much of the report's hopes for improvement relied on the recommended expansion of the safety case regime that covers offshore production facilities in Australian waters, to be applied to onshore major hazards facilities in the state of Victoria (Victoria Longford Royal Commission 1999). However, cases for safety are reviewed and regulated by engineers and inspectors. Not a single recommendation addressed the organization's need to widen its perspective from engineering or systems solutions alone, to where the interaction between organizational dynamics, humans and engineering/systems is recognized as critical to safe operations in a hazardous facility.

In the aviation industry, the impact of stress has been documented in aeroplane crashes such as Asiana Airlines flight 214 where inadequate pilot training and supervision, non-standard communication processes, and flight crew fatigue were identified as contributing factors (NTSB 2014);

and American Airlines 1420 where the flight crew's impaired performance from fatigue and the situational stress associated with their intent to land during severe thunderstorms were contributing factors to the crash (NTSB 1999). There is also a significant need to monitor the psychological health of employees working in high-risk industries, as demonstrated by the Germanwings flight 9525 pilot suicide in 2015 that resulted in the deaths of 150 people (BEA 2016).

While the Germanwings incident brought stark attention to the potential multiple fatality outcomes from the act of a suicidal pilot, the problem is much wider than a single incident. The largest study on depression and suicidal thoughts among commercial pilots found levels almost three times that of the general population (Pasha and Stokes, 2018).

A FOCUS ON MINING

In the continuous production of oil, gas, nuclear power and chemicals (excludes batch production) people oversee the process using computer monitors or by reading site instrumentation. The hazardous product moves from the fluid feeder piping to the facility, through pumps, compressors, heaters, coolers and vessels until it leaves the facility through the discharge piping. The product is always held in pressure-containing engineered assets. These fluids can be relied on to conform to scientific rules based on their flow rate, pressure, temperature and composition.

However, mining is very different, as people are not merely overseeing the process but are a direct component within it. On a mine site, people drill into rock using hand-held and motorized devices, set explosives into small-diameter holes in difficult to reach places, de-scale loose material from underground shafts, drive excavators and haul trucks and operate bulldozers on top of ore stockpiles. Often while equipment is being driven it is necessary to have people in close proximity. Only when the ore reaches the milling and separation process does it finally start to be comparable to other automated industries. The rock itself is not the only hazard; air, mud and/or water are often present in the ore body and, because it is difficult to predict their behaviour, they may burst into the areas in which miners work. In coal mining, methane gas is a constant threat because it is naturally generated and can even self-combust.

Given the importance of the mining sector in providing employment in Australian rural communities in particular, and its profile in both national and international markets, it is timely to investigate new and emerging psychosocial risks within this high-risk industry. For example, the inquiry into the Moura No. 2 mine incident in 1994 noted 'less than adequate responses to information available', 'the abysmal state of communications' and the working relationship between key people was 'less than co-operative' (Queensland Coroner 1996, pp. 28, 34, 55). An inquest into three deaths in two separate car accidents in 15 months found fatigue while driving home after completing a shift in a Bowen Basin mine contributed to all three deaths (Queensland Government 2018). The inquest resulted in a series of Queensland Government-led initiatives from 2007 to 2015, but the initiatives focused solely on fatigue and not on cumulative stress from multiple sources that could be attributed to PSC. The problem of a low PSC in mining is not just a First World problem, for example, PSC factors are one of the key causes of mining disasters in China (Hongxia 2014).

The International Council on Mining and Metals (ICMM 2018a, 2018b, 2018c) is a powerful representative body for mining and metal production, setting social responsibility performance requirements for its members which include 27 of the largest companies (including Anglo, Barrick, BHP Billiton, Glencore, Newmont, Rio Tinto and Vale) and over 30 national or regional industry associations (including those from Australia, Canada, Europe, India, Japan and most countries in South America and Africa). The council itself comprises the chief executive officers of each organization, and is therefore powerful (ICMM 2018a). It is well intentioned and has successfully raised health and safety performance expectations for many of its members (Globescan 2017). However, the ICMM has no direct reference to work stressors in relation to fatal incidents in its two publications on reducing fatalities, one aimed at senior leaders (ICMM 2018d) and the other at managers of operations and site leadership teams (ICMM 2018e).

The ICMM (2018f) showcases BHP's 2015 mental wellness framework which identifies long hours and extended periods of time living away from home as potential stressors, and recognizes that depression, fatigue and stress increase the risk of safety incidents at work, with potentially fatal consequences. However, the BHP framework does not cover potential stressors other than FIFO stressors. A review of the Australian (Minerals Council of Australia 2018), Canadian (The Mining Association of Canada 2018) and US association (National Mining Association 2018) websites demonstrates that mining associations do not cover workplace stress in any depth, perhaps regarding it as an individual company responsibility.

MINING INDUSTRY CONTEXT

For readers outside the social sciences, the terms for PSC stress in hazardous industries may be conceived as follows. Physical stress outcomes can be likened to an asset or equipment failure, such as structural, pressure-containing or rotating parts. Physiological stress outcomes are similar to a system or process failure where the assets have not exceeded their physical limits, but one or more are responding to external influences instead of working in harmony with the remaining assets. Psychological stress outcomes are like programming or signal faults where there is no asset or system failure but illogical or no decisions may result.

The following examples of different or elevated stressors in mining are provided to guide further development of PSC in this industry.

Physical and Physiological Health of Workers

The working environment can include harsh temperatures, vibration and noise that can cause fatigue (Safe Work Australia 2013). While some women in mining are challenged by men in terms of their physical size and stature, others may try to adapt, be tougher or act like men (Abrahamsson et al. 2014) with the potential for overexertion.

Fear of the mined product (for example, uranium), its contaminants (for example, asbestos) or processes (for example, cyanide) creates stress because people are generally more frightened of a threat that they cannot see. For example, at the time of the Three Mile Island nuclear incident in the US in 1979, journalists who had previously worked in war zones found the inability to shelter from an invisible threat to be much worse than avoiding bullets (Stencel 1999). Today, as the shallow mineral resources have been largely consumed, exploration is increasingly turning to deeper ore bodies to find commercial quantities, and as a consequence, rock temperatures will rise owing to the geothermal gradient (Anderson and De Souza 2017), which will create more difficult and extreme working environments in the future.

Psychological Health of Workers

A particular area of concern in Australia is rural workers aged between 25–34 years, who are at high risk of poor psychological health owing to working long hours, competing work and family demands, and who experience low levels of skill discretion (Dollard et al. 2012). As the high-risk mining industry operates in remote locations, it requires employment under the FIFO structure (Cameron et al. 2014) which can impact a worker's emotional health and well-being in several ways, including isolation, lack of autonomy, fatigue and shift work. All of these factors need to be considered in undertaking mining investigations. The FIFO structure requires an employee to be: flown to and from their home residence to a work site for defined time periods; isolated from family and other social supports while away regularly from home; housed within a controlled camp environment; and placed on a compression roster of long working hours and shift work that subjects them to fatigue. A parliamentary report (Western Australian Education and Health Standing Committee 2015) found that the typical FIFO resource worker comes from the highest risk demographic (male aged 18–44) for mental illness and suicide. Given that approximately 100 000 Australian workers are employed under a FIFO structure, and as FIFO workers are up to 70 per cent more likely to suffer mental health problems than the general Australian population (Deceglie 2015), there is a significant need to assess and monitor their psychosocial work conditions in isolated locations and consider policies targeting psychosocial factors to improve worker psychological health and well-being.

Mine workers working in isolated locations, report work–family or work–home conflict associated with spending long periods away from home. McTernan et al. (2016) suggest that social support from co-workers buffers the level of work–family conflict within isolated work groups which highlights the benefits and importance of workplace relationships within the mining industry. As PSC has been confirmed 'as a leading organizational predictor for psychosocial risk factors and employee mental health', it is a specific prevention target of work-related mental health illnesses such as job stress and depression (Bailey et al. 2015a, p. 24). Using national Australian benchmarks, Bailey et al. (2015a) found that low levels of PSC (a score less than 37) predict job strain and poor worker mental health. Therefore, knowing the levels of the PSC can indicate the likelihood of workers developing health-related outcomes such as depression or anxiety, and/or productivity-related outcomes of workplace injuries (Bailey et al. 2015b), absenteeism and presenteeism (Dollard et al. 2016). The regular measurement of psychosocial factors can target risks for prevention and hazards for intervention. For mining workers, psychosocial intervention strategies to address stress and depression may include orchestrating workplace social activities to increase social support and opportunities for social interaction, raising awareness of stress and depression, and implementing mental health first aid training (Becher and Dollard 2016).

ASSESSING PSC IN DISASTER INVESTIGATIONS

As investigators in hazardous industries typically have a technical background such as engineering, drilling or flying, they may not have an adequate perspective to identify the root causes of human behaviour (US

Coast Guard Research and Development Center 2002). Incident analysts may not only be inexperienced in human aspects of disasters, they may also unknowingly ignore causes for which they have no solutions and therefore need both training and specific methodological approaches to show the way (Dien et al. 2012). The management and mitigation of psychosocial risk requires an interdisciplinary approach which also joins the interests of the individual and the organization (larger system) so that both can mutually benefit (Pignata et al. 2014).

If the PSC framework is to assist disaster investigations in hazardous industries with arduous and specific working conditions, it must consider the elevated stresses that may be experienced in comparison with most other workplaces (for example, corporate offices, education, health, retail and the public sector). Some examples are provided in the previous mining context section, but there is a need for a formal industry or academic group to review and develop a comprehensive mining PSC profile survey and measurement criterion for avoiding potential disasters.

REFERENCES

Abrahamsson, L., Segerstedt, E., Mygren, M., Johansson, J., Johannson, B., Edman, I. et al. (2014), *Gender, Diversity, and Work Conditions in Mining*, Luleå: Luleå University of Technology.

Anderson, R., and De Souza, E. (2017), 'Heat stress management in underground mines', *International Journal of Mining Science and Technology*, 27 (4), 651–5.

Australian Psychological Society (APS) (2015), 'Stress and wellbeing in Australia survey 2015: one-third of Australia stressed out', accessed 3 October 2018 at https://www.psychology.org.au/news/media_releases/8Nov2015-pw.

Bailey, T.S., Dollard, M.F. and Richards, P.A.M. (2015a), 'A national standard for psychosocial safety climate (PSC): PSC 41 as the benchmark for low risk of job strain and depressive symptoms', *Journal of Occupational Health Psychology*, 20 (1), 15–26.

Bailey, T.S., Dollard, M.F., McLinton, S.S. and Richards, P.A.M. (2015b), 'Psychosocial safety climate and physical factors in the etiology of MSDs and workplace physical injury compensation claims', *Work and Stress*, 29 (2), 190–211.

Bartone, P.T. (2006), 'Resilience under military operational stress: can leaders influence hardiness?', *Military Psychology*, 18 (suppl.) S131–S148.

Bartone, P.T., Adler, A.B. and Vaitikus, M.A. (1998), 'Dimensions of psychological stress in peacekeeping operations', *Military Medicine*, 163 (9), 587–93.

Becher, H. and Dollard, M.F. (2016), *Psychosocial Safety Climate and Better Productivity in Australian Workplaces. Costs, Productivity, Presenteeism Absenteeism*, Canberra: Safe Work Australia.

Bell, C.R. and Provins, K.A. (1962), 'Effect of high temperature environmental conditions on human performance', *Occupational Medicine*, 4 (4), 202–11.

Bell, J. and Healey, N. (2006), *The Causes of Major Hazard Incidents and How* to Improve Risk Control and Health and Safety Management: A Review of the Existing Literature*, Buxton: Health and Safety Laboratory for the UK Health and Safety Executive.

Bureau d'Enquêtes et d'Analyses (BEA) (2016), 'Final report accident on 24 March 2015 at Prads-Haute-Bléone (Alpes-de-Haute-Provence, France) to the Airbus A320-211 registered D-AIPX operated by Germanwings', accessed 6 September 2018 at https://www.bea.aero/uploads/tx_elyextendttnews/BEA2015-0125.en-LR_03.pdf.

Cameron, R., Lewis, J. and Pfeiffer, L. (2014), 'The FIFO experience: a Gladstone case study', *Australian Bulletin of Labour*, **40** (2), 221–41.

Castro, C.A. and Adler, A.B. (1999), 'OPTEMPO: Effects on soldier and unit readiness', *Parameters*, Autumn, 86–95.

Civil Aviation Safety Authority (CASA) (2009), *Safety Behaviours: Human Factors for Pilots Kit*, Canberra: Civil Aviation Safety Authority and Australian Government.

Croon, M.A., van Veldhoven, M.J.P.M., Peccei, R.E. and Wood, S. (2015), 'Researching individual well-being and performance in context: multilevel mediational analysis for bathtub models', in M.J.P.M. van Veldhoven and R. Peccei (eds), *Well-Being and Performance at Work: the Role of Context*, Hove: Psychology Press, pp. 129–54.

Deceglie, A. (2015), 'FIFO suicides: parliamentary inquiry findings into WA FIFO workers and mental health released', accessed 7 September 2018 at https://www.perthnow.com.au/news/wa/fifo-suicides-parliamentary-inquiry-findings-into-wa-fly-in-fly-out-workers-and-mental-health-released-ng-481e3b66b89fcd418ec4dce9a302e3e3.

Dien, Y., Dechy, N. and Guillaume, E. (2012), 'Accident investigation: from searching direct causes to finding in-depth causes – problem of analysis or/and of analyst?', *Safety Science*, **50** (6), 1398–407.

Dollard, M., Bailey, T., McLinton, S., Richards, P., McTernan, W., Taylor, A. et al. (2012), 'The Australian Workplace barometer: report on psychosocial safety climate and worker health in Australia', Safe Work Australia, accessed 6 September 2018 at http://www.safeworkaustralia.gov.au/sites/swa/about/Publications/Documents/748/The-Australian-Workplace-Barometer-report.pdf.

Dollard, M., Zadow, A., Pignata, S. and Bailey, T. (2016), 'Stress management', in A. Farazmand (ed.), *Global Encyclopedia of Public Administration, Public Policy and Governance*, Springer International, doi:10.1007/978-3-319-31816-5_2509-1.

Dollard, M.F. and Bakker, A.B. (2010), 'Psychosocial safety climate as a precursor to conducive work environments psychological health problems, and employee engagement', *Journal of Occupational and Organizational Psychology*, **83** (3), 579–99.

Frazer, D. (1955), 'Effect of heat stress on serial reaction time in man', *Nature*, **176** (4490), 976–7.

Globescan (2017), *ICMM 2017 Industry Stakeholders Opinion Survey: Final Report*, London: ICMM.

Gopinathan, P., Pichan, G. and Sharma, V.M. (1988), 'Role of dehydration in heat stress-induced variations in mental performance', *Archives of Environmental Health: An International Journal*, **43** (1) 15–17.

Hall, G., Dollard, M. and Coward, J. (2010), 'Psychosocial safety climate: development of the PSC-12', *International Journal of Stress Management*, **17** (4), 353–83.

Health and Safety Executive (HSE) (2004), 'Investigating accidents and incidents: a workbook for employers, unions, safety representatives and safety professionals', HSG245, Health and Safety Executive, Liverpool.

Health and Safety Executive (HSE) (2006), *Investigation of the Links between Psychological Ill-Health, Stress and Safety*, Norwich: HSE Books.

Health and Safety Executive (HSE) (2007), *Managing the Causes of Work-Related Stress: A Step-by-Step Approach Using the Management Standards*, Norwich: Health and Safety Executive.

Hongxia, L. (2014), 'Study on the job stress of miners', *Procedia Engineering*, **84**, 239–46.

IChemE (2017), 'IChemE presidents – Antony Charles Barrell: 1993', accessed 11 October 2018 at www.icheme.org/about_us/icheme presidents/1993 anthony charles barrell.aspx.

Idris, M.A., Dollard, M.F., Coward, J. and Dormann, C. (2012), 'Psychosocial safety climate: conceptual distinctiveness and effect on job demands and worker psychological health', *Safety Science*, **50** (1), 19–28.

International Council on Mining and Metals (ICMM) (2018a), 'About us', accessed 11 October 2018 at http://www.icmm.com/en-gb/about-us.

International Council on Mining and Metals (ICMM) (2018b), 'Member associations', accessed 11 October 2018 at http://www.icmm.com/en-gb/members/mem ber-associations.

International Council on Mining and Metals (ICMM) (2018c), 'Member companies', accessed 11 October 2018 at http://www.icmm.com/en-gb/members/mem ber-companies.

International Council on Mining and Metals (ICMM) (2018d), 'Leadership matters: the elimination of fatalities', accessed 11 October 2018 at http://www.icmm. com/en-gb/publications/health-and-safety/leadership-matters-the-elimination-of -fatalities.

International Council on Mining and Metals (ICMM) (2018e), 'Leadership matters: Managing fatal risk guidance', accessed 11 October 2018 at http://www.icmm. com/en-gb/publications/health-and-safety/leadership-matters-managing-fatal-risk -guidance.

International Council on Mining and Metals (ICMM) (2018f), 'Improving employee mental health and wellbeing in the mining industry', accessed 11 October 2018 at https://www.icmm.com/en-gb/case-studies/improving-employee-mental -health-and-wellbeing.

Leiter, M.P., Bakker, A.B. and Maslach, C. (2014), *Burnout at Work: A Psychological Perspective*, Hove: Psychology Press.

McTernan, W.P., Dollard, M.F., Tuckey, M.R. and Vandenberg, R.J. (2016), 'Enhanced co-worker social support in isolated work groups and its mitigating role on the work–family conflict–depression loss spiral', *International Journal of Environmental Research and Public Health*, **13** (4), 382.

Minerals Council of Australia (2018), website, accessed 12 October 2018 at https:// www.minerals.org.au/.

Mucci, N., Giorgi, G., Roncaioli, M., Fiz Perez, J. and Arcangeli, G. (2016), 'The correlation between stress and economic crisis: a systematic review', *Neuropsychiatric Disease and Treatment*, **12** (April), 983–93.

National Mining Association (2018), website, accessed 12 October 2018 at https:// nma.org/.

National Transportation Safety Board (NTSB) (1999), 'Aircraft accident report PB2001-910402 NTSB/aar-01/02 Dca99ma060 Runway overrun during landing

American Airlines flight 1420 McDonnell Douglas md-82, n215aa, Little Rock, Arkansas, June 1, 1999', accessed 6 September 2018 at https://www.ntsb.gov/investigations/AccidentReports/Reports/AAR0102.pdf.

National Transportation Safety Board (NTSB) (2014), 'Descent below visual glide-path and impact with seawall, Asiana Airlines flight 214, Boeing 777-200ER, HL7742, San Francisco, California, July 6, 2013', accessed 6 September 2018 at https://www.ntsb.gov/investigations/AccidentReports/Reports/AAR1401.pdf.

Pasha, T. and Stokes, P.R. (2018), 'Reflecting on the Germanwings disaster: a systematic review of depression and suicide in commercial airline pilots', *Frontiers in Psychiatry*, **9** (March), 1–8.

Pepler, R. (1963), 'Performance and wellbeing in heat temperature – its measurement and control', *Science and Industry*, **3** (3), 319–36.

Pignata, S., Biron, C. and Dollard, M.F. (2014), 'Managing psychosocial risks in the workplace: Prevention and intervention', in M. Peeters, J., de Jonge and T. W. Taris (eds), *People at Work: An Introduction to Contemporary Work Psychology*, Hoboken, NJ: Wiley Blackwell, pp. 393–413.

Queensland Coroner (1996), 'Warden Inquiry: report on an accident at Moura No 2 underground mine on Sunday 7 August 1994', Queensland.

Queensland Government (2018), 'Inquest into the death of Graham Brown, Malcolm MacKenzie and Robert Wilson', Brisbane.

Reason, J. (2000), 'Human error: models and management', *British Medical Journal*, **320** (7237), 768–70.

SafeWork Australia (2013), *How to Manage Risks Associated with Fatigue*, Canberra: Safe Work Australia.

Sanders, R. (2015), *Chemical Process Safety: Learning from Case Histories*, 4th edn, Oxford: Butterworth Heinemann/Elsevier.

Sexton, B.J., Thomas, E.J. and Helmreich, R.L. (2000), 'Error, stress, and teamwork in medicine and aviation', *British Medical Journal*, **320** (7237), 745–9.

Skakon, J., Nielsen, K., Borg, V. and Guzman, J. (2010), 'Are leaders' well-being, behaviours and style associated with the affective wellbeing of their employees? A systematic review of three decades of research', *Work and Stress*, **24** (2), 107–39.

Snook, B., Haines, A., Taylor P.J., and Bennell, C. (2007), 'Criminal profiling belief and use: a study of Canadian police officer opinion', *Canadian Journal of Police and Security Services*, **5** (3–4), 1–11.

Stencel, M. (1999), 'A nuclear nightmare in Pennsylvania', *Washington Post*, 27 March.

Texas Medical Institute of Technology – Safetyleaders (2018), 'Safetyleaders Superpanel', accessed 11 October 2018 at http://www.safetyleaders.org/superpanel/superpanel_james_reason.html.

The Mining Association of Canada (2018), website, accessed 12 October 2018 at http://mining.ca/.

US Coast Guard Research and Development Center (2002), '2nd International Workshop on Human Factors in Offshore Operations (HFW2002): human factors in incident investigation and analysis', Houston, Texas, pp. 1–141.

Victoria Longford Royal Commission (1999), *The Esso Longford Gas Plant Accident, Report of the Longford Royal Commission*, Melbourne: Government Printer for the State of Victoria.

Western Australian Education Health Standing Committee (2015), *The Impact of FIFO Work Practices on Mental Health: Final Report*, Report No. 5, June,

Legislative Assembly, Parliament of Western Australia, accessed 6 September 2018 at https://www.menshealthwa.org.au/wp-content/uploads/2016/02/The-Impact-of-FIFO-Work-Practices-on-Mental-Health.pdf.

Zadow, A.J., Dollard, M.F., McLinton, S.S., Lawrence, P. and Tuckey, M.R. (2017), 'Psychosocial safety climate, emotional exhaustion, and work injuries in healthcare workplaces', *Stress and Health*, **33** (5), 558–69.

Zohar, D. (1980), 'Safety climate in industrial organizations: theoretical and applied implications', *Journal of Applied Psychology*, **65** (1), 96–102.

13. Aggressive and criminal behavior of police officers

Philip Matthew Stinson Sr

Police crime refers to the criminal behavior of sworn law enforcement officers. This chapter focuses on the criminological conceptualization and research on police crime within the United States. Little is known about the crimes of police officers because policing is decentralized, with more than 18 000 state and local law enforcement agencies in 50 states and the District of Columbia employing approximately 600 000 total sworn officers, and because there are no nationwide data collected by the federal government available to researchers. In the absence of any meaningful official statistics on police crime, researchers have resorted to a variety of methodologies such as surveys, field studies, quasi-experiments, internal agency records, and content analysis of news articles and court records. The availability of more comprehensive data on police crime could improve policing and police legitimacy through comparison of agency-specific police crime rates and the development of policies to deter criminal behavior by sworn law enforcement officers.

THE POLICE SUBCULTURE

It has long been recognized that the police operate in a social environment of secrecy that fosters a thriving subculture based on an unwritten set of norms permeated by systemic patterns of police violence and police corruption. Vollmer, writing for the Wickersham Commission in 1931, noted that 'it is an unwritten law in police departments that police officers must never testify against their brother officers' (Wickersham Commission 1931, p. 48). Also in the 1930s, Key recognized that there was an informal rule of silence within police departments across the United States (Key 1935). Westley extensively studied the police department in Gary, Indiana, during the 1950s and found that there was a code of secrecy among police officers that served the dual purpose of being a social bond between the city's police officers and was used as a shield against outsiders by police who treated the public as their enemy (Westley 1956). Stoddard (1968) also recognized an informal social code within police departments, arguing that newly hired police officers are socialized through a group deviance process of blue-coat crime that introduces them into an unlawful informal code of secrecy that perpetuates illegal police conduct as a matter of routine. Stoddard defined this informal code as a functioning social system whose norms and practices are contrary to the law. Such informal codes of secrecy within police departments quickly socialize new officers into an us-versus-them social structure that serves as the linchpin of the police subculture that views the public with extreme distrust.

Van Maanen (1973, 1975) identified four sequential stages of police socialization in his ethnographic study of police recruits during and after their police academy training. The stages represent how each new police officer is initiated into the organizational setting of working in a large municipal police department: choice (the pre-entry state of police sociali-zation where the recruit decides to become a police officer), introduction (the admittance state of police socialization), encounter (the change stage of police socialization) and metamorphosis (the continuance stage of police socialization). Van Maanen later admitted that his status as a participant-observer in the ethnographic research led to his own involve-ment in the police subculture in a variety of ways, such as helping other police recruits cheat on examinations while at the police academy, carrying unauthorized equipment while on patrol (for example, backup gun, sap gloves and an unapproved type of ammunition), observing veteran police officers stealing from a warehouse where an open door was found late at night, lying to police supervisors to cover up misconduct of fellow officers, and drinking beer with officers while on duty working a midnight shift (Van Maanen 1988).

The police socialization process involves distinct phases that inevitably lead to adoption of a new worldview occurring over several years at the beginning of each officer's law enforcement career. According to Van Maanen, individuals are drawn to police work because they want to work for an elite organization where they can do meaningful work. It is typically a deliberate career choice that attracts individuals with a strong sense of right and wrong. Police academies are full-time training programs where newly hired officers are introduced to the formal structure of the police organization. Police academy training programs typically last anywhere from two to six months, depending on state-specific training and certifica-tion requirements. The most crucial stage of socialization into the police subculture occurs after a police recruit completes their academy training and is assigned to work under the direct supervision and mentoring of a field training officer. Field training officers often will encourage the trainee to forget what they learned in the police academy in favor of the real ways that a police officer works on the street. This is the encounter stage of police socialization where each new police officer learns 'how to survive on the job . . . how to walk, how to stand, and how to speak and how to think and what to say and see' (Van Maanen 1973, p. 412). According to Van Maanen, socialization into the police subculture is completed with a metamorphosis when an officer realizes several years into his or her law enforcement career that much of police work is mundane and unsatisfying. Many officers become disenchanted with the social structure of policing at the metamorphosis stage of their police socialization, and often simply try to get through each work shift without making waves or getting hurt. It is not uncommon for police officers to quit their job after three to five years working in law enforcement after becoming frustrated with the day-to-day realities of working within a police organization.

Stinson et al. (2010) argue that there is another final stage in the police socialization process: the exit strategy. It occurs late in an officer's law enforcement career, often within three years of retirement eligibility. The exit strategy posits that long-term police officers are often unprepared to return to civilian life without the power that comes with the legal authority, badge and gun of a sworn officer. The exit strategy for some police officers is manifested when they commit a crime late in their law enforcement career. Stinson et al. (2016) found that 13 percent of police crime arrest cases in years 2005–11 involved an officer who had 18 or more years of service as a sworn officer at the time of their arrest. It makes little sense that officers who presumably served throughout their career without committing crimes would suddenly engage in criminal activity on the cusp of retirement. One possible explanation, from a life course criminology perspective, is that there is something about the anticipated loss of bonds and attachments closely tied to a police officer's self-identification as a police officer that leads an officer to a turning point in life where he or she starts to engage in criminal behaviors (cf. Laub and Sampson 2003). Future research should more closely examine factors as to why so many police officers are getting arrested in the final years of their law enforcement careers.

CONCEPTUALIZATION OF POLICE CRIME

The study of police crime has been hampered by conceptual confusion in large part because scholars have considered crimes perpetrated by sworn law enforcement officers within the broader context of police corruption, police misconduct or police deviance (Stinson 2015). Police corruption involves acts of organized occupational deviance committed in violation of recognized norms within policing. Some acts of police corruption are crimes for which an officer could be criminally prosecuted, while other acts of police corruption are not crimes but considered unethical because the acts are committed for an improper purpose in violation of departmental policies. Police misconduct involves acts committed by a sworn law enforcement officer in violation of the administrative policies of an officer's employing law enforcement agency (Punch 2000). Police corruption and police misconduct are both forms of police deviance because they involve either an abuse of authority or occupational misconduct. Police crime involves the criminal behavior of nonfederal sworn law enforcement officers who, by virtue of their employment with a state or local law enforcement agency, are sworn to uphold the law and empowered with the general power of arrest by statutory authority (Stinson 2009). Wilson (1963, p. 190) defines police criminality as 'illegally using public office for private gain without the inducement of a bribe, whereas acts of corruption do involve the acceptance of bribes'. Sherman (1978) does not distinguish between police corruption and police crime in his study of police corruption as a form of organizational deviance exhibited within police departments. Punch (2000, pp. 302–3) distinguishes police 'crimes committed by criminals in uniform' from other acts of police misconduct in the nature of violations of administrative rules and policies that are investigated and sanctioned internally within an officer's employing law enforcement agency. Ross (2001) presents a multidimensional taxonomy of police crime based on whether an act was violent, motivated by profit, perpetrated by an officer on behalf of the police organization, and/or committed by an officer in their individual capacity or in their official capacity as a sworn law enforcement officer. Building on each of these conceptualizations, Stinson's (2009, 2015) typology of police crime posits most crime committed by sworn law enforcement officers is alcohol related, drug related, profit motivated, sex related and/or violence related. These five types of police crime are each dichotomous variables and are not mutually exclusive categories of a single variable. There are many possibilities within the conceptualization of the typology of police crime. Some drug-related police crime, for example, is also profit-motivated police crime. Similarly, most sex-related police crimes are also violence-related offenses. Police crime is a multifaceted phenomenon that makes it very different from virtually all other crimes.

Central to the conceptualization of police crime is whether crimes committed by police officers while off duty should be included within the conceptual and operational definitions of police crime. Some scholars emphasis the occupational aspect and focus only on those criminal acts that occur under the guise of police authority while an officer is on duty (Foster 1966; Stoddard 1968; Barker and Carter 1994). Others have argued that police crime should only encompass crimes committed during the course of an officer's normal work activities (Barker 1978; Ross 2001) or that off-duty crimes do not include any aspect of a police officer's official position (Kappeler et al. 1998). Fyfe and Kane (2006), however, make a compelling case for the inclusion of off-duty crimes committed by sworn officers within the conceptualization of police crime. First, the occupation of policing provides unique criminal offending opportunities that can be taken advantage of while an officer is either on or off duty. Second, police officers know that they are unlikely to be arrested for any crimes they might commit while either on or off duty because sworn law enforcement officers are generally exempt from law enforcement – meaning that police officers do not like to arrest other police officers and often look the other way when confronted with a colleague who has committed a crime (Reiss 1971; Stinson et al. 2014c). Third, most state and local law enforcement agencies encourage or require their sworn officers to carry a handgun while off duty and grant full law enforcement powers to off-duty sworn officers. There is no bright line distinction between actions taken while an officer is technically on or off duty because of the 'knowledge, gun, and badge that comes with being a police officer' (Mollen Commission 1994, p. 30).

Alcohol-Related Police Crime

Alcohol-related problems in policing have long been recognized by scholars (Dishlacoff 1976; Hurrell Jr. et al. 1984; Violanti et al. 1985; Dietrich and Smith 1986). Violanti et al. (2011) identified numerous factors that promote drinking to excess among police officers, including officer demographics, stress and the police subculture. Young adult males are over-represented among the ranks of police officers, and that at least partially explains the demographics (Stinson 2015). Alcohol-use disorders are most prevalent among adult males, age 18–24 (National Institute on Alcohol Abuse and Alcoholism 2008), and young adult males are significantly more likely to drive a motor vehicle while intoxicated than other age groups (Chou et al. 2006). Numerous studies attribute excessive consumption of alcoholic beverages by police officers to job-related stresses such as chronic problematic street encounters with the public and a perceived lack of organizational support from officers' employing law enforcement agencies (Ayres et al. 1992; Anshel 2000; Kohan and O'Connor 2002). Leino et al. (2011) argue that repeated exposure to violence in policing may increase excessive alcohol consumption by officers who have inadequate coping mechanisms. The police subculture has been described as a drinking culture that includes frequent off-duty social gatherings that involve the consumption of alcoholic beverages (Violanti et al. 1985). Macdonald et al. (1999) found that the existence of an occupational drinking subculture was associated with the development and onset of drinking problems among employees. Other studies have focused on the personal lives of police officers and the relationship to alcoholism among officers. Violanti et al. (2011) found that police officers' failed interpersonal relationships often resulted in excessive alcohol consumption.

Few empirical studies address alcohol-related police crime. Fyfe and Kane (2006) include off-duty driving under the influence within the category of off-duty public order crimes in their study of career-ending misconduct within the New York Police Department. Using the same dataset, Kane and White (2009) describe numerous cases that involved the drunken behavior of New York City police officers in personal disputes, drunk driving incidents and bar-room fights. Another study analyzed 782 cases of police officers from more than 500 agencies across the United States who were each arrested for driving while intoxicated offenses (Stinson et al. 2014c). More than half of the police drunk-driving arrest cases involved drunken officers who were involved in a motor vehicle crash, and more than 13 percent of the arrest cases involved an officer who was arrested while on duty driving a police vehicle.

Drug-Related Police Crime

Research on drug-related police crime generally addresses three primary themes, including the etiology of drug-related police misconduct, classifications of drug-related police misconduct and the prevalence of illicit drug use by police officers (Stinson et al. 2016). Policing scholars often attribute drug-related police corruption to factors related to the occupational organizational structure and the nature of socialization into the police subculture (for example, Stoddard 1968). Patrol officers have ample on-the-job opportunities to engage in drug-related police misconduct owing to constant exposure to drug dealers and drug users, the availability of narcotics and other illicit drugs, and a general lack of direct contact with police supervisors during work shifts (Kraska and Kappeler 1988). The Mollen Commission (1994) hypothesized that the constant exposure to opportunities for drug-related corruption in some New York City police precincts worked as erosion on other police officers who then developed a high tolerance for widespread corruption throughout the New York Police Department. Adherence to the norms of the police subculture through secrecy, loyalty to fellow officers, and cynicism about police work and the criminal justice system all work to protect officers who engage in drug-related police crime (US General Accounting Office 1998). Carter (1990) posits that there are two primary types of police drug corruption. The first involves police officers who are motivated by illegitimate goals such as personal profit and takes the form of accepting bribes, extortion and robbing drug dealers they come into contact with through their police work. The second type of drug corruption in Carter's typology of police drug corruption is sometimes referred to as noble cause corruption (Crank et al. 2006; Crank and Caldero 2010). This involves officers who are motivated by ostensibly legitimate organizational goals tied to the criminal arrest and conviction of drug users and drug dealers, but takes the form of illegitimate law enforcement practices including police perjury, planting of evidence, and violations of constitutional criminal procedure that are commonplace at many state and local law enforcement agencies across the United States (see, for example, Dershowitz 1994).

The prevalence of illicit drug use among police officers in the United States is unknown and few studies have examined the phenomenon (Mieczkowski 2002). Kraska and Kappeler's (1988) study of one medium-sized police department found that 20 percent of the officers surveyed admitted to smoking marijuana while on duty at least twice a month, and 10 percent of the officers surveyed admitted to using non-prescribed controlled substances including barbiturates, hallucinogens and/or stimulants while they were on duty. Random drug testing of police officers suggests that the drugs of choice for drug-using police officers are cocaine and marijuana (Mieczkowski and Lersch 2002; Lersch and Mieczkowski 2005; Smalley 2006). Cocaine is the most prevalent drug in drug-related police crime arrest cases (Stinson et al. 2013), followed by (in descending order) marijuana, oxycodone, hydrocodone, amphetamine and methamphetamine, heroin, unclassified narcotics, anabolic steroids not including testosterone, unclassified depressants, benzodiazepines and testosterone (Stinson et al. 2016).

Profit-Motivated Police Crime

Profit-motivated police crimes involve the use of police authority for an officer's personal enrichment (Stinson et al. 2018). Ross's (2001) taxonomy included economically motivated criminal offenses by officers as one of four primary types of police crime. Profit-motivated police crime often involves acts of corruption characterized by a profit-driven cycle that is often also drug related (Carter 1990). The Knapp Commission (1972, p. 283) differentiated between 'grass eaters' (police officers who engaged in petty acts of corruption) and 'meat eaters' (police officers who engaged in serious forms of organized police corruption). In their investigation of

New York City police in the early 1970s, the Knapp Commission found that many police officers engaged in petty acts of corruption, but that relatively few officers engaged in the more serious meat-eater forms of police corruption. Two decades later, however, the Mollen Commission (1994) found that greed is the primary motivation behind most police corruption, and that many police officers in New York City were engaged in drug-related organized crime activities of the more serious meat-eater variety of corruption. In their study of career-ending misconduct in the New York Police Department, Kane and White (2013) found that profit-motivated police crime often included acts of bribe-taking, burglary, extortion, illegal gambling, insurance fraud, petty larceny, receiving stolen property and robbery. Similarly, Stoddard (1968) found that many police officers in one Indiana police department routinely engaged in on-duty crimes such as bribery, extortion, perjury, shakedowns (that is, robberies during traffic stops, searches and other street encounters) and thefts.

Stinson and colleagues studied profit-motivated police crime arrest cases during 2005–11 at 782 state and local law enforcement agencies across the United States. Their findings indicate that profit-motivated police crime occurs in a variety of contexts across the life-course of police careers, with more than 20 percent of the profit-motivated police crime arrest cases involving an officer with 18 or more years of service at the time of arrest (Stinson et al. 2018). Many of the profit-motivated crimes are committed by officers acting in their official capacity and often on duty when their crimes were committed (Stinson et al. 2012, 2018). Officers arrested for profit-motivated police crimes are often mid-rank field supervisors or high-ranking managers and administrators, although the majority of officers arrested for profit-motivated crimes are nonsupervisory street-level officers, deputies, troopers and detectives (Stinson et al. 2010, 2018). Crimes by female officers are most often profit-motivated offenses (Stinson et al. 2015). In one study, Stinson et al. (2018) examined predictors of criminal conviction following arrest for profit-motivated offenses. Arrested officers were most likely to be convicted of a profit-motivated police crime if an officer's crime involved a drug-related shakedown or theft during an off-duty robbery, or if the profit-motivated crime was also violence related (Stinson et al. 2018).

Sex-Related Police Crime

The organizational structure of state and local law enforcement agencies and the police subculture provide ample opportunities for on-duty officers to engage in acts of police sexual misconduct. Police officers often spend most of their shifts on patrol working alone, and can go many hours without coming into contact with fellow officers or police supervisors. Many police–citizen interactions occur in late-night hours during encounters with low visibility that are out of public view. Barker (1978) surveyed police officers in one American city and found that consensual sex between on-duty officers and adult females was fairly common, especially in police vehicles. Many police officers have met girls and women who actively seek out sexual encounters with on-duty uniformed police officers, but most on-duty sexual encounters are not consensual because they are sexual acts that are coerced from a vulnerable person – often under the threat of arrest – by a police officer who carries a gun and badge. Some police officers are consumed by inappropriate behaviors that constitute police sexual misconduct, such as seeking on-duty opportunities to view nude females, sexual harassment of criminal suspects and crime victims, and sexual encounters with underage girls (Sapp 1994). Persons who identify as lesbian, gay, bisexual, transgender and queer (or questioning) and others (LGBTQ+) are especially at risk of being victimized by a variety of types of police sexual misconduct, including police violence against transgender women and gay men, sexual assault by police officers against transgender men, police harassment of lesbians and gay men, as well as discrimination, harassment and retaliation by police officers against LGBTQ+ police officers (Copple and Dunn 2017).

Kraska and Kappeler (1995) conceptualize police sexual violence in the context of gender bias and systematic differential treatment of girls and women in the criminal justice system. Police sexual violence encompasses a continuum of inappropriate police behaviors that range from voyeurism and other unobtrusive acts at one end of the continuum, to rape and sexual assault at the other end of the continuum. Acts of police sexual violence are not rare or isolated events at state and local law enforcement agencies across the United States (Stinson et al. 2014b). Stinson et al. found 771 cases of 555 sworn officers, each of whom were arrested during the years 2005–08 for sex-related police crimes (Stinson et al. 2014a). The arrested officers were employed by 449 state and local law enforcement agencies located in 349 counties and independent cities in 44 states and the District of Columbia. Included within the arrest cases were officers charged with forcible rape ($n = 174$), forcible fondling ($n = 177$), statutory rape ($n = 66$), forcible sodomy ($n = 62$). The vast majority of the victims in the sex-related police crime arrest cases were female (91.4 percent), and more than half of the victims (52.9 percent) were age 15 or younger in cases where the exact age of the victim was known to the researchers. The odds of an officer being convicted for a sex-related police crime were greatest in arrest cases with a child victim and in arrest cases where an officer was charged with a criminal offense involving forcible fondling.

Violence-Related Police Crime

Police violence includes any justified or unjustified use of physical force, including deadly force, against a citizen (Sherman 1980). In the United States, police officers are legally justified in using the amount of force that is reasonably and proportionately necessary to effect an arrest or otherwise protect public safety. Police violence is unjustified when it involves any amount of force by a police officer that cannot be accounted for under the auspices of lawful necessity in the line of duty as a sworn law enforcement officer (Westley 1970). Unjustified force includes violence-related police crimes such as intimidation or coercion under the threat of police violence or police sexual violence. Acts of police violence are often hidden crimes that are not labeled as criminal offenses because everyone recognizes that policing is often violent. Police officers often encounter violence in the normal course of their patrol duties in responding to calls for service and street encounters with the public; rarely is violence by the police recognized as police misconduct or law-breaking behavior. Victims of police violence are often left with no recourse because their stories typically lack corroboration and their complaints are viewed as untrustworthy. Past research examining factors that influence police use of force have focused on various predictors, including situational, individual, organizational and/ or community variables. The studies have largely found that situational factors are the primary variables that influence police decisions to use coercive non-deadly force (Terrill 2003; Skogan and Frydl 2004).

A police officer may be criminally prosecuted if he or she uses excessive force (any level of force that is unjustified in the course of an officer's official actions). Researchers have struggled to determine the true nature and extent of police violence, in large part because the federal government has not yet collected or disseminated any meaningful data on police coercive force, unjustified police violence and violence-related police crime (Stinson 2018). Stinson et al. (2016) analyzed 3328 criminal cases of 2586 nonfederal sworn law enforcement officers who were each arrested during 2005–11. The officers were employed by 1445 state and local law enforcement agencies located in 805 counties and independent cities in 49 states (all except Wyoming) and the District of Columbia. Some of the officers ($n = 407$) had more than one case because they had more than one victim and/or were arrested more than once for a violence-related police crime during the study years. Among the most serious offenses charged in the violence-related cases were simple assault ($n = 870$, 26.4 percent), aggravated assault ($n = 570$, 17.1 percent) and murder or non-negligent manslaughter ($n = 104$, 3.1 percent).

One of the most troubling forms of aggressive police behavior is in the area of officer-involved domestic violence. The occupationally derived etiology of officer-involved domestic violence includes authoritarianism, problem drinking behaviors and repeated exposure to work-related violence. Each of these factors leads to spillover effect from work to family life at home (Mullins and McMains 2000; Johnson et al. 2005). Johnson (1991) found that 40 percent of officers surveyed admitted they had behaved violently toward their spouse at least once in the preceding six months. Similarly, Neidig and colleagues found that 41 percent of male officers surveyed admitted at least one act of physical aggression toward their wives in the prior year, and 8 percent of those male officers indicated that the physical aggression had involved severe incidents of choking, strangling and/or the actual or threatened use of a gun or knife against their spouse (Neidig et al. 1992a, 1992b).

Officer-involved domestic violence is a serious problem, in part, because police officers in the United States are required to carry a gun, and federal law prohibits anyone convicted of a felony or qualifying misdemeanor crime involving domestic violence from owning or possessing firearms and ammunition (see Lautenberg Amendment 1996). Stinson and Liederbach (2013) examined 324 cases in which police officers were arrested for crimes arising out of officer-involved domestic violence incidents during 2005–07. The arrested officers were employed by 226 state and local law enforcement agencies across the United States. One-third of the victims were the current spouse of the arrested officer, and approximately one-fourth of the victims were children. Other victims included current and former girlfriends or boyfriends, former spouses, other relatives and several on-duty police officers who were assaulted when they responded to the call for service. Most of the officers arrested in officer-involved domestic violence cases were charged with assault, but some were charged with offenses as varied as forcible rape, intimidation and harassment, murder or non-negligent manslaughter, violation of a protection order, burglary and vandalism. The simple odds of criminal conviction in an officer-involved domestic violence arrest case were greatest if the arrested officer's crime involved a sex-related offense, use of a personally owned firearm or resulted in a fatal injury to the victim.

REFERENCES

Anshel, M.H. (2000), 'A conceptual model and implications for coping with stressful events in police work', *Criminal Justice and Behavior*, **27** (3), 375–400, doi:10.1177/0093854800027003006.

Ayres, R.M., Flanagan, G.S. and Ayres, M.B. (1992), *Preventing law Enforcement Stress: The Organization's Role*, No. NCJ 141934, Washington, DC: US Department of Justice, Office of Justice Programs, Bureau of Justice Assistance.

Barker, T. (1978), 'An empirical study of police deviance other than corruption', *Journal of Police Science and Administration*, **6** (3), 264–72.

Barker, T. and Carter, D.L. (1994), 'A typology of police deviance', in T. Barker and D.L. Carter (eds), *Police Deviance*, 3rd edn, Cincinnati, OH: Anderson, pp. 3–11.

Carter, D.L. (1990), 'Drug-related corruption of police officers: a contemporary typology', *Journal of Criminal Justice*, **18** (2), 85–98, doi:10.1016/0047-2352(90)90028-A.

Chou, S.P., Dawson, D.A., Stinson, F.S., Huang, B., Pickering, R.P., Zhou, Y. et al. (2006), 'The prevalence of drinking and driving in the United States, 2001–2002: results from the national epidemiological survey on alcohol and related conditions', *Drug and Alcohol Dependence*, **83** (2), 137–46, doi:10.1016/j.drugalcdep.2005.11.001.

Copple, J.E. and Dunn, P.M. (2017), *Gender, Sexuality, and 21st Century Policing: Protecting the Rights of the LGBTQ+ Community*, Washington, DC: US Department of Justice, Office of Community Oriented Policing Services.

Crank, J.P. and Caldero, M.A. (2010), *Police Ethics: The Corruption of Noble Cause*, 3rd edn, Cincinnati, OH: Anderson.

Crank, J.P., Flaherty, D. and Giacomazzi, A. (2006), 'The noble cause: an empirical assessment', *Journal of Criminal Justice*, **35** (1), 103–16, doi:10.1016/j.jcrimjus.2006.11.019.

Dershowitz, A.M. (1994), 'Controlling the cops; accomplices to perjury', *New York Times*, 2 May, p. A17.

Dietrich, J.F. and Smith, J. (1986), 'The nonmedical use of drugs including alcohol among police personnel: a critical literature review', *Journal of Police Science and Administration*, **14** (4), 300–306.

Dishlacoff, L. (1976), 'The drinking cop', *Police Chief*, **43** (1), 32–9.

Foster, G.P. (1966), 'Police administration and the control of police criminality: A case study approach', PhD dissertation, University of Southern California, Los Angeles.

Fyfe, J.J. and Kane, R.J. (2006), *Bad Cops: A Study of Career-Ending Misconduct among New York City Police Officers*, NCJ No. 215795, Washington, DC: US Department of Justice, National Institute of Justice.

Hurrell, Jr, J.J., Pate, A., Kliesmet, R., Bowers, R.A., Lee, S. and Burg, J.A. (1984), *Stress among Police Officers*, Cincinnati, OH: US Department of Health and Human Services, Public Health Service, Centers for Disease Control, National Institute for Occupational Safety and Health, Division of Biomedical and Behavioral Science.

Johnson, L.B. (1991), 'Statement of Leonor Boulin Johnson', in *On the Front Lines: Police Stress and Family Well-Being: Hearings before the Select Committee on Children, Youth, Families, House of Representatives, 102 Congress, First Session*, Washington, DC: US Government Printing Office, pp. 32–48.

Johnson, L.B., Todd, M. and Subramanian, G. (2005), 'Violence in police families: work–family spillover', *Journal of Family Violence*, **20** (1), 3–12, doi:10.1007/s10896-005-1504-4.

Kane, R.J. and White, M.D. (2009), 'Bad cops: a study of career-ending misconduct among New York City police officers', *Criminology and Public Policy*, **8** (4), 737–69, doi:10.1111/j.1745-9133.2009.00591.x.

Kane, R. J., and White, M. D. (2013), *Jammed Up: Bad Cops, Police Misconduct, and the New York City Police Department*, New York: New York University Press.

Kappeler, V.E., Sluder, R.D. and Alpert, G.P. (1998), *Forces of Deviance:*

Understanding the Dark Side of Policing, 2nd edn, Long Grove, IL: Waveland Press.

Key, V.O. (1935), 'Police graft', *American Journal of Sociology*, **40** (5), 624–36, doi:10.1086/216900.

Knapp Commission (1972), *Commission to Investigate Allegations of Police Corruption and the City's Anti-Corruption Procedures: The Knapp Commission Report on Police Corruption*, New York: George Braziller.

Kohan, A. and O'Connor, B.P. (2002), 'Police officer job satisfaction in relation to mood, well-being, and alcohol consumption' *Journal of Psychology*, **136** (3), 307–19, doi:10.1080/00223980209604158.

Kraska, P.B. and Kappeler, V.E. (1988), 'Police on-duty drug use: a theoretical and descriptive examination', *American Journal of Police*, **7** (1), 1–28.

Kraska, P.B. and Kappeler, V.E. (1995), 'To serve and pursue: exploring police sexual violence against women', *Justice Quarterly*, **12** (1), 85–111, doi:10.1080/07418829500092581.

Laub, J.H. and Sampson, R.J. (2003), *Shared Beginnings, Divergent Lives: Delinquent Boys to Age 70*, Cambridge, MA: Harvard University Press.

Lautenberg Amendment (1996), 'Gun ban for individuals convicted of a misdemeanor crime of domestic violence', Pub.L. No. P.L. 104–208, s. 658.

Leino, T., Eskelinen, K., Summala, H. and Virtanen, M. (2011), 'Work-related violence, debriefing and increased alcohol consumption among police officers', *International Journal of Police Science and Management*, **13** (2), 149–57, doi:10.1350/ijps.2011.13.2.229.

Lersch, K.M., and Mieczkowski, T. (2005), 'Drug testing sworn law enforcement officers: one agency's experience', *Journal of Criminal Justice*, **33** (3), 289–97, doi:10.1016/j.jcrimjus.2005.02.008

Macdonald, S., Wells, S. and Wild, T.C. (1999), 'Occupational risk factors associated with alcohol and drug problems', *American Journal of Drug and Alcohol Abuse*, **25** (2), 351–69, doi:10.1081/ADA-100101865.

Mieczkowski, T. (2002), 'Drug abuse, corruption, and officer drug testing: an overview in K.M. Lersch (ed.), *Policing and Misconduct*, Upper Saddle River, NJ: Prentice Hall, pp. 157–92.

Mieczkowski, T. and Lersch, K.M. (2002), 'Drug-testing police officers and police recruits: the outcome of urinalysis and hair analysis compared', *Policing: An International Journal of Police Strategies & Management*, **25** (3), 581–601, doi:10.1108/13639510210437041.

Mollen Commission (1994), *Commission to Investigate Allegations of Police Corruption and the Anti-Corruption Procedures of the Police Department: Commission Report: Anatomy of Failure: A Path for Success*, New York: City of New York.

Mullins, W.C. and McMains, M.J. (2000), 'Impact of traumatic stress on domestic violence in policing', in D.C. Sheehan (ed.), *Domestic Violence by Police Officers: A Compilation of Papers Submitted to the Domestic Violence by Police Officers Conference at the FBI Academy, Quantico, VA*, Washington, DC: US Government Printing Office, pp. 257–68.

National Institute on Alcohol Abuse and Alcoholism (2008), 'Alcohol and other drugs', *Alcohol Alert*, No. 76, US Department of Health and Human Services, National Institutes of Health and National Institute on Alcohol Abuse and Alcoholism, Bethesda, MD, accessed 14 May 2019 at http://pubs.niaaa.nih.gov/publications/AA76/AA76.htm.

Neidig, P.H., Russell, H.E. and Seng, A.F. (1992a), 'Interspousal aggression in law enforcement families: a preliminary investigation', *Police Studies: The International Review of Police Development*, **15** (1), 30–38.

Neidig, P.H., Seng, A.F. and Russell, H.E. (1992b), 'Interspousal aggression in law enforcement personnel attending the FOP biennial conference', *National FOP Journal*, Fall–Winter, 25–8.

Punch, M. (2000), 'Police corruption and its prevention', *European Journal on Criminal Policy and Research*, **8** (3), 301–24, doi:10.1023/A:1008777013115.

Reiss, A.J. (1971), *The Police and the Public*, New Haven, CT: Yale University Press.

Ross, J.I. (2001), 'Police crime & democracy: demystifying the concept, research, and presenting a taxonomy', in S. Einstein and M. Amir (eds), *Policing, Security and Democracy: Special Aspects of Democratic Policing*, Huntsville, TX: Sam Houston State University, Office of International Criminal Justice, pp. 177–200.

Sapp, A.D. (1994), 'Sexual misconduct by police officers', in T. Barker and D.L. Carter (eds), *Police Deviance*, Cincinnati, OH: Anderson, pp. 187–99.

Sherman, L.W. (1978), *Scandal and Reform: Controlling Police Corruption*, Berkeley, CA: University of California Press.

Sherman, L.W. (1980), 'Causes of police behavior: the current state of quantitative research', *Journal of Research in Crime and Delinquency*, **17** (1), 69–100, doi:10.1177/002242788001700106.

Skogan, W. and Frydl, K. (2004), *Fairness and Effectiveness in Policing: The Evidence*, Washington, DC: National Academies Press.

Smalley, S. (2006), '75 officers failed city drug tests: cocaine use most prevalent, raising concern', *Boston Globe*, 30 July, p. A1.

Stinson, P.M. (2009), 'Police crime: a newsmaking criminology study of sworn law enforcement officers arrested, 2005–2007', PhD dissertation, Indiana University of Pennsylvania, Indiana, PA.

Stinson, P.M. (2015), 'Police crime: the criminal behavior of sworn law enforcement officers', *Sociology Compass*, **9** (1), 1–13, doi:10.1111/soc4.12234.

Stinson, P.M. (2018), 'The federal government doesn't track police violence – but I do', *The Atlantic*, 11 September, accessed 30 November 2018 at https://www.theat lantic.com/ideas/archive/2018/09/amber-guyger-fallout-how-common-is-police -crime/569950/.

Stinson, P.M. and Liederbach, J. (2013), 'Fox in the henhouse: a study of police officers arrested for crimes associated with domestic and/or family violence', *Criminal Justice Policy Review*, **24** (5), 601–25, doi:10.1177/0887403412453837.

Stinson, P.M., Brewer, S.L., Mathna, B.E., Liederbach, J. and Englebrecht, C.M. (2014a), 'Police sexual misconduct: arrested officers and their victims', *Victims & Offenders*, **10** (2), 117–51, doi:10.1080/15564886.2014.939798.

Stinson, P.M., Liederbach, J. and Freiburger, T.L. (2010), 'Exit strategy: an exploration of late-stage police crime', *Police Quarterly*, **13** (4), 413–35, doi:10.1177/1098611110384086.

Stinson, P.M., Liederbach, J. and Freiburger, T.L. (2012), 'Off-duty and under arrest: a study of crimes perpetuated by off-duty police', *Criminal Justice Policy Review*, **23** (2), 139–63, doi:10.1177/0887403410390510.

Stinson, P.M., Liederbach, J., Brewer, S.L. and Mathna, B.E. (2014b), 'Police sexual misconduct: a national scale study of arrested officers', *Criminal Justice Policy Review*, **26** (7), 665–90, doi:10.1177/0887403414526231.

Stinson, P.M., Liederbach, J., Brewer, S.L. and Todak, N.E. (2014c), 'Drink, drive, go to jail? A study of police officers arrested for drunk driving', *Journal of Crime and Justice*, **37** (3), 356–76, doi:10.1080/0735648X.2013.805158.

Stinson, P.M., Liederbach, J., Buerger, M. and Brewer, S.L. (2018), 'To protect and collect: a nationwide study of profit-motivated police crime', *Criminal Justice Studies*, **31** (3), 310–31, doi:10.1080/1478601X.2018.1492919.

Stinson, P.M., Liederbach, J., Brewer, S.L., Schmalzried, H.D., Mathna, B.E. and Long, K.L. (2013), 'A study of drug-related police corruption arrests', *Policing: An International Journal of Police Strategies & Management*, **36** (3), 491–511, doi:10.1108/PIJPSM-06-2012-0051.

Stinson, P.M., Liederbach, J., Lab, S.P. and Brewer, S.L. (2016), 'Police integrity lost: a study of law enforcement officers arrested', No. NCJ 249850, US Department of Justice, Office of Justice Programs and National Institute of Justice, Washington, DC, accessed 30 July 2018 at https://www.ncjrs.gov/App/publications/abstract. aspx?ID=272010.

Stinson, P.M., Todak, N.E. and Dodge, M. (2015), 'An exploration of crime by policewomen', *Police Practice and Research*, **16** (1), 79–93, doi:10.1080/156142 63.2013.846222.

Stoddard, E.R. (1968), 'The informal code of police deviancy: a group approach to blue-coat crime', *Journal of Criminal Law, Criminology and Police Science*, **59** (2), 201–13, doi:10.2307/1141940.

Terrill, W. (2003), 'Police use of force and suspect resistance: the micro process of the police-suspect encounter', *Police Quarterly*, **6** (1), 51–83, doi:10.1177/10986 11102250584.

US General Accounting Office (1998), *Report to the Honorable Charles B. Rangel, House of Representatives: Law Enforcement: Information on Drug-Related Police Corruption*, Washington, DC: US General Accounting Office.

Van Maanen, J. (1973), 'Observations on the making of policemen', *Human Organization*, **32** (4), 407–18, doi:10.17730/humo.32.4.13h7x81187mh8km8.

Van Maanen, J. (1975), 'Police socialization: a longitudinal examination of job attitudes in an urban police department', *Administrative Science Quarterly*, **20** (2), 207–28, doi:10.2307/2391695.

Van Maanen, J. (1988), *Tales of the Field: On Writing Ethnography*, Chicago, IL: University of Chicago Press.

Violanti, J.M., Marshall, J.R. and Howe, B. (1985), 'Stress, coping, and alcohol use: the police connection', *Journal of Police Science and Administration*, **13** (2), 106–10.

Violanti, J.M., Slaven, J.E., Charles, L.E., Burchfiel, C.M., Andrew, M.E. and Homish, G.G. (2011), 'Police and alcohol use: a descriptive analysis and associations with stress outcomes', *American Journal of Criminal Justice*, **36** (4), 344–56, doi:10.1007/s12103-011-9121-7.

Westley, W.A. (1956), 'Secrecy and the police', *Social Forces*, **34** (3), 254–7, doi:10.2307/2574048.

Westley, W.A. (1970), *Violence and the Police: A Sociological Study of Law, Custom, and Morality*, Cambridge, MA: MIT Press.

Wickersham Commission (1931), *United States National Committee on Law Observance and Enforcement: Report on the Police*, Washington, DC: US Government Printing Office.

Wilson, J.Q. (1963), 'The police and their problems: a theory', *Public Policy*, **12**, 189–216.

14. Workplace stress and firefighter health and safety

Todd D. Smith, Mari-Amanda Dyal and David M. DeJoy

INTRODUCTION

Protecting and promoting the health of firefighters is of great importance, especially because firefighters risk their lives daily (and nightly) performing critical operations to protect the public, businesses, government and more. Firefighter injuries, illnesses and fatalities are problematic in the United States and internationally. In this chapter, we present current data highlighting the veracity of these incidents and trends associated with incidents and causal factors. A review of these incidents identify that firefighting is stressful, demanding and exhausting work. Often the inherent job demands associated with firefighting cannot be fully abated, but there are opportunities to control or counter excessive job demands though appropriate job resources (Smith and Dyal 2016). Countering these job demands should curtail firefighter health impairment and should help with preventing firefighter injuries, illnesses and fatalities (Smith and Dyal 2016).

This chapter, through a theoretical framework associated with the job demands–resources model, highlights physical and psychosocial job demands associated with firefighting. We focus on physical job demands, workload, work-role expectations and role conflict, work–family conflict and passive leadership. As noted above, job resources may be utilized to control or counter excessive job demands. In the chapter, with a focus on means to protect and promote worker health, job resources to include recovery, transformational leadership, support and safety climate are presented and suggested.

FIREFIGHTER HEALTH, SAFETY AND WELL-BEING

Firefighters are often injured, become ill and/or die at rates higher than many other workers (Walton et al. 2003; Lee et al. 2004; Marsh et al. 2018). Firefighters historically have been hospitalized more often that many other workers (Lee et al. 2004) and their nonfatal injury rate is significantly higher than other workers in the United States (Marsh et al. 2018). Marsh et al. (2018) also report that between 2003 and 2014, 351 800 firefighters were treated in emergency departments for nonfatal injuries. These injuries were sustained on the fireground (38 percent), during training (7 percent) and during patient medical treatment or care (7 percent), and were mostly associated with strains and sprains (Marsh et al. 2018).

According to Haynes and Molis (2017) there were 62 085 firefighter injuries in the United States in 2016. This was fewer than the previous year, but still suggests a significant safety and health problem. Similar to the findings by Marsh et al. (2018), Haynes and Molis (2017) reported that most of the injuries sustained in 2016 occurred on the fireground (39 percent) and were associated with overexertion or strain (27 percent). Thirty-one percent of all the 2016 injuries suffered by firefighters in the United States resulted in lost work time (Haynes and Molis 2017).

As regards firefighter line-of-duty fatalities, there have been some reductions over the past decade within the United States; however, during this time, the average number of deaths was still 75 per year (Fahy et al. 2018). Current data from the National Fire Protection Association indicates 60 firefighters died in the line of duty in the United States in 2017 (Fahy et al. 2018). This is fewer than the reported 93 line-of-duty fatalities suffered in 2016 (USFA 2018).

In 2017, as in many previous years, sudden cardiac death accounted for the majority of the deaths (Fahy et al. 2018). Sudden cardiac deaths have long been problematic in the fire service, accounting for a large percentage of fatalities (Kales et al. 2003; Kunadharaju et al. 2011; Soteriades et al. 2011). Most on-duty firefighter fatalities associated with coronary heart disease are triggered by the demands of the work performed and occur in firefighters with underlying risk factors (Kales et al. 2003). Firefighters are more likely to die of a heart attack while on duty than are those in other strenuous occupations (USFA 2002).

Firefighter injuries and fatalities are not limited to the United States. In the United Kingdom, firefighter injury rates are high (Watterson 2015) and are generally higher than most other occupations when compared in Health and Safety Executive statistics. Firefighter fatalities are also problematic in the United Kingdom. Fourteen firefighters were killed in the line of duty between 2004 and 2013 (Watterson 2015).

Brushlinsky et al. (2017) present additional data associated with firefighter fatalities and injuries for 32 countries between 2011 and 2015. These data are limited to the countries that reported their data, and are not reflective of all injuries and fatalities throughout the world, but provide some additional insights into the extent of firefighter injuries and fatalities internationally. The highest average number of fatalities per year occurred in the United States, where the average number of firefighter fatalities was 70 per year between 2011 and 2015 (Brushlinsky et al. 2017). Other countries reporting double-digit fatalities per year during this period included Japan (62 firefighter fatalities per year), Russia (15 firefighter fatalities per year) and France, which averaged 10 firefighter fatalities per year (Brushlinsky et al. 2017). During the same period, Austria reported an average of three firefighter fatalities per year; however, their average number of firefighter injuries per year totaled 1045 (Brushlinsky et al. 2017). For the other countries mentioned above, on average, 133 French firefighters were injured each year, 319 Russian firefighters were injured each year and 2457 Japanese firefighters were injured each year (Brushlinsky et al. 2017). The highest mean count for injury for the period was in the United States. Brushlinsky et al. (2017) report that within the United States, between 2011 and 2015, 67 361 American firefighters were injured each year. Burgess et al. (2014) also determined that injury rates were higher in the United States when compared with those in Japan, Australia, Canada and the United Kingdom.

Beyond line-of-duty injuries and fatalities, it is difficult to determine the long-term effects of workplace exposures on occupational health related injuries, illnesses and fatalities, including those associated with suicide (Fahy et al. 2018). Behavioral health issues among fire service members are now garnering increased attention since current firefighters are experiencing high levels of stress, depression, anxiety, burnout and post-traumatic stress disorder as a result of job demands and work tasks that expose them to violence, trauma and life or death situations (Jahnke et al. 2014). The emergence of these behavioral health outcomes has further increased the risk of injury and fatalities, including suicidal ideation and suicides (Antonellis and Thompson 2012; Stanley et al. 2015; Henderson et al. 2016).

JOB DEMANDS–RESOURCES MODEL

Over the years, there have been numerous models and frameworks associated with workplace stress and strain with a focus on advancing worker health. These models have been presented in numerous review articles and book chapters. Some of the most notable theories that pertain to workplace or occupational stress include person–environment (P-E) fit, job demands–control model, conservation of resources theory and effort–reward imbalance theory. See Ganster and Perrewé (2011) for details associated with these theories and models.

One of the more prominent models currently is the job demands–resources (JD-R) model, which was first presented in the early 2000s (Demerouti et al. 2001) and has since become a popular model for researchers and practitioners in the area of stress, burnout and health impairment. It has also been expanded to focus on safety and health protection (Nahrgang et al. 2011). The JD-R model implies that a balance between negative and positive job characteristics can influence worker health outcomes (Schaufeli and Bakker 2004; Bakker and Demerouti 2007). Job demands are the factors that influence health impairment, and job resources are the factors that foster engagement and the control of health impairment. Health impairment is attributed to an imbalance consisting of high job demands and low job resources, which may lead to negative health outcomes, such as burnout (Schaufeli and Bakker 2004) and safety outcomes related to injury, accidents and adverse events (Nahrgang et al. 2011).

According to the JD-R model, risk factors, particularly stressors, are categorized as job demands and protective factors are categorized as job resources (Bakker and Demerouti 2007). Job demands are inevitable in the workplace and may function as physical or psychosocial stressors (Carayon et al. 1999). Worker health outcomes or impairment may occur if the demands are excessive, particularly when workers do not have appropriate tools or resources to cope with or control the stressors. Individual differences or trait characteristics play a part in how the demands are appraised.

Resources are crucial to minimizing and controlling damaging physical and mental health impairment outcomes in the workplace. The second process associated with the JD-R model is motivational and suggests that imperative job resources bolster work engagement (Bakker and Demerouti 2007), which empowers workers to counter high job demands to protect and promote health. Nonexistent or diminished job resources prevent workers from coping with excessive job demands, which leads to reduced motivation and health impairment, such as burnout (Demerouti et al. 2001). Job resources can include physical, psychological, social or organizational aspects of the job.

WORK-ASSOCIATED STRESSORS AND EFFECTS ON FIREFIGHTER HEALTH AND SAFETY

Job demands include the physical and psychosocial aspects of the job that necessitate sustained physical and/or psychological effort. Consequently, job demands are associated with certain physiological and/or psychological outcomes (Bakker et al. 2003), such as increased worker burnout and diminished job performance (Bakker and Demerouti 2007). These outcomes can be significant in high hazard occupations such as firefighting, where diminished job or task performance can result in injuries and fatalities. The fire service presents very complex job demands. In this section of the chapter, we present job demands or stressors in the context of firefighting operations, which include physical job demands, workload pressure, work-role expectations, work–family conflict and passive leadership. We describe these job demands and their relationships with health impairment, including physical health impairment (including injury) and/ or psychological health impairment.

Physical Job Demands

There are numerous physical job demands, which are often presented in the context of ergonomic exposures or stressors. These include bending, standing, lifting, carrying, pushing, pulling, forceful exertions, static or sustained postures, awkward postures and repetition. Physical activities that result in energy expenditure, such as walking, running, climbing, and so on are also included (Hakanen et al. 2005). Many of these job demands require significant effort and the most demanding jobs require greater energy expenditure in the context of firefighting (Gledhill and Jamnik 1992). These physical job demands are often associated with injury outcomes when demands are excessive, but there is evidence that they are also associated with health impairment, including burnout (Nahrgang et al. 2011).

Firefighting is a very physically demanding job and there are numerous physical hazards encountered with fire service operations. Smith and Dyal (2016, p. 448) state:

> Physical demands in firefighting are acute, require endurance, and push bodily systems to extremes. Repeated activation of these systems without proper safeguards can lead to physical and mental fatigue. Fatigue at this level delays reaction time, impairs health behaviors, and creates a sense of indifference for safety procedures and protocols, which serves as a negative defense mechanism against high physical demands. (Basińska and Wiciak, 2012; Takahashi et al., 2011)

In the context of ergonomic exposures and physical job demands, we generally find that fire service job tasks are not repetitive and job tasks vary significantly (Conrad et al. 2000). Although the tasks are not repetitive, they are physically demanding and often result in strains, sprains and other related musculoskeletal disorders. These types of disorders occur frequently in the fire service and are associated with overexertion. These disorders often occur when firefighters are providing medical treatment. Firefighters, especially those performing emergency medical technician work operations, are at significant risk of musculoskeletal disorders (Conrad et al. 2000).

Physical activity job demands are often encountered during exercise, wellness and physical activity training programs. These activities account for a large number of injuries within the fire service (Szubert and Sobala 2002; Poplin et al. 2012; Katsavouni et al. 2014; Vaulerin et al. 2016). Ironically, these programs are meant to improve firefighter health and are meant to prevent injuries. Poplin et al. (2012, p. 231) best capture the irony of the situation in the following statement:

> The purpose of physical exercise is to prepare one for their job and to condition a person to perform those job tasks with the utmost amount of efficiency, so that injuries are prevented. Therefore, it is somewhat of a paradox that physical exercise, which aims to prevent injuries (and other adverse health outcomes), is actually the most frequent cause of injury.

Workload

Workload is often addressed in the context of job demands. Workload generally signifies the amount of work completed and the pace at which it is completed under time pressure (Euwema and Bakker 2009). Workload has been prominently featured in research associated with occupational safety and health, particularly in the context of safety climate (for example, Lawton and Parker 1998; Brown et al. 2000; Flin et al. 2000; Guldenmund 2000; Seo 2005). Production pressure, time pressure, external work pacing, and excessive work hours have been associated with both unsafe behavior and accidents.

Lusa et al. (2002) found that very long work hours (more than 70 hours per week) were associated with a significant increase in the risk of accidents among firefighters. Also, high time pressure has been associated with increased firefighter fatalities (Rosmuller and Ale 2008). Time pressure is a key attribute of many, if not most, firefighting situations, and direct fire suppression activities are often initiated before sufficient personnel and equipment are at the scene, which has been a contributing factor to firefighter fatalities (Kunadharaju et al. 2011).

In the context of health outcomes, excessive workload has been associated with heat stress and cardiovascular risk factors. In particular, researchers determined that firefighting operations subsequently increase heart rate and core body temperature (Horn et al. 2011) and that there is an increased risk of heat stress with long-duration firefighting, even with extended break times (Horn et al. 2013). This risk is present within both structural firefighting and wildland firefighting. Cuddy and Ruby (2011) determined that the risk of heat exhaustion was elevated during wildland firefighting with high workload and high temperature situations, even with adequate hydration. Of great significance is that the workload associated with firefighting produces significant cardiovascular changes, which increases cardiovascular risk during firefighting operations (Fernhall et al. 2012).

Work-Role Expectations

Work-role expectations refer to beliefs workers have about what tasks persons in their specific jobs are expected to perform and under what conditions (Hackman 1992). These perceptions are derived from information communicated to them during initial orientation and training and, later, from expectations communicated on the job by co-workers and supervisors (Ashforth et al. 1998). Training and socialization processes for high-hazard occupations are typically very elaborate and well organized; newcomers quickly develop a good understanding of what it takes to perform and integrate into the work context. For firefighters, these work-role expectations are developed primarily during their training at a fire academy and during their probationary service, when they are assigned to a functioning company or other work unit. In addition, many newcomers enter the occupation with well-developed expectations about what it means to be a firefighter (Myers 2005). The well-established and positive external public image of firefighting undoubtedly contributes to, and reinforces, the establishment of these expectations (Hatch and Schultz 1997).

Expectations, however, can be altered and role conflict may arise, resulting in strain and associated outcomes. Job-role conflict and ambiguity has long been an area of research in the context of stress, work and organizations (Kahn et al. 1964; Rizzo et al. 1970). Within the fire service, as noted above, firefighters generally learn their work role during their initial training within a fire service academy. When they move into the second stage of socialization, which is less structured and occurs during the firefighters' probationary service within their fire company, firefighters learn how things are 'really done' in the field (Hackman 1992; Myers 2005; Bauer et al. 2007), which is often in contrast to what is learned within their initial training and established standard operating procedures. Given their rank within a paramilitary-style organization, probationary firefighters often do not contest what they are told to do, even if it is wrong and counter to their training, resulting in conflict. What is often deemed as more important within their roles is listening to superiors and being a trusted and cohesive member of the group. Trust and group cohesion are widely considered key attributes of effective firefighting units (DeJoy et al. 2017). Although high levels of trust and group cohesion often facilitate team performance and safety, these same characteristics can also produce detrimental effects. Strong group affiliation might lead to excessive risk-taking and/or inappropriate dismissal of problems affecting individual team members (Weick and Roberts 1993; Jones and Roelofsma 2000). High interdependence can make groups susceptible to social influence and group think. Shared mental models may lead to group members thinking similarly and confidently, but incorrectly.

Many firefighters, especially career firefighters, identify strongly with being a firefighter. It is not unusual for firefighters to say that all they ever wanted to be is a firefighter. This identity often begins to develop through family and friends prior to entry into the occupation (Myers 2005). This identity is then further strengthened and reinforced during training and assimilation. Part of this job-role identity involves believing that firefighters have a duty to face danger and take risks to help others, that they are able to function under conditions of high stress, and that they are members of an elite and tightly knit occupational group with certain privileges and obligations (Bellrose and Pilisuk 1991; Myers 2005). This identity is further reinforced by the strong positive public image of firefighters and firefighting. The image is of heroism and self-sacrifice for the good of others. These expectations may create excessive stress and conflict, particularly when it results in unacceptable risks that may further expose firefighters to injury and increased work roles and activities that may result in health impairment. Further, these expectations may result in additional inter-role conflict, such as work–family conflict.

Work–Family Conflict

Interest in the work–family interface has expanded significantly over the past few decades (Allen and Martin 2017). The work–family interface is generally comprised of work–family conflict and family–work conflict, also referred to as work–family interference and family–work interference. Most researchers currently support the notion of the bidirectional relationships and multiple constructs present within the work–family interface. In general, work–family conflict implies that participation in the work role interferes or becomes incompatible with participation in the family role, and family–work conflict implies that participation in the family role interferes or becomes incompatible with work roles. Current thinking about these aspects of inter-role conflict emerged from the work of Greenhaus and Beutell (1985). Beyond directionality, researchers often differentiate the various types of conflict, including time-based conflict, strain-based conflict and behavior-based conflict (Greenhaus and Beutell 1985; Carlson et al. 2000). Allen and Martin (2017) indicate that time-based conflict occurs when time spent on responsibilities associated with one role interferes with the completion of responsibilities in another role, and that strain-based conflict occurs when pressures from one role make it challenging to complete requirements of another role. Further, Allen and Martin (2017) indicate that behavior-based conflict occurs when the behaviors necessary for one role are incompatible with the required behaviors in the other role.

The work–family conflict literature is robust. There is a substantial body of research that has examined the associations between work–family conflict and health outcomes, including health impairment associated with both mental and physical health (Allen et al. 2000; Judge et al. 2006; Panatik et al. 2011; Poms et al. 2016; Allen and Martin 2017; Davis et al. 2017). Despite this robust literature associated with work–family conflict, little work has been conducted in the fire service until recently. Shreffler et al. (2011) found that occupational stress was associated with work–family conflict in a sample of male firefighters, who were fathers. Smith et al., in multiple studies (Smith et al. 2017, 2018), found that work–family conflict is associated with firefighter burnout outcomes, which can ultimately impact safety outcomes, including safety behaviors associated with personal protective equipment practices, general safe work practices, and safety reporting and communication behavior within the fire service (Smith et al. 2018).

Passive Leadership

There is also a robust literature on the many benefits of transformational leadership, particularly as it relates to psychological well-being (Arnold 2017) and enhanced worker health and safety outcomes (Barling et al. 2002; Zohar 2002; Kelloway et al. 2006; de Koster et al. 2011; Clarke 2013; Smith et al. 2016; Arnold 2017). Supportive leadership, such as transformational leadership, has been delineated as a job resource in the context of the JD-R model (Nahrgang et al. 2011; Smith and Dyal 2016). Together with knowledge, social support and positive safety climate, supportive leadership, as a measure of job resources, was found to be associated with engagement, compliance and satisfaction (Nahrgang et al. 2011). In contrast though, we have learned that passive leadership can be detrimental to health and safety (Kelloway et al. 2006; Smith et al. 2016) and general health and well-being (Mullen and Kelloway 2011; Barling and Frone 2017). Much of the research associated with poor leadership has been in the context of stress and health impairment (Offerman and Hellman 1996; Kelloway et al. 2004).

At the time of writing, the literature associated with passive leadership in the context of the fire service is not well established. A knowledge gap exists in that the extant literature does not provide ample evidence of the relationships between passive leadership and health impairment. There is some evidence that passive leadership is detrimental to safety in the fire service. Smith et al. (2016) found that passive leadership diminishes perceptions of safety climate, especially when fire service leaders fail to intervene in safety issues and concerns, when they fail to take action until things go wrong and when they fail to make decisions important to firefighter safety and health. These diminished perceptions of safety climate are a concern given the significant relationship between safety climate and safety behaviors in the fire service (Smith et al. 2016). Reductions in safety-climate perceptions would thus be associated with reductions in safety behavior outcomes, which would be problematic for the fire service (Smith et al. 2016).

PROTECTING AND PROMOTING FIREFIGHTER HEALTH AND SAFETY

As previously noted, improving firefighter health and safety outcomes is of vast importance as firefighter injuries, illnesses and fatalities remain problematic. Strategic approaches are needed to counter excessive job demands that result in firefighter health impairment. As indicated within the JD-R model, essential job resources, particularly those that protect and promote firefighter health should be incorporated into safety and workplace health promotion (WHP) efforts. Providing and maintaining essential resources as a safety and health strategy may counter firefighter health impairment (Smith and Dyal 2016).

We next present job resources aimed at curtailing excessive firefighter job demands. These resources should help firefighters achieve work goals, foster engagement, be feasible and efficacious, exhibit an integrated approach to health protection and promotion and should counter excessive job demands. These resources are based on the work of Smith and Dyal (2016) and include recovery, transformational leadership, support and safety climate. Implications and content have been expanded to address broader current thinking and job demands that were not previously considered.

Recovery

Recovery is a necessary job resource in firefighting needed to counter high job demands, particularly those associated with physical job demands and workload, but also emotional demands as Katsavouni et al. (2016) found an increased risk of occupational health problems as a result of both psychological and physical stress. Recovery provides a break from job demands so that they do not become excessive and result in health impairment. Recovery results in positive mental health and well-being, and may serve to reduce inter-role conflict, minimizing work–family conflict. Recovery while on shift is difficult as firefighters have little downtime. Between fire-extinguishing activities, they are required to perform other job duties such as training, community service projects, exercise programs, station-related activities and emergency calls (Basińska and Wiciak 2012). Furthermore, most firefighters are not able to fully recover when they are off shift as they often have second jobs. These jobs vary in job demands but can lead to strain, especially since many firefighters take on additional employment as a firefighter with a different fire department. In the context of safety and WHP, programs and policies should be implemented to ensure recovery while on duty, particularly during firefighting operations as rehabilitation and while off duty. Together with recovery, programs that bolster resilience may be beneficial to counter excessive job demands.

Transformational Leadership

To curtail the negative influences associated with passive leadership, transformational leadership strategies and tactics should be integrated at multiple levels within fire service organizations. These efforts should especially be carried out at the workgroup level by direct supervisors who lead firefighters. These tactics differ from the authoritarian leadership style usually exemplified within the fire service. Transformational leadership tactics should include: being a good role model; communicating a vision of positive health, safety and well-being; motivating firefighters to think of means to bolster health and safety; motivating firefighters to perform safety citizenship or extra-role behaviors protecting others and not only themselves; and, showing concern for the safety, health and well-being of subordinates. In the context of safety, transformational leadership is associated with safety and health protection (Barling et al. 2002; Zohar

2002; Kelloway et al. 2006), including within the fire service (Smith et al. 2016). Apart from safety, transformational leadership is also associated with overall health and well-being (Arnold 2017).

Support

Support provides a means to cope with stress. Supervisor support for safety and health is important as support has been associated with firefighter health and safety outcomes. Support is associated with reduced stress (Regehr et al. 2003; Cowman et al. 2004; Yarnal et al. 2004; Oginska-Bulik 2005; Bacharach and Bamberger 2007; Varvel et al. 2007), improvements in wellness (Yarnal et al. 2004), reduced depression (Regehr et al. 2003), reduced anxiety (Bacharach and Bamberger 2007) and reductions in over-all negative health outcomes (Oginska-Bulik 2005; Tuckey and Hayward 2011). These factors are all significantly related to cardiovascular health.

There is ample evidence associated with the health promotion aspects of support within the fire service, but a knowledge gap exists regarding the relationship between support and safety outcomes in the fire service. DeJoy et al. (2017) and Smith and DeJoy (2014) illustrate its importance to safety as a factor of safety climate, but other findings associated with fire service members are not evident in the extant literature. Outside the fire service, existing research illustrates the positive associations that could likely be obtained within fire service organizations. Support has been associated with safe work behaviors (Parker et al. 2001; Torp et al. 2005), voicing safety concerns (Hofmann and Morgeson 1999; Tucker et al. 2008) and higher levels of psychological safety (May et al. 2004). It also buffered the negative effects of working in a garage with high job demands (Torp et al. 2005).

Safety Climate

Safety climate is associated with shared perceptions among members of an organization concerning the importance of workplace safety (Flin et al. 2000; Zohar 2003; DeJoy et al. 2004). The most important aspects of safety climate appear to be management commitment to safety and perceptions of policies, procedures and practices (Griffin and Neal 2000; Barling et al. 2002; Zohar 2003; Hahn and Murphy 2008; Smith and DeJoy 2014). Safety climate perceptions have been directly linked to safety behavior outcomes, including safety compliance behaviors and safety participation behaviors of firefighters (Smith and DeJoy 2014; Smith et al. 2016) and were deemed protective against injury (Smith and DeJoy 2014). From a multi-level perspective, fire service organizations should aim to address the most important aspects associated with fire service safety climate. These include supervisor support for safety, vertical cohesion (between firefighter and company officer) and horizontal cohesion (between firefighters) at the work-group level (DeJoy et al. 2017). At the organizational level, the focus should be on management commitment to safety, formalized and enacted safety programs, policies and procedures, quality human resource programs and effective incident command (DeJoy et al. 2017).

SUMMARY AND CONCLUSIONS

As exhibited, firefighter injuries, illnesses and fatalities remain a significant safety and health problem within the United States and internationally. These work-related injuries, illnesses and fatalities are often directly associated with workplace stressors, particularly the physical and psychosocial job demands encountered by firefighters during fire service operations. There are numerous theoretical models associated with occupational stressors, but one of the more recent models, the job demands–resources model provides an opportunity for us to address firefighter health and safety. As regards the health impairment process of the model, we identified job demands that if not controlled or managed, would likely impair firefighter health. These job demands included physical job demands, workload, work-role expectations, particularly role conflict, work–family conflict and passive leadership. In the context of the model, we also presented job resources that can be enacted to curtail health impairment. Here, we focused on recovery, transformational leadership, support and safety climate. Ultimately, these job resources should serve as targets for interventions to protect and promote firefighter health by practitioners.

REFERENCES

Allen, T.D. and Martin, A. (2017), 'The work–family interface: a retrospective look at 20 years of research in JOHP', *Journal of Occupational Health Psychology*, **22** (3), 259–72.

Allen, T.D., Herst, D.L., Bruck, C.S. and Sutton, M. (2000), 'Consequences associated with work-to-family conflict: a review and agenda for future research', *Journal of Occupational Health Psychology*, **5** (2), 278–308, doi:10.1037/1076-8998.5.2.278.

Antonellis, P.J. and Thompson, D. (2012), 'A firefighter's silent killer: suicide', *Fire Engineering*, **165** (12), 69–76.

Arnold, K.A. (2017), 'Transformational leadership and employee psychological wellbeing: a review and directions for future research', *Journal of Occupational Health Psychology*, **22** (3), 381–93.

Ashforth, B.E., Saks, A.M. and Lee, R.T. (1998), 'Socialization and newcomer adjustment: the role of organizational context', *Human Relations*, **51** (7), 897–926.

Bacharach, S.B. and Bamberger, P.A. (2007), '9/11 and New York City firefighters' post hoc unit support and control climates: a context theory of the consequences of involvement in traumatic work-related events', *Academy of Management Journal*, **50** (4), 849–68.

Bakker, A.B. and Demerouti, E. (2007), 'The job demands-resources model: state of the art', *Journal of Managerial Psychology*, **22** (3), 309–28.

Bakker, A.B., Demerouti, E. and Schaufeli, W.B. (2003), 'Dual processes at work in a call centre: an application of the job demands-resources model', *European Journal of Work & Organizational Psychology*, **12** (4), 393–417.

Barling, J. and Frone, M.R. (2017), 'If only my leader would just do something! Passive leadership undermines employee well-being through role stressors and psychological resource depletion', *Stress and Health*, **33** (3), 211–22.

Barling, J., Loughlin, C. and Kelloway, E.K. (2002), 'Development and test of a model linking safety-specific transformational leadership and occupational safety', *Journal of Applied Psychology*, **87** (3), 488–96.

Basińska, B.A. and Wiciak, I. (2012), 'Fatigue and professional burnout in police officers and firefighters', *Internal Security*, **4** (2), 265–73.

Bauer, T.N., Bodner, T., Erdogan, B., Truxillo, D.M. and Tucker, J.S. (2007), 'Newcomer adjustment during organizational socialization: a meta-analytic review of antecedents, outcomes, and methods', *Journal of Applied Psychology*, **92** (3), 707–21.

Bellrose, C.A. and Pilisuk, M. (1991), 'Vocational risk tolerance and perceptions of occupational hazards', *Basic and Applied Social Psychology*, 12 (3), 303–23.

Brown, K.A., Willis, P.G. and Prussia, G.E. (2000), 'Predicting safe employee behavior in the steel industry: development and test of a sociotechnical model', *Journal of Operations Management*, **18** (4), 445–65.

Brushlinsky, N.N., Ahreus, M., Sokolov, S.V. and Wagner, P. (2017), 'World fire statistics. Center of Fire Statistics (CFS) of the International Association of Fire and Rescue Services: Russia', accessed 25 September 2018 at https://www.ctif.org/sites/default/files/ctif_report22_world_fire_statistics_2017.pdf.

Burgess, J.L., Duncan, M., Mallett, J., LaFleur, B., Littau, S. and Shiwaku, K. (2014), 'International comparison of fire department injuries', *Fire Technology*, **50** (5), 1043–59.

Carayon, P., Smith, M.J. and Haims, M.C. (1999), 'Work organization, job stress, and work-related musculoskeletal disorders', *Human Factors*, **41** (4), 644–63.

Carlson, D.S., Kacmar, K.M. and Williams, L.J. (2000), 'Construction and initial validation of a multidimensional measure of work–family conflict', *Journal of Vocational Behavior*, **56** (2), 249–76.

Clarke, S. (2013), 'Safety leadership: a meta-analytic review of transformational and transactional leadership styles as antecedents of safety behaviours', *Journal of Occupational and Organizational Psychology*, **86** (1), 22–49.

Conrad, K.M., Lavender, S.A., Reichelt, P.A. and Meyer, F.T. (2000), 'Initiating an ergonomic analysis: a process for jobs with highly variable tasks', *AAOHN Journal*, **48** (9), 423–9.

Cowman, S.E., Ferrari, J.R. and Liao-Troth, M. (2004), 'Mediating effects of social support on firefighters' sense of community and perceptions of care', *Journal of Community Psychology*, **32** (2), 121–6.

Cuddy, J.S. and Ruby, B.C. (2011), 'High work output combined with high ambient temperatures caused heat exhaustion in a wildland firefighter despite high fluid intake', *Wilderness & Environmental Medicine*, **22** (2), 122–5.

Davis, K.D., Gere, J. and Sliwinski, M.J. (2017), 'Investigating the work–family conflict and health link: repetitive thought as a mechanism', *Stress and Health*, **33** (4), 330–38.

De Koster, R.B.M., Stam, D. and Balk, B.M. (2011), 'Accidents happen: the influence of safety-specific transformational leadership, safety consciousness, and hazard reducing systems on warehouse accidents', *Journal of Operations Management*, **29** (7–8), 753–65, doi:10.1016/j.jom.2011.06.005.

DeJoy, D.M., Schaffer, B.S., Wilson, M.G., Vandenberg, R.J. and Butts, M.M. (2004), 'Creating safer workplaces: assessing the determinants and role of safety climate', *Journal of Safety Research*, **35** (1), 81–90.

DeJoy, D.M., Smith, T.D. and Dyal, M. (2017), 'Safety climate and firefighting: focus group results', *Journal of Safety Research*, **62** (September), 107–16, doi:10.1016/j.jsr.2017.06.011.

Demerouti, E., Bakker, A., Nachreiner, F. and Schaufeli, W. (2001), 'The job demands-resources model of burnout', *Journal of Applied Psychology*, **86** (3), 499–512.

Euwema, M.C. and Bakker, A.B. (2009), 'Explaining employees' evaluations of organizational change with the job-demands resources model', *Career Development International*, **14** (6), 594–613.

Fahy, R.F., LeBlanc, P.R. and Molis, J.L. (2018), 'Firefighter fatalities in the United States – 2017', report, National Fire Protection Association, Quincy, MA.

Fernhall, B., Fahs, C.A., Horn, G., Rowland, T. and Smith, D. (2012), 'Acute effects of firefighting on cardiac performance', *European Journal of Applied Physiology*, **112** (2), 735–41.

Flin, R., Mearns, K., O'Connor, P. and Bryden, R. (2000), 'Measuring safety climate: identifying the common features', *Safety Science*, **34** (1–3), 177–92.

Ganster, D.C. and Perrewé, P.L. (2011), 'Theories of occupational stress', in J.C. Quick and L.E. Tetrick (eds), *Handbook of Occupational Health Psychology*, Washington, DC: American Psychological Association, pp. 37–53.

Gledhill, N. and Jamnik, V.K. (1992), 'Characterization of the physical demands of firefighting', *Canadian Journal of Sport Sciences*, **17** (3), 207–13.

Greenhaus, J.H. and Beutell, N.J. (1985), 'Sources of conflict between work and family roles', *Academy of Management Review*, **10** (1), 76–88, doi:10.5465/AMR.1985.4277352.

Griffin, M.A. and Neal, A. (2000), 'Perceptions of safety at work: a framework for linking safety climate to safety performance, knowledge, and motivation', *Journal of Occupational Health Psychology*, **5** (3), 347–58, doi:10.1037/1076-8998.5.3.347.

Guldenmund, F.W. (2000), 'The nature of safety culture: a review of theory and research', *Safety Science*, **34** (1), 215–57.

Hackman, J.R. (1992), 'Group influences on individuals in organizations', in M.D. Dunnette and L.M. Hough (eds), *Handbook of Industrial and Organizational Psychology*, vol. 3, Palo Alto, CA: Consulting Psychologists Press, pp. 199–268.

Hahn, S.E. and Murphy, L.R. (2008), 'A short scale for measuring safety climate', *Safety Science*, **46** (7), 1047–66.

Hakanen, J.J., Bakker, A.B. and Demerouti, E. (2005), 'How dentists cope with their job demands and stay engaged: the moderating role of job resources', *European Journal of Oral Sciences*, **113** (6), 479–87.

Hatch, M. J. and Schultz, M. (1997), 'Relations between organizational culture, identity and image', *European Journal of Marketing*, **31** (5–6), 356–65.

Haynes H.J.G and Molis, J.L. (2017), 'Firefighter injuries in the United States – 2016', National Fire Protection Association, Quincy, MA.

Henderson, S.N., Van Hasselt, V.B., LeDuc, T.J. and Couwels, J. (2016), 'Firefighter suicide: understanding cultural challenges for mental health professionals', *Professional Psychology: Research and Practice*, **47** (3), 224–30.

Hofmann, D.A. and Morgeson, F.P. (1999),' Safety-related behavior as a social exchange: the role of perceived organizational support and leader–member exchange', *Journal of Applied Psychology*, **84** (2), 286–96.

Horn, G.P., Blevins, S., Fernhall, B. and Smith, D.L. (2013), 'Core temperature and heart rate response to repeated bouts of firefighting activities', *Ergonomics*, **56** (9), 1465–73.

Horn, G.P., Gutzmer, S., Fahs, C.A., Petruzzello, S.J., Goldstein, E., Fahey, G.C. et al. (2011), 'Physiological recovery from firefighting activities in rehabilitation and beyond', *Prehospital Emergency Care*, **15** (2), 214–25, doi:10.3109/1090312 7.2010.545474.

Jahnke, S.A., Gist, R., Poston, W.S.C. and Haddock, C.K. (2014), 'Behavioral health interventions in the fire service: stories from the firehouse', *Journal of Workplace Behavioral Health*, **29** (2), 113–26.

Jones, P.E. and Roelofsma, P.H. (2000), 'The potential for social contextual and group biases in team decision-making: biases, conditions and psychological mechanisms', *Ergonomics*, **43** (8), 1129–52.

Judge, T.A., Ilies, R. and Scott, B.A. (2006), 'Work–family conflict and emotions: effects at work and at home', *Personnel Psychology*, **59** (4), 779–814.

Kahn, R.L., Wolfe, D.M., Quinn, R.P., Snoek, J.D. and Rosenthal, R.A. (1964), *Organizational Stress: Studies in Role Conflict and Ambiguity*, Oxford: John Wiley.

Kales, S.N., Soteriades, E.S., Christoudias, S.G. and Christiani, D.C. (2003), 'Firefighters and on-duty deaths from coronary heart disease: a case control study', *Environmental Health: A Global Access Science Source*, **2** (1), 14.

Katsavouni, F., Bebetsos, E., Antoniou, P., Malliou, P. and Beneka, A. (2014), 'Work-related risk factors for low back pain in firefighters. Is exercise helpful?', *Sport Sciences for Health*, **10** (1), 17–22.

Katsavouni, F., Bebetsos, E., Malliou, P. and Beneka, A. (2016), 'The relationship between burnout, PTSD symptoms and injuries in firefighters', *Occupational Medicine*, **66** (1), 32–7.

Kelloway, E.K., Mullen, J. and Francis, L. (2006), 'Divergent effects of transformational and passive leadership on employee safety', *Journal of Occupational Health Psychology*, **11** (1), 76–86.

Kelloway, E.K., Sivanathan, N., Francis, L. and Barling, J. (2004), 'Poor leadership', in J. Barling, E.K. Kelloway and M.R. Frone (eds), *Handbook of Workplace Stress*, Thousand Oaks, CA: Sage, pp. 89–112.

Kunadharaju, K., Smith, T. and DeJoy, D. (2011), 'Line-of-duty deaths among U.S. firefighters: an analysis of fatality investigations', *Accident Analysis and Prevention*, **43** (3), 1171–80.

Lawton, R. and Parker, D. (1998), 'Individual differences in accident liability: a review and integrative approach', *Human Factors*, **40** (4), 655–71.

Lee, D.J., Fleming, L.E., Gomez-Marin, O. and Leblanc, W. (2004), 'Risk of hospitalization among firefighters: the National Health Interview Survey, 1986–1994', *American Journal of Public Health*, **94** (11), 1938–9.

Lusa, S., Häkkänen, M., Luukkonen, R. and Viikari-Juntura, E. (2002), 'Perceived physical work capacity, stress, sleep disturbance and occupational accidents among firefighters working during a strike', *Work & Stress*, **16** (3), 264–74.

Marsh, S.M., Gwilliam, M., Konda, S., Tiesman, H.M. and Fahy, R. (2018), 'Nonfatal injuries to firefighters treated in US emergency departments, 2003–2014', *American Journal of Preventive Medicine*, **55** (3), 353–60.

May, D.R., Gilson, R.L. and Harter, L.M. (2004), 'The psychological conditions of meaningfulness, safety and availability and the engagement of the human spirit at work', *Journal of Occupational and Organizational Psychology*, **77** (1), 11–37.

Mullen, J. and Kelloway, E.K. (2011), 'Occupational health and safety leadership', in J. Campbell Quick and L. Tetrick (eds), *Handbook of Occupational Health Psychology*, 2nd edn, Washington, DC: APA, pp. 357–72.

Myers, K.K. (2005), 'A burning desire', *Management Communication Quarterly*, **18** (3), 344–84.

Nahrgang, J.D., Morgeson, F.P. and Hofmann, D.A. (2011), 'Safety at work: a meta-analytic investigation of the link between job demands, job resources, burnout, engagement, and safety outcomes', *Journal of Applied Psychology*, **96** (1), 71–94, doi:10.1037/a0021484.

Offermann, L.R. and Hellmann, P.S. (1996), 'Leadership behavior and subordinate stress: a 360□ view', *Journal of Occupational Health Psychology*, **1** (4), 382–90.

Oginska-Bulik, N. (2005), 'Emotional intelligence in the workplace: exploring its effects on occupational stress and health outcomes in human service workers', *International Journal of Occupational Medicine and Environmental Health*, **18** (2), 167–75.

Parker, S.K., Axtell, C.M. and Turner, N. (2001), 'Designing a safer workplace: importance of job autonomy, communication quality, and supportive supervisors', *Journal of Occupational Health Psychology*, **6** (3), 211–28.

Panatik, S.A.B., Badri, S.K.Z., Rajab, A., Rahman, H.A. and Shah, I.M. (2011), 'The impact of work–family conflict on psychological well-being among school teachers in Malaysia', *Procedia-Social and Behavioral Sciences*, **29**, 1500–1507.

Poms, L.W., Fleming, L.C. and Jacobsen, K.H. (2016), 'Work–family conflict, stress, and physical and mental health: a model for understanding barriers to and opportunities for women's well-being at home and in the workplace', *World Medical & Health Policy*, **8** (4), 444–57.

Poplin G.S., Harris R.B., Pollack K.M., Peate W.F. and Burgess J.L. (2012), 'Beyond the fireground: injuries in the fire service', *Injury Prevention*, **18** (4), 228–33.

Regehr, C., Hill, J., Knott, T. and Sault, B. (2003), 'Social support, self-efficacy and trauma in new recruits and experienced firefighters', *Stress and Health*, **19** (4), 189–93.

Rizzo, J.R., House, R.J. and Lirtzman, S.I. (1970), 'Role conflict and ambiguity in complex organizations', *Administrative Science Quarterly*, **15** (2), 150–63.

Rosmuller, N. and Ale, B.J.M. (2008), 'Classification of fatal firefighter accidents in the Netherlands: time pressure and aim of the suppression activity', *Safety Science*, **46** (2), 282–90.

Schaufeli, W.B. and Bakker, A.B. (2004), 'Job demands, job resources, and their relationship with burnout and engagement: a multi-sample study', *Journal of Organizational Behavior*, **25** (3), 293–315, doi:10.1002/job.248.

Seo, D.C. (2005), 'An explicative model of unsafe work behavior', *Safety Science*, **43** (3), 187–211.

Shreffler, K.M., Meadows, M.P. and Davis, K.D. (2011), 'Firefighting and fathering: work–family conflict, parenting stress, and satisfaction with parenting and child behavior', *Fathering*, **9** (2), 169–88.

Smith, T. and DeJoy, D. (2014), 'Safety climate, safety behaviors and line-of-duty injuries in the fire service', *International Journal of Emergency Services*, **3** (1), 49–64.

Smith, T.D., DeJoy, D.M., Dyal, M.A. and Huang, G. (2017), 'Impact of work pressure, work stress and work–family conflict on firefighter burnout', *Archives of Environmental & Occupational Health*, online 27 November, doi:10.1080/193 38244.2017.1395789.

Smith, T.D. and Dyal, M.A. (2016), 'A conceptual safety-oriented job demands and resources model for the fire service', *International Journal of Workplace Health Management*, **9** (4), 443–60.

Smith, T., Eldridge, F. and DeJoy, D. (2016), 'Safety-specific transformational and passive leadership influences on firefighter safety climate perceptions and safety behavior outcomes', *Safety Science*, **86** (July), 92–7.

Smith, T.D., Hughes, K., DeJoy, D.M. and Dyal, M.A. (2018), 'Assessment of relationships between work stress, work–family conflict, burnout and firefighter safety behavior outcomes', *Safety Science*, **103** (March), 287–92.

Soteriades, E.S., Smith, D.L., Tsismenakis, A.J., Baur, D.M. and Kales, S.N. (2011), 'Cardiovascular disease in US firefighters: a systematic review', *Cardiology in Review*, **19** (4), 202–15.

Stanley, I.H., Hom, M.A., Hagan, C.R. and Joiner, T.E. (2015), 'Career prevalence and correlates of suicidal thoughts and behaviors among firefighters', *Journal of Affective Disorders*, **187** (November), 163–71.

Szubert, Z. and Sobala, W. (2002), 'Work-related injuries among firefighters: sites and circumstances of their occurrence', *International Journal of Occupational Medicine and Environmental Health*, **15** (1), 49–55.

Takahashi, M., Iwasaki, K., Sasaki, T., Kubo, T., Mori, I. and Otsuka, Y. (2011), 'Worktime control-dependent reductions in fatigue, sleep problems, and depression', *Applied Ergonomics*, **42** (2), 244–50, doi:10.1016/j.apergo.2010.06.006.

Torp, S., Grøgaard, J.B., Moen, B.E. and Bråtveit, M. (2005), 'The impact of social and organizational factors on workers' use of personal protective equipment: a multilevel approach', *Journal of Occupational and Environmental Medicine*, **47** (8), 829–37.

Tucker, S., Chmiel, N., Turner, N., Hershcovis, M.S. and Stride, C.B. (2008), 'Perceived organizational support for safety and employee safety voice: the mediating role of coworker support for safety', *Journal of Occupational Health Psychology*, **13** (4), 319–30.

Tuckey, M.R. and Hayward, R. (2011), 'Global and occupation-specific emotional resources as buffers against the emotional demands of fire-fighting', *Applied Psychology: An International Review*, **60** (1), 1–23.

United States Fire Administration (USFA) (2002), 'Firefighter fatality retrospective study', USFA and Federal Emergency Management Agency (FEMA), Emmitsburg, MD.

Varvel, S.J., He, Y., Shannon, J.K., Tager, D., Bledman, R.A., Chaichanasakul, A. et al. (2007), 'Multidimensional, threshold effects of social support in firefighters: is more support invariably better?', *Journal of Counseling Psychology*, **54** (4), 458–65.

Vaulerin, J., d'Arripe-Longueville, F., Emile, M. and Colson, S.S. (2016), 'Physical exercise and burnout facets predict injuries in a population-based sample of

French career firefighters', *Applied Ergonomics*, **54** (May), 131–5, doi:10.1016/j. apergo.2015.12.007.

Walton, S.M., Conrad, K.M., Furner, S.E. and Samo, D.G. (2003), 'Cause, type, and workers' compensation costs of injury to fire fighters', *American Journal of Industrial Medicine*, **43** (4), 454–8.

Watterson, A. (2015), 'Briefing: firefighter fatalities at fires in the United Kingdom: 2004–2013: voices from the fireground', report, Fire Brigade's Union, Kingston upon Thames, accessed 18 September 2018 at https://www.fbu.org.uk/publication/briefing-firefighter-fatalities-fires-uk-2004-2013-voices-fireground.

Weick, K.E. and Roberts, K.H. (1993), 'Collective mind in organizations: heedful interrelating on flight decks', *Administrative Science Quarterly*, **38** (3), 357–81.

Yarnal, C.M., Dowler, L. and Hutchinson, S. (2004), 'Don't let the bastards see you sweat: masculinity, public and private space, and the volunteer firehouse', *Environment and Planning A: Economy and Space*, **36** (4), 685–99.

Zohar, D. (2002), 'The effects of leadership dimensions, safety climate, and assigned priorities on minor injuries in work groups', *Journal of Organizational Behavior*, **23** (1), 75–92.

Zohar, D. (2003), 'Safety climate: conceptual and measurement issues', in J.C. Quick and L.E. Tetrick (eds), *Handbook of Occupational Health Psychology*, Washington, DC: APA, pp. 123–42.

PART IV

Building Safety Climates, Promoting Worker
Health and Changing Workplace Cultures

15. Types of safety cultures and best practice suggestions

Sharon Clarke

Safety culture reflects core values and assumptions about organizational safety (Reason 1997; Clarke 1999, 2000; Guldenmund 2000, 2007). Based on a review of the safety culture and climate literature, and drawing on theories of organizational culture, Guldenmund (2000) describes safety culture as comprising three layers: core assumptions, which are implicit and pre-conscious in nature; espoused values and attitudes, which are explicit and conscious; and, artefacts, which are visible and reflect the underlying culture, such as organizational statements, use of safety equipment and employee behavioral norms. A key aspect of safety culture is that it is shared across the organization, and reflected in organizational safety policies, procedures and practices, such as incident reporting and sharing of safety information (Reason 1997). There is considerable evidence linking safety culture to organizational safety, though most has derived from retrospective analyses of major accidents (for example, Reason 1997; Hopkins 2008, 2012). These analyses have demonstrated how the influence of safety culture is manifest in failures throughout the organization (for example, communication, managing and organizing, leadership, and behavioral norms) in the context of a major accident. However, these analyses also indicate that such failures often exist within organizations for long periods of time, and are not readily recognized by the organization due to the prevailing safety culture.

TYPES OF SAFETY CULTURE

To better understand the nature of safety culture, and assist organizations to identify and improve their culture, researchers have developed typologies of safety culture. One of the earliest and most influential was developed by Westrum (1993). It describes three types of safety culture: pathological, bureaucratic and, generative. These types are distinguished by the way in which organizations manage safety-related information (for example, whether new ideas are actively crushed, create problems or are welcomed). In generative organizations, speaking up is encouraged, safety-related information is actively sought, failures thoroughly investigated and lessons learnt.

Based on Westrum's (1993) typology, Parker et al. (2006) developed a safety culture maturity model that describes different types of safety culture, and proposes that the model maps how safety culture is developed and matures over time. Derived from in-depth interviews with 26 oil and gas company executives, the typology of five safety cultures comprises: pathological, reactive, calculative, proactive and generative. This typology is based on 11 dimensions, and so draws on an expanded set of cultural indicators, including: benchmarking, audits and reviews, incident and accident reporting, hazard reporting, work planning and contractor management. Pathological organizations are reluctant compliers with legal requirements and value getting the job done over safety considerations; reactive organizations meet the legal requirements, and respond to failures, but do little more than required by law; calculative organizations operate according to the rules and are concerned with data collection, benchmarking, tabulating and monitoring accidents and audit data, and having procedures in place to ensure safety; proactive organizations are concerned to be the best in the industry and wish to benchmark themselves against their competitors, with a wish to understand why incidents happen and learn lessons from them, and they see audits as necessary; generative organizations benchmark outside the industry, focus on continuous improvement and encourage speaking up at all levels in the organization. The model forms the basis of a self-assessment tool for organizations to identify strengths and weaknesses, and so take actions to enable movement up the levels of maturity (Parker et al. 2006). These ideas have led to a number of subsequent, similar safety culture maturity models with applications in different sectors, including construction and healthcare, as well as the oil and gas industry.

A recent review of these models concluded that, while popular, there has been little published validity evidence supporting a link between safety culture maturity and safety performance (Filho and Waterson 2018). For example, although a basic tenet of this type of model is that safety culture matures over time from low to high levels, there is little evidence to suggest this is how safety culture actually develops or changes over time. Filho and Waterson (2018) identify one publication in their review that reports both reliability and validity information, but note that none report predictive validity, in terms of safety-related outcomes. Thus, while appealing to organizations, there are few published examples of organizations that have successfully matured through the application of safety culture improvement efforts leading to better safety performance outcomes. Nevertheless, there is evidence to suggest that organizations with more sophisticated safety management systems, high-performance human resource (HR) practices and greater investment in safety initiatives (which might be described as having a higher level of cultural maturity) are more receptive to safety interventions, such as training (for example, Burke et al. 2008).

Researchers (for example, Petitta et al. 2017a; 2017b) have responded to the need for greater attention to exploring the link between safety culture and safety performance in organizations. Petitta et al. developed a model of safety culture, which describes five dimensions of organizational culture (autocratic, bureaucratic, clan-patronage, technocratic and cooperative) within a safety context, based on the ways in which safety rules are developed and implemented. The five types comprise: autocratic (safety behaviors are indicated by the boss); bureaucratic (rules and safety procedures are proposed by the organization); clan-patronage (there is one set of safety rules when employees are with outsiders, but different practices within the group); technocratic (safety practices are followed only if this does not impede the achievement of the result and the progress of the work); and cooperative (all members of the organization are proactively involved in the diffusion and adoption of safety practices). In an empirical study, Petitta et al. (2017a) found that accident under-reporting, at an organizational level (N = 28), was positively correlated with all types of culture (except the autocratic culture); given the small sample size, there were significant correlations for technocratic and clan-patronage only, but a moderate non-significant positive correlation was also reported for the cooperative culture (r = .34 ns). The only culture which was associated with reduced accident under-reporting at an organizational level was the autocratic culture, although this was non-significant (r = −.18 ns). At the individual level, accident under-reporting was not significantly correlated with any of the culture types, except a positive correlation with technocratic culture, suggesting that accident under-reporting by individuals was most associated with cultures which endorsed getting the job done more strongly than it endorsed following safety procedures. In addition, strong endorsement of following safety rules (by the boss) was the only significant cultural influence on accident under-reporting at an organizational level. Correlations between the different types of culture also suggest the possibility of disparate values coexisting within an organization (for example, the cooperative culture was significantly correlated with autocratic, bureaucratic and clan-patronage at an organizational level), so that organizations can attach significant value to both reliability and flexibility. Indeed, this would be consistent with high-reliability organizations (HROs) which manage to balance the need for reliability with flexibility.

The study signals the need for further research in this area, given some exploratory findings, but with a small sample of organizations and being restricted by cross-sectional self-reported data. For example, none of the safety culture types were (significantly) negatively associated with accident under-reporting (only autocratic culture at a non-significant level), even though a reporting culture is a necessary aspect of an effective safety culture (Reason 1997). Speaking-up (providing safety-related information) is also an indicator of an effective safety culture (such as a generative culture, which encourages and rewards speaking-up; Westrum 1993), yet the autocratic culture does not seem to reflect such an open and responsive safety culture, which would appear closer to the cooperative culture (despite this type of culture being unrelated to accident under-reporting).

In addition, in defining key aspects of safety culture, theorists have rarely addressed the challenges associated with managing safety in global environments. Reader et al. (2015) argue that the safety culture construct has been largely developed in Western cultures, and does not adequately reflect the influence of national cultural values. They developed a safety culture model for the air traffic control industry, which they found operated effectively across different European countries, but safety culture perceptions differed depending on national cultural dimensions. Noort et al. (2016) focused on the influence on uncertainty avoidance (UA) as a dimension of national culture; they expected high UA to be associated with a reduced tendency to speak up or admit errors, and feelings of chronic unease (Fruhen et al. 2014), and thereby affect safety culture. They found that high UA was associated with lower perceptions of safety culture, and so suggested that UA may shape safety practices in an organization and also influence employees' attitudes towards safety practices. These findings have significant implications for organizations that operate in a global environment, particularly those that operate across national boundaries. Companies that attempt to export their corporate safety culture may experience unanticipated problems, where it is assumed that cultural values will translate to different national cultures, or be interpreted in the same way.

Reason (1997) describes the need for constant vigilance and not forgetting to be afraid as necessary for an effective safety culture. Fruhen et al. (2014) define a manager's chronic unease as a strain reaction to the management of risks where managers experience uneasiness about the management of safety risks and to what extent this might inform their managerial actions. The authors argue that it is possible to have a healthy approach to risk management that does not result in significant emotional distress, but that managers should take their (sometimes non-specific) unease about safety risks into account in making risk-related decisions. Thus, particularly in high-risk contexts, there is a balance to be achieved in being vigilant, while avoiding an over-emphasis on identifying potential threats to system safety.

SAFETY CULTURE IN SAFETY-CRITICAL ORGANIZATIONS

High-reliability organizations are organizations that operate in high-risk environments but with a very low level of safety-related incidents. Similar to the generative organization, HROs have an effective safety culture that focuses on continuous improvement and being mindful about safety (Weick et al. 1999). However, HROs operate on specific safety principles, such as in the way that they are organized (for example, simultaneous centralization and decentralization), which allows them to balance their routine operations (focus on centralized control and standard operating procedures), with responses both to known emergencies and to unknown threats (focus on decentralized control, where expertise and local experience takes precedence).

In safety-critical organizations, Grote (2007, 2015) argues that the way in which organizations manage uncertainty determines the role of safety culture; where the approach may focus on minimizing uncertainty versus competent coping with uncertainty (Grote 2007): minimization of uncertainty depends heavily on planning and automation to reduce degrees of freedom, so that employees need to follow predetermined rules and regulations, with little opportunity for autonomy; coping with uncertainty as an alternative approach, allows greater autonomy for employees to manage uncertainty on a localized basis. Most high-risk systems focus on standard operating procedures and strict rule-following (which is seen as the benchmark of reliability and safety), with reliance on a shared understanding of safety culture through a knowledge of the rules and procedures at team level. However, there can be negative consequences of such an approach as it can lead to an over-reliance on rules, including rote rule-following, or a lack of adaptation in non-routine situations when this is needed. Grote (2007) argues for an approach based on loose coupling, as in HROs, where a different form of organization and planning is advocated to maintain a high level of safety.

The implications of how uncertainty is managed within organizations, depends on the level of uncertainty likely to be encountered by organizations. In safety-critical contexts, where uncertainty is high, there is a need to organize in such a way that teams are able to adapt effectively, and have the autonomy at all levels to do so, including operational autonomy (that is, control over how rules are followed at an operational level, such as determining how goals are achieved). At low levels of uncertainty, however, operational autonomy can be counter-productive and have negative effects on safety in a context where predetermined rules define the safest means of operating, and deviations from rules are likely to lead to accidents and injuries (Grote 2007). Some forms of autonomy have a positive effect on safety regardless of the level of uncertainty, such as higher level autonomy (that is, involvement in decision making about how rules are implemented at their operational level), and others are critical for safety in high uncertainty (for example, local control) but have little effect on safety in low uncertainty (Grote 2007).

This line of work focuses on autonomy in the context of rule-following, but it should be recognized that there are other types of safety behavior, in addition to compliance, that are critical for promoting an effective safety culture (for example, safety participation), in which autonomy may play a role (for example, employees have greater freedom to make safety suggestions, or be proactive in relation to safety). In addition, organizations may face situations that fluctuate in the degree of uncertainty (that is, even safety-critical organizations face both certain and uncertain situations, and so must have a degree of flexibility to manage uncertainty when it arises). For example, HROs have the flexibility to reorganize in the face of known and unknown risks. Grote (2015) discusses different ways in which organizations may manage uncertainty and how they might switch between different modes of operation.

Literature on HROs has suggested that this type of organization engages in a process of continuous change and improvement. In contrast, Griffin et al. (2016) describe the idea of dynamic safety capability which reflects how organizations may make more radical changes in response to major shifts in their industry or more broadly. They define dynamic safety capability as 'an organization's capacity to generate, reconfigure, and adapt operational routines to sustain high levels of safety performance in environments characterized by change and uncertainty' (Griffin et al. 2016, p.249). Furthermore, they argue that this approach requires more emphasis on safety culture as it involves preparing an organization for future change and adaptation. This theoretical framework is consistent with the literature on HROs, which emphasizes a focus on vigilance and anticipating change (Weick et al. 1999), and the principles of resilience engineering (Hollnagel et al. 2011; Hollnagel, 2014). Thus, dynamic safety capability provides a framework for further research to explore how these processes operate in practice within teams and how radical changes at an organizational level translate into improved safety performance for organizations.

SAFETY CULTURE AND THE DEVELOPMENT OF SAFETY CLIMATE

Based on an integration of systems theory and psychological approaches to safety, and incorporating ideas from the HRO literature, Casey et al. (2017) developed a safety culture model that describes how types of safety climate are associated with different control strategies. They argue that the layer metaphor of safety culture (for example, Guldenmund 2000) provides an overly static view of safety culture, and does not reflect the interactions between layers, or the interactions between elements nested within layers. They describe organizational (that is, HR practices, policies and procedures), social (that is, leadership and teamwork) and human (that is, expertise, motivation and energy of people) capital as built upon the core assumptions that organizational members share about the meaning of safety. Safety climate is described as located within social capital, and defined as 'perceptions of behavioral norms and espoused values around safety, aggregated at different levels of the organization (for example, team, department, company)' (Casey et al. 2017, p. 345). Safety culture is conceptualized as a distal control mechanism, which is embedded in organizational functions, and acts to optimize the balance between productivity and efficiency versus safety. Safety climate is conceptualized as a proximal control mechanism, which can be more readily modified through 'specific organizational, supervisor, and co-worker practices' (Casey et al. 2017, p. 346). The model maintains a social exchange perspective, in which employees act safely as they believe they are obligated to do so, in return for their managers' perceived commitment to safety.

Safety culture is defined in terms of two dimensions: reliability versus flexibility and promotion versus prevention. Reliability reflects top-down control using rules and procedures to cope with low-uncertainty routine operations, while flexibility comprises bottom-up adaptive responses in highly uncertain non-routine situations, where employees have autonomy in the form of local control. Promotion is a focus on growth and continual development, and prevention is a focus on minimizing risk from economic, social and environmental threat. The interaction of these two dimensions is used to define four types of control strategy. While it is noted that these control strategies are best used in balance and combination, Casey et al. (2017) describe different types of safety climate that are associated with each of the control strategies: defend (reliability and prevention); adapt (flexibility and prevention); leverage (reliability and promotion); and, energize (flexibility and promotion).

While consistent with previous conceptualizations of the relationship between safety culture and safety climate (for example, Guldenmund

2000), this model suggests that culture and climate are closely aligned and that culture is something of a singularity, which does not necessarily align with Reason's (1997) conceptualization of multiple safety cultures within an organization, or with the view that creating uncertainty is a better way of ensuring safety in complex systems, than attempting to reduce or control it (Grote 2015). Nevertheless, these theoretical developments are needed to prompt further empirical research to explore the nature of safety culture and its relationship with safety climate.

Very few studies have measured both safety climate and safety culture within an empirical study, and can provide evidence of the relationship between them. One exception is a study by Petitta et al. (2017b), which tested a multi-level model in which organizational safety culture and safety climate both acted as moderators of the relationship between supervisor enforcement and safety compliance (as well as a direct effect being modeled). In this study, perceptions of safety culture and safety climate were aggregated to the organizational level (N = 32). Both safety climate and supervisor enforcement were found to have a significant direct effect on safety compliance under all safety cultures. Autocratic and bureaucratic cultures had a positive effect on safety compliance, while cooperative and clan-patronage cultures had no significant effect; technocratic culture had a significant negative effect on safety compliance. In addition to testing direct effects, the study examined the moderation effects of safety culture. In some cultures (such as autocratic and bureaucratic cultures which place value on rule following), the effects of safety climate and supervisor enforcement were nullified (possibly as these were not needed to ensure safety compliance). A clan-patronage culture had no effect on safety compliance where there is positive safety climate and safety enforcement by supervisors; similarly, the negative effects of a technocratic culture were nullified by a positive safety climate and supervisory enforcement. Perhaps because cooperative safety culture is more likely to generate a positive safety climate, safety compliance was encouraged through safety climate and supervisory actions (but safety compliance must be enforced through safety climate and supervisory enforcement as this cannot be achieved through a cooperative climate alone). As regards the relationship between cultures and climate, positive safety climates were most likely to develop in autocratic, bureaucratic and cooperative cultures (these were all positively correlated at an organizational level, and at an individual level), and negative safety climate was most likely to develop in a technocratic culture (negatively correlated at organizational and individual level).

Given these interesting explorative findings, further research is needed to extend the consideration of safety behaviors beyond safety compliance, and look at the combined effects of safety culture and safety climate on a range of other safety behaviors (such as safety citizenship behaviors). The existing research has also taken a limited view of supervisory leadership behaviors (measuring supervisory enforcement of rule following), but might be extended to other types of leader behaviors. For example, Grill et al. (2017) found that while rule-orientated leadership was effective in enhancing safety behaviors and reducing accidents, it was particularly effective when rules and procedures had been established in a collaborative manner (participative leadership moderated the link between rule-orientated leadership and safety outcomes). Participative leadership in this study reflected the participative decision-making dimension of empowering leadership.

SAFETY CULTURE AND TEAM SAFETY CLIMATE

Much of our understanding of how teams operate reliably and safely derives from research on teams in high-risk contexts, such as in aviation, air traffic control, process industries and medicine (including healthcare more generally). In these contexts, understanding safety at a team level is necessary, as work activities are interdependent, and teams must cooperate and coordinate effectively to resolve situations safely (for example, surgical teams). However, all types of organizations depend increasingly on teams to achieve their organizational goals. Despite safety climate being a group-level variable (given that it is defined in terms of shared perceptions of the priority of safety), the psychological research has placed more emphasis on understanding individual safety behavior than team processes in relation to safety. For example, based on a review of the psychological safety literature (Beus et al. 2016), the integrative model of this research conflates group and organizational levels into contextual factors acting upon the individual. However, the review highlighted group-level mechanisms as missing links and suggested that individual-level antecedents might be considered at the group level (for example, shared mental models and team cohesion).

There has been limited empirical research looking at team-level processes, but the few studies in this area indicate a significant role for team-level variables, such as team cohesion. Yagil and Luria (2010) examined the effect of social relationships with colleagues on employees' safety compliance; they found that this relationship was moderated by safety climate, such that the relationship was strengthened when safety climate was low. Therefore, when employees perceived that social bonds with team members were stronger (for example, co-workers give sound advice) this led to higher safety compliance when perceptions of safety climate were low. The findings suggested that team cohesion boosted safety compliance in the absence of positive safety climate. Similarly, Fruhen and Keith (2014) measured team-level task cohesion (related to perceptions of colleagues' safety attitudes) and social cohesion (related to social relationships outside work); they found that task cohesion had a significant negative correlation with accident occurrence (such that cohesive teams reported fewer accidents), but social cohesion had no significant relationship with accident occurrence. Thus, team-level variables, particularly team cohesion, have significant implications for safety-related outcomes, but have lacked systematic, theory-driven research efforts.

Contextual antecedents of safety climate, including the influence of organizational policies and practices, and leadership (for example, Christian et al. 2009; Clarke 2010), are largely conceptualized to influence safety behaviors at an individual level, and emphasize psychological mechanisms, such as social exchange between the leader and the individual employee. However, in an era of safety management characterized by the Safety-II resilient engineering approach (Hollnagel et al. 2011; Hollnagel 2014), there might be greater focus on building capacity and resilience within teams, and recognizing the need for empowering leadership to encourage proactivity and engagement with safety. Research also needs to include team-level processes and behaviors, such as shared mental models, team learning, co-worker support, resilience and team knowledge. Although conceptualized as a group-level variable, team safety climate is most often operationalized as aggregated across individuals to represent shared perceptions, rather than measured using team-level referents.

Indicators of an effective safety culture, such as speaking up, safety initiative and employee participation, have been explored in the general psychological literature (for example, work on organizational citizenship behavior, employee voice and empowering leadership). This work suggests that these factors are facilitated by a supportive team climate, and can be conceptualized at a team level, although there has been less research attention within the safety literature.

EMPOWERING LEADERSHIP AND SAFETY

There is a considerable body of research which has emphasized the importance of safety leadership in terms of developing positive safety climate and encouraging safety behaviors. A meta-analysis of the safety leadership literature (Clarke 2013) showed that both transformational (for example, inspiring and motivating leader behaviors) and active transactional (for example, proactive monitoring) leadership was associated with safety behaviors. The latter type of leader behaviors, which emphasize vigilance, close monitoring and intervening to avoid errors, were most associated with effective leadership in safety-critical contexts (Willis et al. 2017). Similarly, Martínez-Córcoles and Stephanou (2017) found that active transactional leadership was associated with safety climate perceptions and safety behaviors (safety compliance, safety participation and risky behaviors) in a sample of military parachutists (that is, in a high-risk military context).

Few studies have examined other forms of leadership, such as participative or empowering leadership. Participative leadership, which encourages participative decision making, has been shown to be particularly effective in combination with rule-orientated leadership (Grill et al. 2017). In addition, there have been studies that have examined the influence of empowering leaders in a safety context, including at team level. For example, Martínez-Córcoles et al. (2013) examined empowering team leadership, defined as the following leader behaviors: lead by example (for example, demonstrating commitment to safety); encourage team members to express their opinions and be involved in decision making (for example, promoting speaking up and employee voice); coaching (for example, encouraging team members to engage in problem solving); informing team members about the organization's mission and vision; and, showing concern and providing support for employees (for example, engaging in safety-related dialogue). The study found that empowering team leadership was related to safety compliance, safety participation, and (low) risk-taking behavior in a high-reliability context (nuclear power plants). These studies demonstrate a potential role for this form of leadership in safety-critical contexts, particularly in terms of empowerment through high-level autonomy (Grote 2007), such as encouraging employee involvement in decision making and speaking up. In the general leadership literature, empowering leadership has been linked to team effectiveness, by promoting team-level processes, including the development of shared mental models, team learning and coordination over time (Lorinkova et al. 2013).

SAFETY CITIZENSHIP BEHAVIORS, SPEAKING UP AND EMPLOYEE VOICE

Safety research has supported the importance of safety behaviors beyond compliance, broadly defined as extra-role or organizational citizenship behaviors (OCBs), which contribute to organizational safety by providing feedback and suggestions for safety improvements. Studies have mainly focused on individual-level processes, such as social exchange between organization and the individual worker, although some studies have referred to team-level constructs, such as team safety climate, they rarely take a team-level perspective. For example, Curcuruto and Griffin (2018) examined the role of team safety climate (measured at the individual level, reflecting safety commitment and safety expectations) and two forms of safety citizenship (proactive and prosocial). They argued that organizational support for safety participation was an important antecedent (based on social exchange perspective), and found that affective commitment (toward the organization) mediated the effects on prosocial behaviors, while psychological ownership (of safety promotion, that is, extent to which individuals feel personal investment in organizational safety initiatives) mediated the effects on proactive behaviors. Similarly, Reader et al. (2017) focused on employees' perceived organizational support (POS) as an antecedent of employee safety citizenship behaviors (SCBs) from a social exchange perspective (where organizational support is signaled by organizational initiatives to promote employee health). The study supported a serial mediation model in which POS led to commitment to the organization which in turn increased SCBs.

Nevertheless, SCBs are also likely to play a significant role in developing and supporting team safety climate. For example, prosocial (affiliative-orientated behaviors) may facilitate team cohesion, and proactive (change-orientated behaviors) may build collective voice and increase team knowledge. At a team level, the interaction between affiliative and challenging OCBs should also be investigated in the safety context, as research has shown that challenging OCBs at moderate levels has a positive effect on team effectiveness, but at high levels (especially when not accompanied by affiliative OCBs) they can have a negative effect on team effectiveness (MacKenzie et al. 2011). Such findings have significant implications for organizational safety as the safety literature has largely focused on the positive consequences of SCBs, including both challenging and affiliative orientated behaviors, but not necessarily within a team context.

Although the general psychological literature has emphasized prosocial behavior as having positive outcomes, including increased cooperation, it also has downsides (Bolino and Grant 2016). For example, other-orientated individuals may offer assistance when it is not needed, or to their own detriment (Bolino et al. 2015). Prosocial behavior can lead to exhaustion, which in turn leads to increased unethical behaviors or rule breaking in an effort to promote the interests of their own team (Bolino and Klotz 2015). Co-workers' reactions to being offered help should also be considered: co-workers may reject offers of help (for example, owing to diminished image, reciprocity obligation, self-reliance, co-worker mistrust or co-worker incompetence) leading to lowered task performance, and fewer OCBs (Thompson and Bolino 2018). Although some researchers have considered the potential advantages and disadvantages of safety pro-activity in a safety-critical context (Curcuruto and Griffin 2016), further research is warranted, particularly when considering the team context within which SCBs take place.

As in the general literature on organizational citizenship behaviors, little work in the safety literature has examined group-level voice behavior; as noted by Frazier and Bowler (2015), the real value of citizenship behavior is its performance in the aggregate, given that work-group speaking up is more likely to impact on organizational policies and practice. There are risks associated with speaking up (for example, upsetting others; Detert and Burris 2007), and so the social context of speaking up needs to be considered. Frazier and Bowler (2015) showed that the effects of supervisor undermining of group voice was fully mediated by perceptions of group-level voice climate; in addition, group voice behavior partially mediated the effect of voice climate on group performance (supervisor ratings). In a safety context, a recent study examined the effects of group-level voice behavior on team safety performance (Li et al. 2017). Team safety performance was based on audit data reflecting team safety compliance (for example, wearing personal protective equipment). Perceptions of team voice behavior included: prohibitive voice (that is, being vigilant to hazards and bringing them to the attention of the team) and promotive voice (that is, surfacing new ideas that can improve team work-practices). The study found that a team-level process of team monitoring (De Jong and Elfring 2010) acted as mediator between prohibitive team voice and safety performance. Team monitoring was also directly associated with higher subsequent safety performance.

RULE VIOLATIONS, ADAPTATION AND JOB CRAFTING

Most theorizing and research on reliability and flexibility has taken place at an organizational level, with the focus on the relationship with safety culture, and organizational efforts to improve work safety. However, this tends to be based on an assumption that all cultural drivers are top down, without reflecting on the implications of safety culture initiatives to encourage bottom-up processes of safety improvement, such as encouraging safety participation and speaking up about safety. Nevertheless, given the disincentives for speaking up, and the effort involved in safety proactivity, employee-driven adaptations within teams may evolve over time. While originating at an individual level, adaptive processes and procedures are likely to be shared and quickly established as behavioral norms within teams.

Research has shown that employees' actions frequently deviate from safety rules and procedures (Lawton 1998). While these rule deviations are rarely intended to be malevolent, they are often viewed as potentially damaging non-compliance by organizations. Rule violations occur for many reasons, sometimes because employees take short cuts, or are under pressure to put production before safety (Lawton 1998; Reason et al. 1998; Hansez and Chmiel 2010) or use a workaround when experiencing burnout at work (Halbesleben 2010). However, rule violations can be adaptive ways of working that are both more efficient and potentially safer than following the predetermined procedure (Jones et al. 2018). In a qualitative study with pharmacists, Jones et al. (2018) found that pharmacists would violate safety procedures which they felt were unnecessarily time-consuming in order to fit in with the social norms of their colleagues, or when they felt that the safety procedures prevented the running of an efficient service for patients. Pharmacists would make decisions to circumvent procedures when they felt that this was in the patient's best interest, even when this could have serious disciplinary consequences for themselves. Thus, working by their own professional standards and meeting patient needs would lead pharmacists to adapt rules and procedures as they saw fit. Having the training and shared safety knowledge within teams is critical to the development of better ways of working that not only improve efficiency but also protect safety.

Halbesleben (2010) found that emotional exhaustion (an element of burnout) in nurses was predictive of increased safety workarounds, where workarounds were ways of working around a blockage in workflow, such as a safety procedure. In this study, increased safety workarounds led to a subsequent increase in the frequency and severity of occupational injuries over time. However, research on job crafting has shown that employees often take the initiative to make adaptations (or actively craft) their work tasks to manage job demands and increase resources to enhance their well-being, and job performance, through increased work engagement (Petrou et al. 2012). Job crafting enables employees to gain personal meaning and enhanced control over their work by exercising their skills, knowledge and expertise (Wrzesniewski and Dutton 2001). Unlike extra-role behaviors or personal initiative (for example, safety citizenship behaviors or speaking up), which aim to support the organization, job crafting is a bottom-up process in which the focus is on improving employee well-being. There is evidence that this adaptation takes place both at an individual and team level (Tims et al. 2013). Mäkikangas et al. (2017) found that team job crafting was associated with a supportive team climate, team cohesion and participative leadership. Team job crafting is more likely to take place within positive and open cultures (such as a cooperative culture; Petitta et al. 2017a), and where employees have greater autonomy at a local level (Grote 2007). In the context of workplace safety, job crafting could have both positive and negative consequences. For example, using expertise might lead to successful safety workarounds, but could also lead to dangerous practices for the sake of stretching the boundaries. A recognition that adaptations to work practices are likely to occur (for example, Petrou et al. 2012 found that job crafting takes place on a daily basis), and that local control has little potential for ill effects on safety (Grote 2007), might encourage organizations to invest further in their employees' training so that they have the expertise to craft successfully. Hollnagel (2014) advocates such an approach from a resilience engineering perspective, but there is as yet little research that has systematically investigated the impact of adaptations either at an individual or team level to support this approach. Given the pace of change in contemporary organizations, other changes within organizations, such as new equipment, technologies and ways of working, adaptations may be adopted by employees before rules and procedures can change, highlighting the need for further research.

BEST PRACTICE SUGGESTIONS

There is often an ambition for organizations to change their safety culture to ensure organizational safety. However, it needs to be recognized that this is a slow process, which will not provide ready results on the ground. Safety culture interventions need to be implemented from the top down as company values are embedded in its history and are deeply rooted. Transformational change can occur; for example, owing to the installation of a new chief executive officer or as the response to a major accident. However, even this can be a slow process (for example, as discussed by Hopkins, the progress of the changes implemented in BP as responses to the Texas City and the Deep Water Horizon accidents, and the failure to learn from previous major incidents over a number of years; Hopkins 2008, 2012). Nevertheless, changes to more localized and proximal safety climates or through changes to leadership styles, can still be effective, and potentially have more powerful effects on employees' safety behavior than cultural improvements (Petitta et al. 2017b). Safety climate interventions can have broader positive effects, such as improvements in work-related attitudes, and the promotion of employee health and well-being (Clarke 2010), which lead to longer-term improvements in organizational functioning. A positive safety climate also makes individual-level interventions, such as safety training, more effective (Burke et al. 2008).

Safety interventions have focused on improving safety through changes to leader behaviors (for example, Zohar and Polachek 2017), which have an important effect on safety behaviors, including safety participation. However, there is a paucity of potentially effective safety interventions that might focus on encouraging speaking up, developing leaders to be more participative and empowering; and creating greater autonomy in involvement in decision making at higher levels. Thus, organizations should focus on implementing safety interventions aimed at improving safety climate, particularly at a team level, safety leadership and safety participation as best practice. Further work is needed to provide evidence-based advice to organizations implementing interventions with the aim to move towards a Safety-II approach to organizational safety.

FURTHER RESEARCH

Scholars have concluded that we have reached a mature stage in research on safety climate (Hofmann et al. 2017). Nevertheless, despite an understanding that safety climate is important for workplace safety, we still know little about the relationship between safety culture and safety climate, or how these two constructs develop over time. In particular, there is a need for researchers to focus more on team-level safety climate. This is likely to be critical for ensuring workplace safety, especially given the increased focus on teams as the key unit within organizations. Most research work has continued to take an individual-level approach and focus on individual safety behaviors, rather than team-level behaviors. While researchers now have considerable knowledge of the role of safety leadership, we still need further work to understand the role of empowering leadership and autonomy in the context of safety. As organizations increasingly develop as more agile and flexible, and integrate more novel technology and ways of working, there is a significant challenge for organizational safety research to keep up with these developments.

REFERENCES

Beus, J.M., McCord, M.A. and Zohar, D. (2016), 'Workplace safety: a review and research synthesis', *Organizational Psychology Review*, **6** (4), 352–81, doi:10.1177/2041386615626243.
Bolino, M.C. and Grant, A.M. (2016), 'The bright side of being prosocial at work, and the dark side, too: a review and agenda for research on other-oriented motives, behavior, and impact in organizations', *Academy of Management Annals*, **10** (1), 599–670, doi:10.1080/19416520.2016.1153260.

Bolino, M.C. and Klotz, A.C. (2015), 'The paradox of the unethical organizational citizen: the link between organizational citizenship behavior and unethical behavior at work', *Current Opinion in Psychology*, 6 (March), 45–9, doi:10.1016/j. copsyc.2015.03.026.

Bolino, M.C., Hsiung, H.H., Harvey, J. and LePine, J.A. (2015), '"Well, I'm tired of tryin'!" Organizational citizenship behavior and citizenship fatigue', *Journal of Applied Psychology*, 100 (1), 56–74, doi:10.1037/a0037583.

Burke, M.J., Chan-Serafin, S., Salvador, R., Smith, A. and Sarpy, S.A. (2008), 'The role of national culture and organizational climate in safety training effectiveness', *European Journal of Work and Organizational Psychology*, 17 (1), 133–52, doi:10.1080/13594320701307503.

Casey, T., Griffin, M.A., Flatau Harrison, H. and Neal, A. (2017), 'Safety climate and culture: integrating psychological and systems perspectives', *Journal of Occupational Health Psychology*, 22 (3), 341–53, doi:10.1037/ocp0000072.

Christian, M.S., Bradley, J.C., Wallace, J.C. and Burke, M.J. (2009), 'Workplace safety: a meta-analysis of the roles of person and situation factors', *Journal of Applied Psychology*, 94 (5), 1103–27, doi:10.1037/a0016172.

Clarke, S. (1999), 'Perceptions of organizational safety: implications for the development of safety culture', *Journal of Organizational Behavior*, 20 (2), 185–98, doi:10.1002/(SICI)1099-1379(199903)20:2<185::AID-JOB892>3.0.CO;2-C.

Clarke, S. (2000), 'Safety culture: under-specified and overrated?', *International Journal of Management Reviews*, 2 (1), 65–90, doi:10.1111/1468-2370.00031.

Clarke, S. (2010), 'An integrative model of safety climate: linking psychological climate and work attitudes to individual safety outcomes using meta-analysis', *Journal of Occupational and Organizational Psychology*, 83 (3), 553–78, doi:10.1348/096317909X452122.

Clarke, S. (2013), 'Safety leadership: a meta-analytic review of transformational and transactional leadership styles as antecedents of safety behaviours', *Journal of Occupational and Organizational Psychology*, 86 (1), 22–49, doi:10.1111/j.20448325.2012.02064.x.

Curcuruto, M. and Griffin, M.A. (2016), 'Safety proactivity in organizations: the initiative to improve individual, team and organizational safety', in S.K. Parker and U.K. Bindl (eds), *Proactivity at Work: Making Things Happen in Organizations*, New York: Routledge, pp. 105–37.

Curcuruto, M. and Griffin, M.A. (2018), 'Prosocial and proactive "safety citizenship behaviour" (SCB): the mediating role of affective commitment and psychological ownership', *Safety Science*, 104 (April), 29–38, doi:10.1016/j.ssci.2017.12.010.

De Jong, B.A. and Elfring, T. (2010), 'How does trust affect the performance of ongoing teams? The mediating role of reflexivity, monitoring, and effort', *Academy of Management Journal*, 53 (3), 535–49, doi:10.5465/amj.2010.51468649.

Detert, J.R. and Burris, E.R. (2007), 'Leadership behavior and employee voice: is the door really open?', *Academy of Management Journal*, 50 (4), 869–84, doi:10.5465/amj.2007.26279183.

Filho, A.P.G. and Waterson, P. (2018), 'Maturity models and safety culture: a critical review', *Safety Science*, 105 (June), 192–211, doi:10.1016/j.ssci.2018.02.017.

Frazier, M.L. and Bowler, W.M. (2015), 'Voice climate, supervisor undermining, and work outcomes: a group-level examination', *Journal of Management*, 41 (3), 841–63, doi:10.1177/0149206311434533.

Fruhen, L.S. and Keith, N. (2014), 'Team cohesion and error culture in risky work environments', *Safety Science*, 65 (June), 20–27, doi:10.1016/j.ssci.2013.12.011.

Fruhen, L., Flin, R. and McLeod, R. (2014), 'Chronic unease for safety in managers: a conceptualisation', *Journal of Risk Research*, **17** (8), 969–79, doi:10.1080/13669877.2013.822924.

Griffin, M.A., Cordery, J. and Soo, C. (2016), 'Dynamic safety capability: how organizations proactively change core safety systems', *Organizational Psychology Review*, **6** (3), 248–72, doi:10.1177/2041386615590679.

Grill, M., Pousette, A., Nielsen, K., Grytnes, R. and Törner, M. (2017), 'Safety leadership at construction sites: the importance of rule oriented and participative leadership', *Scandinavian Journal of Work, Environment & Health*, **43** (4), 375–84, doi:10.5271/sjweh.3650.

Grote, G. (2007), 'Understanding and assessing safety culture through the lens of organizational management of uncertainty', *Safety Science*, **45** (6), 637–52, doi:10.1016/j.ssci.2007.04.002.

Grote, G. (2015), 'Promoting safety by increasing uncertainty – implications for risk management', *Safety Science*, 71 (part B), 71–9, doi:10.1016/j.ssci.2014.02.010.

Guldenmund, F.W. (2000), 'The nature of safety culture: a review of theory and research', *Safety Science*, **34** (1), 215–57, doi:10.1016/S0925-7535(00)00014-X.

Guldenmund, F.W. (2007), 'The use of questionnaires in safety culture research: an evaluation', *Safety Science*, **45** (6), 723–43, doi:10.1016/j.ssci.2007.04.006.

Halbesleben, J.R.B. (2010), 'The role of exhaustion and workarounds in predicting occupational injuries: a cross-lagged panel study of health care professionals', *Journal of Occupational Health Psychology*, **15** (1), 1–16, doi:10.1037/a0017634.

Hansez, I. and Chmiel, N. (2010), 'Safety behavior: job demands, job resources, and perceived management commitment to safety', *Journal of Occupational Health Psychology*, **15** (3), 267–78, doi:10.1037/a0019528.

Hofmann, D.A., Burke, M.J. and Zohar, D. (2017), '100 years of occupational safety research: from basic protections and work analysis to a multilevel view of workplace safety and risk', *Journal of Applied Psychology*, **102** (3), 375–88, doi:10.1037/apl0000114.

Hollnagel, E. (2014), *Safety-I and Safety-II: The Past and Future of Safety Management*, Farnham: Ashgate.

Hollnagel, E., Paries, J., Woods, D.D. and Wreathall, J. (2011), *Resilience Engineering in Practice: A Guidebook*, Burlington, VT: Ashgate.

Hopkins, A. (2008), *Failure to Learn: The BP Texas City Refinery Disaster*, Sydney: CCH Australia.

Hopkins, A. (2012), *Disastrous Decisions: The Human and Organisational Causes of the Gulf of Mexico Blowout*, Sydney: CCH Australia.

Jones, C.E.L., Phipps, D.L. and Ashcroft, D.M. (2018), 'Understanding procedural violations using Safety-I and Safety-II: the case of community pharmacies', *Safety Science*, **105** (June), 114–20, doi:10.1016/j.ssci.2018.02.002.

Lawton, R. (1998), 'Not working to rule: understanding procedural violations at work', *Safety Science*, **28** (2), 77–95, doi:10.1016/S0925-7535(97)00073-8.

Li, A.N., Liao, H., Tangirala, S. and Firth, B.M. (2017), 'The content of the message matters: the differential effects of promotive and prohibitive team voice on team productivity and safety performance gains', *Journal of Applied Psychology*, **102** (8), 1259–70, doi:10.1037/apl0000215.

Lorinkova, N.M., Pearsall, M.J. and Sims, H.P. (2013), 'Examining the differential longitudinal performance of directive versus empowering leadership in teams', *Academy of Management Journal*, **56** (2), 573–96, doi:10.5465/amj.2011.0132.

MacKenzie, S.B., Podsakoff, P.M. and Podsakoff, N.P. (2011), 'Challenge-oriented organizational citizenship behaviors and organizational effectiveness: do challenge-oriented behaviors really have an impact on the organization's bottom line?', *Personnel Psychology*, **64** (3), 559–92, doi:10.1111/j.1744-6570.2011.01219.x.

Mäkikangas, A., Bakker, A.B. and Schaufeli, W.B. (2017), 'Antecedents of daily team job crafting', *European Journal of Work and Organizational Psychology*, **26** (3), 421–33, doi:10.1080/1359432X.2017.1289920.

Martínez-Córcoles, M. and Stephanou, K. (2017), 'Linking active transactional leadership and safety performance in military operations', *Safety Science*, **96** (July), 93–101, doi:10.1016/j.ssci.2017.03.013.

Martínez-Córcoles, M., Gracia, F.J., Tomás, I., Peiró, J.M. and Schöbel, M. (2013), 'Empowering team leadership and safety performance in nuclear power plants: a multilevel approach', *Safety Science*, **51** (1), 293–301, doi:10.1016/j.ssci.2012.08.001.

Noort, M.C., Reader, T.W., Shorrock, S. and Kirwan, B. (2016), 'The relationship between national culture and safety culture: implications for international safety culture assessments', *Journal of Occupational and Organizational Psychology*, **89** (3), 515–38, doi:10.1111/joop.12139.

Parker, D., Lawrie, M. and Hudson, P. (2006), 'A framework for understanding the development of organisational safety culture', *Safety Science*, **44** (6), 551–62, doi:10.1016/j.ssci.2005.10.004.

Petitta, L., Probst, T. and Barbaranelli, C. (2017a), 'Safety culture, moral disengagement, and accident under-reporting', *Journal of Business Ethics*, **141** (3), 489–504, doi:10.1007/s10551-015-2694-1.

Petitta, L., Probst, T.M., Barbaranelli, C. and Ghezzia, V. (2017b), 'Disentangling the roles of safety climate and safety culture: multi-level effects on the relationship between supervisor enforcement and safety compliance', *Accident Analysis and Prevention*, **99** (February), 77–89, doi:10.1016/j.aap.2016.11.012.

Petrou, P., Demerouti, E., Peeters, M.C.W., Schaufeli, W.B. and Hetland, J. (2012), 'Crafting a job on a daily basis: contextual correlates and the link to work engagement', *Journal of Organizational Behavior*, **33** (8), 1120–41, doi:10.1002/job.1783.

Reader, T.W., Mearns, K., Lopes, C. and Kuha, J. (2017), 'Organizational support for the workforce and employee safety citizenship behaviors: a social exchange relationship', *Human Relations*, **70** (3), 362–85, doi:10.1177/0018726716655863.

Reader, T.W., Noort, M.C., Shorrock, S. and Kirwan, B. (2015), 'Safety sans Frontieres: an international safety culture model', *Risk Analysis*, **35** (6), 770–89, doi:10.1111/risa.12327.

Reason, J., Parker, D. and Lawton, R. (1998), 'Organizational controls and safety: the varieties of rule-related behavior', *Journal of Occupational and Organizational Psychology*, **71** (4), 289–304, doi:10.1111/j.2044-8325.1998.tb00678.x.

Reason, J.T. (1997), *Managing the Risks of Organizational Accidents*, Aldershot: Ashgate.

Thompson, P.S. and Bolino, M.C. (2018), 'Negative beliefs about accepting coworker help: Implications for employee attitudes, job performance, and reputation', *Journal of Applied Psychology*, **103** (8), 842–66, doi:10.1037/apl0000300.

Tims, M., Bakker, A.B., Derks, D. and van Rhenen, W. (2013), 'Job crafting at the team and individual level: implications for work engagement and performance', *Group and Organization Management*, **38** (4), 427–54, doi:10.1177/1059601113492421.

Weick, K.E., Sutcliffe, K.M. and Obstfeld, D. (1999), 'Organizing for high reliability: processes of collective mindfulness', in R.S. Sutton and B.M. Staw (eds), *Research in Organizational Behavior*, vol. 21, Stamford, CT: JAI Press, pp. 81–123.

Westrum, R. (1993), 'Cultures with requisite imagination', in J. Wise, P. Stager and J. Hopkin (eds), *Verification and Validation in Complex Man-Machine Systems*, New York: Springer, pp. 401–16.

Willis, S., Clarke, S. and O'Connor, E. (2017), 'Contextualizing leadership: transformational leadership and management-by-exception-active in safety-critical contexts', *Journal of Occupational and Organizational Psychology*, **90** (3), 281–305, doi:10.1111/joop.12172.

Wrzesniewski, A. and Dutton, J.E. (2001), 'Crafting a job: revisioning employees as active crafters of their work', *Academy of Management Review*, **26** (2), 179–201, doi:10.5465/amr.2001.4378011.

Yagil, D. and Luria, G. (2010), 'Friends in need: the protective effect of social relationships under low-safety climate', *Group & Organization Management*, **35** (6) 727–50, doi:10.1177/1059601110390936.

Zohar, D. and Polachek, T. (2017), 'Using event-level data to test the effect of verbal leader behavior on follower leadership perceptions and job performance: a randomized field experiment', *Group & Organization Management*, **42** (3), 419–49, doi:10.1177/1059601115619079.

16. The role of safety culture and safety leadership on safety-related outcomes

Çakıl Agnew and Laura Fruhen

INTRODUCTION

Following the Chernobyl disaster in 1986, concern over safety in hazardous industries attracted wider attention to research that could help improve the understanding of human behaviour and performance in relation to risks and safety. Although earlier research mostly focused on individual factors as sources of error vulnerability, organizational factors such as safety culture and safety leadership were identified as the main cause of major accidents (Neal et al. 2000; Borys et al. 2009). In line with the theoretical developments emphasizing organizational, and leader commitment to safety as one of the key factors of safety culture, a number of organizations adopted safety culture assessments in their safety management systems in order to evaluate the state of safety, as well as to identify the weaknesses and strengths regarding their safety culture.

The Accident causation theories in high-risk industries progressed via the technical period, human error, social-technical and, finally, the organizational culture period where the interaction between the human performance and the environment was acknowledged (see Wiegmann et al. 2004). Pioneer work on safety culture was conducted in hazardous industries such as aviation, manufacturing, military and nuclear power. Later, despite the discrepancies between the industry and the healthcare, the medical domain also started to seek guidance from the industry and adopted various safety approaches around safety culture, and leadership (Somekh 2005). These two concepts are closely connected yet make key and unique contributions to workplace safety. This chapter presents theories and findings on safety culture and safety leadership, and how they relate to a number of safety outcomes in various work settings.

WHAT IS SAFETY CULTURE?

One of the most widely cited definitions of safety culture is provided by the International Atomic Energy Agency (IAEA) following the Chernobyl disaster. It defines safety culture as:

> The product of individual and group values, attitudes, perceptions, competencies and patterns of behavior that determine the commitment to, and the style and proficiency of, an organization's health and safety management. Organizations with a positive safety culture are characterized by communications founded on mutual trust, by shared perceptions of the importance of safety and by confidence in the efficacy of preventive measures. (ACSNI 1993, p. 23)

Previously, in a workplace safety investigation in a manufacturing company, Zohar (1980, p. 96) introduced the term 'safety climate' as reflecting 'employees' perceptions about the relative importance of safe conduct in their occupational behavior'. Since then, the definitions of both safety culture and safety climate have been subject to numerous theoretical discussions and the terms often used interchangeably. For example, Clarke's (2006a) meta-analysis showed that when compared with safety attitudes and dispositions, safety perceptions had a more predictive value in relation to work accidents.

In the wider organizational literature, the concept of culture is generally taken to mean something less tractable and more complex than climate. Schein (1990) proposed that climate is only a surface manifestation of culture and that culture manifests itself in deeper levels of unconscious assumptions. Similarly, Cox and Cox (1996) linked culture to personality, which is generally more pervasive and stable, and climate to mood, that is, subject to short-term fluctuations. Again, this was also reflected in the assessment of safety culture. For example, Denison (1996) argued that compared with safety climate – a snapshot – of the prevailing safety culture that can be measured via surveys (Mearns et al. 1997), in order to assess safety culture, qualitative methods should be preferred as surveys cannot fully represent the underlying safety culture.

In addition to conceptual arguments, both methodological issues, such as the level of measurement, and the predictive power of safety culture in relation to safety outcomes have been raised (Griffin and Neal, 2000; Zohar and Luria 2005; Clarke 2006c; Christian et al. 2009). Given the conceptual and methodological differences that have been discussed in the safety literature, we discuss both the theoretical and applied issues of safety culture and safety climate in different contexts in the following sections. Due to the scope of this chapter, and in order to maintain consistency with the cited literature, we use both the terms safety culture and safety climate based on the mentioned studies.

SAFETY CULTURE–SAFETY OUTCOMES RELATIONSHIP

When it comes to investigating the link between safety culture and safety outcomes, there are a number of factors that have been discussed in the literature. One of the main arguments is based on the nature of study design. In particular, study design issues concern whether studies are cross-sectional or longitudinal, what hierarchical level the concept is measured at, and what type of outcome or criterion is used.

The majority of the research in the literature is conducted via cross-sectional surveys assessing all variables at a single point in time. Although cross-sectional studies are less costly and more time efficient, it is not possible to offer a cause–effect direction but, instead, a correlational relationship (Mann 2003). In this respect, in an earlier meta-analysis by Clarke (2006b), the direction of the causation between safety climate and future safety-related outcomes have been demonstrated. Contrary to this, in their meta-analysis, Beus et al. (2010) distinguished between individual-level and organizational (group-level) safety climates. In their meta-analysis, injury rates were found to be a stronger predictor of future safety perceptions than safety climate perceptions for future injury rates (both at unit level). However, it was not possible to investigate the same causal relationship for individual-level safety climate perceptions owing to lack of data.

Although, the majority of the research investigated safety climate at the individual level (Griffin and Neal 2000), previous literature demonstrated safety climate as a multi-level construct (Zohar 2000) where policies and procedures are established at organizational level and executed at unit level (Zohar and Luria 2005). Based on their multi-level model of safety climate (Zohar and Luria 2005), it was argued that policies and procedures adapted at organizational level could not be an objective measure of safety priorities. However, the execution of these procedures and policies at group level give a more realistic frame of reference of safety climate for employees to refer to when deciding which safety behaviours are more rewarding with regards to competing goals (that is, production versus safety). In their influential work, Zohar and Luria (2005) further argued that since policies and procedures are set at the organizational level, they cannot capture every aspect of practice. Therefore a discrepancy between written policies and procedures and their execution is inevitable and the influence of top management commitment to safety on workers' safety behaviours would be rather limited. Zohar and Luria (2005) further differentiated between the effect of safety climate for routine and non-routine tasks, and underlined the crucial role of supervisory practices especially for the non-routine tasks, when speed is valued over safety with consistent feedback, which might lead to the employee perceptions of lower levels of safety climate. However, when the task is routine and highly formalized (with specific and rigid procedures), the supervisory discretion is suggested not to have the same effect on climate perceptions since there would be less discrepancy at execution level when compared with non- routine and less-formalized tasks.

In connection with the issue of measurement level, in a meta-analysis by Christian et al. (2009), the hierarchical structure of safety-related perceptions was investigated in relation to safety behaviours and safety outcomes. Their results demonstrate the association between safety perceptions at each level of analysis (individual, group and organizational), and revealed a stronger association between the organizational/group-level safety climates and safety outcomes compared with psychological safety climate (individual-level perceptions.

Another argument in the safety literature relates to the conceptualizing of the criterion. In order to demonstrate the predictive validity of safety climate, a number of different safety-related outcomes have been adopted in the industry such as accident and injury rates and safety behaviours of the workers. However, while some studies measure these accidents and injuries via archival records or self-report measures, others use self-report measures. Similarly, safety behaviour data were either collected via self-report items or observations. Such discrepancies across studied clouded the predictions. In the following sections, we provide an overview of the links of safety culture and climate with various safety outcomes.

SAFETY CULTURE: SAFETY BEHAVIOURS

The relationship between safety culture and safety behaviours are often assessed via observations or self-report measures. In a study with production workers, group-level safety climate was found to predict observed safety behaviours measured prospectively (Zohar and Luria 2005). These findings were also replicated in the healthcare sector where both unit- and organizational-level climates were predictors of future safety behaviours of the workers (Zohar et al. 2007). Similarly, in a heavy manufacturing company, Johnson (2007) reported links between individual-level safety climate and observed future safety behaviours.

When measured through self-report items, different conceptualizations of safety behaviours are adopted in the literature. such as safety participation, compliance, unsafe behaviours and co-workers' safety behaviours. For example, in a study flight attendants (Kao et al. 2009), positive associations were found between the management's commitment to safety and safety participation, but not compliance. Similarly, safety climate has been shown to be linked to lower rates of unsafe behaviours of Italian blue-collar workers (Cavazza and Serpe 2009). Although Clarke (2006c) found an association between safety climate and unsafe behaviours in the UK manufacturing sector, no such link was observed for accident history. In another study in the US (Seo 2005), safety climate was shown to be the best predictor of unsafe behaviours compared with perceived risk, barriers, hazard level and work pressure.

As regards the definition of safety-related behaviours, one of the most frequently adopted conceptualizations was based on Neal et al.'s (2000) work, which identified two components of safety behaviours of workers as safety compliance and participation. Safety compliance was defined as behaviours rooted in workers' rule obedience, whereas safety participation was related to engaging in activities to promote safety in an organization instead of just following the written safety rules (Clarke 2006b). Based on their definition (Neal et al. 2000), Griffin and Neal (2000) demonstrated the links between, safety climate, safety motivation, knowledge and the self-report safety behaviours in the healthcare sector. Subsequently, in a longitudinal study on healthcare workers, Neal and Griffin (2006) also showed the influence of safety climate on safety participation, but no effect was observed on safety compliance.

SAFETY CULTURE: ACCIDENTS AND INJURIES

Next to its links with employee behaviours, safety culture and climate have also been shown to affect the bottom line of workplace safety: accidents and injuries. In the literature, a more positive safety climate has been reported to be related to fewer experienced injuries (Probst et al. 2008). Although, a number of studies reported direct links between safety climate and injury or accident rates (Zohar and Luria 2004), other research indicates safety climate as a more distal predictor of safety-related worker injuries. For example, in a manufacturing company, safety behaviours were found to mediate the safety climate and injury rate relationship (Cooper and Phillips 2004). In a cross-sectional retrospective survey study on a sample of pilots, flight officers and other aircrew in the US Navy, while a positive association was reported for minor and intermediate accidents and future safety climate perceptions, no such effect of major accidents was found (Desai et al. 2006).

Compared to objective data obtained from archival records, injury rates were also assessed through self-report items. For example, safety perceptions were related to self-reported worker injuries experienced in the past six months (Zacharatos et al. 2005). Huang et al. (2006) also collected injury history (in the past six months) data with a self-report single-item measure in a number of industrial settings (transportation, service, construction and manufacturing) in the US, and reported indirect links between safety climate and self-reported injury experience.

Similarly, in the healthcare sector, Gimeno et al. (2005) showed that lower levels of safety climate at the individual level coupled with lower rates of safety compliance were related with past worker injuries. While a direct effect of safety climate on back injuries was found, no such effect was reported for needle-stick injuries (Mark et al. 2007).

SAFETY CULTURE: PATIENT SAFETY

Although the number of adverse events affecting patients has captured a great deal of attention in the healthcare sector, not many studies investigated the causes of negative patient safety outcomes in relation to safety climate factors. Rates of errors (medication, treatment, and so on), incidents, failures and severity rates can be measured in healthcare settings via incident reports or case record reviews. However, the challenges of acquiring accurate patient adverse event data (Williams et al. 2008) have been reflected in the literature, with relatively few safety climate studies using patient outcome data. While studies considering safety culture and climate in the healthcare sector are limited, their findings are consistent with those from other sectors and illustrate that the concept also has a key role for patient safety. For example, supervisor safety-related activities were found to be effective on reducing the treatment errors (Naveh et al. 2005). Subsequently, they also demonstrated a curvilinear relationship between treatment errors and safety performance where overly detailed, or too many, written rules did not assure higher levels of safety performance (Katz-Navon et al. 2005).

Safety climate and its consequences were also investigated simultaneously in relation to patients and workers. For example, Hofmann and Mark (2006) demonstrated a link between safety climate, and patient safety outcomes (urinary tract infections and medication errors) from archival records.

Overall, a number of different safety-related outcomes have been adopted while examining the predictive power of safety climate. However, the decision to select a criterion measure requires careful attention. Challenges associated with reporting near-misses, injuries and incidents might have consequences on the validity of the company records of accidents and injuries. However, self-reported measures of safety performance were also associated with a number of problems, such as biased recall, common method bias and social desirability (Cooper and Phillips 2004).

In summary, attention should be drawn to adopting incident rates as an outcome criterion. For example, in order to evaluate the effectiveness of any intervention designed to improve overall patient safety, incident rates might be considered as a more problematic outcome variable (Ross et al. 2010) owing to the possibility of an under-reporting tendency among healthcare staff.

APPLICATIONS ACROSS INDUSTRIES

The research summarized above illustrates that safety culture can be applied to various industries and settings. High-reliability industries, such as nuclear power and aviation, have led the way for safety culture and its development. Other sectors, such as healthcare, have built on the progress that had been made in the aviation industry and successfully adopted safety culture in their sector. In fact, the medical domain sought guidance from other high-risk industries, especially following the investigations into tragedies in British hospitals (Kennedy 2001) and fatal drug errors (Toft 2001) that identified a weak safety culture as a causal factor. Both sectors are highly complex and potentially risky; however, a number of notable differences require adaptation of the concept. Unlike healthcare staff, pilots experience the same consequences of fatal error as their passengers. Also, an aviation accident might produce mass casualties (Wachter 2004), whereas in healthcare multiple simultaneous fatalities are very rare and, although workers do get injured, this will not generally happen when the patient is affected by an adverse event. Another difference between these two industries is attitudes to error; compared with healthcare, human error is now perceived as more inevitable in aviation, which designs special training programmes to reinforce this thinking (Helmreich et al. 1999). In contrast, there is a great emphasis on perfectionism in healthcare, in that error has been considered as a weakness of the individual (Leape 1994). This stress on error-free performance in healthcare might contribute to the difficulty of communicating errors or learning from mistakes.

Therefore, although healthcare can benefit from industrial safety experiences, it is not always suitable to adopt techniques directly from industry owing to the nature of the particular occupation in the two sectors. Medical workers are dealing with unstable states of patients, and the decision-making process in the medical domain is highly distributed between distinct roles (caring versus curing), whereas only two pilots are in charge while flying a plane and focused on the specific task (Randell

2003). Finally, as healthcare involves the risk of both patient and worker injury, the well-being of staff may have an effect on the safety of the patient, making healthcare safety more complicated compared with other industries (Flin 2007).

These differences between industry sectors illustrate some of the unique issues that different workplace settings may experience when considering their safety culture. As such, the work conducted in the healthcare sector illustrates how important it is to consider the uniqueness of an industry or sector when it comes to adopting a safety culture framework.

Despite the discrepancies, based on the safety literature from both industrial and healthcare settings, it appears that utilizing two complementary methods rather than a single approach (Cox and Cheyne 2000; Battles and Lilford 2003) is advisable to understand the state of safety culture within workplaces. For example, similar to the approach used in air traffic management (see Eurocontrol 2008; Mearns et al. 2013), as part of a larger quantitative study on 1866 healthcare workers in Scotland (Agnew et al., 2013), a workshop was designed to provide feedback to the healthcare management and provide opportunities to discuss the areas of strength and improvement in addition to future recommendations (Agnew and Flin 2014).

While the measurement of safety culture is key to most organizations' initiatives around the concept, it is important to consider that measuring safety culture is akin to taking the temperature of a workplace. Inviting employees to complete surveys should be regarded as the start of a process aimed to strengthening and building a better safety culture. The actions that follow from this assessment are what will shape the safety culture of the future. Part of these actions need to be open and honest feedback to the workforce and workforce involvement in identifying what steps can be taken going forward.

SAFETY LEADERSHIP

Leaders influence their followers' attitudes, experience of the workplaces and behaviour, and this impact also applies to safety and safety culture. The role of leadership for safety is evident in investigation reports into major accidents in various industries, such as aviation and energy generation (for example, the Deepwater Horizon oil spill in 2010 and the Fukushima nuclear accident in 2011). These reports often refer to leadership at the very highest levels of organizations, the senior management, highlighting this group's key influence on safety in organizations. These reports also illustrate how leadership in relation to safety permeates all layers of an organization. It has been argued that leadership in organizations can turn out differently depending on the organizational level at which it is exerted (Andriessen 1978; Flin 2003). Despite the key role of senior and middle managers and the distinct nature of leadership at different organizational levels, research into safety leadership is predominantly conducted at the supervisory level (Pilbeam et al. 2016; see also Flin 2003).

Since the early 2000s, leadership has received attention in the safety literature as an antecedent of employee perceptions of culture and climate, attitudes and behaviours (Clarke 2013). A meta-analysis by Nahrgang et al. (2011) indicates that among a range of other workplace factors, safety leadership (that is, styles of leadership – for example, transformational, relationships between leaders and workers, trust and supervisor support for safety) explained a 15.5 per cent variance in accidents and injuries, a 26.2 per cent variance in adverse events (that is, near misses, safety events, and errors), as well as a 20.3 per cent variance in unsafe behaviour (unsafe behaviours, absence of safety citizenship behaviours, and negative health and safety). These results therefore identify leadership to be a key factor when it comes to safety in organizations.

Despite this recognition and research into safety leadership, to date no agreed upon definition of the concept exists. The available conceptualizations and definitions of safety leadership are derived from each study's focus on a particular theoretical model or view on (safety) leadership. However, for the purpose of this chapter we put forward a general definition of safety leadership as the process of influence and guidance that one individual, probably in a leadership position, exerts over another individual or a group of individuals regarding their attitudes and behaviours towards safety issues. As a consequence of the lack of an agreed upon definition, the safety literature has adopted various approaches towards leadership. From these approaches in different studies, we identify two main distinctive conceptual dimensions that describe four distinct approaches to safety leadership (see Figure 16.1). The first dimension describes the actions (the 'how') of safety leadership, that is, the specific behaviours through which leaders exert their influence on safety. Along this dimension, studies vary in the extent to which they approach safety leadership with a focus on interpersonal relationships (for example, Hofmann and Morgeson 1999; Barling et al. 2002; Kelloway et al. 2006) or with a focus on managerial activities such as decision-making and in particular priorities (for example, Zohar and Luria 2004: Dahl and Olsen 2013; Fruhen et al. 2014). The second dimension relates to the content (the 'what') of safety leadership, and particularly concerns the specificity of the concept in relation to safety. Safety leadership may be approached as a safety-specific concept, or with a focus on leadership more generally (for an overview, see Zohar 2003; Hofmann and Morgeson 2004). As noted by Clarke (2013) the

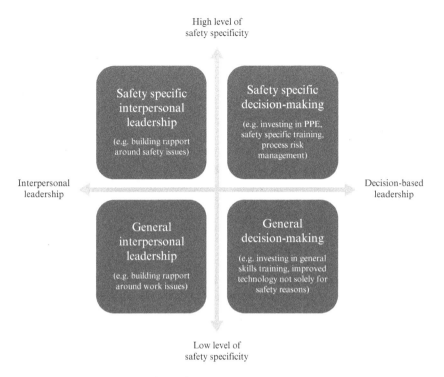

High level of
safety specificity

Safety specific
interpersonal
leadership

(e.g. building rapport
around safety issues)

Safety specific
decision-making

(e.g. investing in PPE,
safety specific training,
process risk
management)

Interpersonal
leadership

Decision-based
leadership

General
interpersonal
leadership

(e.g. building rapport
around work issues)

General
decision-making

(e.g. investing in general
skills training, improved
technology not solely for
safety reasons)

Low level of
safety specificity

Note: PPE = personal protective equipment.

Figure 16.1 Dimensions of safety leadership

extent to which safety leadership needs to be thought of as a safety specific concept is not clear.

Scholars have acknowledged both these dimensions as relevant to leaders' impact on organizational safety and safety culture (for example, Zohar 2003; Griffin and Talati 2014) and have addressed these to varying degrees. However, only limited research exists that considers both these dimensions' relative importance for leaders' impact on safety (for example, Zohar 2002).

DIMENSION 1: THE ACTIONS OF SAFETY LEADERSHIP

Actions of safety leadership are distinguishable based on their focus on interpersonal relationships, as defined in classic models of leadership (for example, Bass 1985; Graen and Uhl-Bien 1995), and conceptualizations of decision-making and classic managerial activities (for example, Hambrick 1989; Mintzberg 1975; Zaccaro and Horn 2003). This proposed grouping resonates with the distinction between initiating structure and consideration as key aspects of a leader's behaviour (Yukl 1998). We argue that both leadership (that is, building relationships and motivating) and managerial activities (that is, the decisions that leaders make about safety issues) contribute to leaders' overall impact on safety. The influence of leaders on safety can only be fully understood by taking both of these aspects into account. For example, followers are likely to be motivated to follow procedures and to act safely based on their relationship with a leader, while also observing their leader's own decisions and actions as indicators of how important safety truly is to the leader.

Interpersonal Leadership

Interpersonal leadership is rooted in the relationships that leaders develop with their team members and is usually captured via traditional leadership theories. This relationship development is a more direct form of safety leadership, and we label it interpersonal safety leadership (also labelled relationship-orientated leadership; Clarke 2013). Clarke (2013, p. 23) defines this type of leadership as:

> enacted through a process of social exchange with subordinates, whereby mutual trust and liking is developed; in exchange for leaders' concern for their safety and well-being [. . .].

Research focusing on interpersonal safety leadership in particular employs two leadership models: the full-range leadership model and leader member exchange. The full-range model of leadership (Bass 1985) defines leadership along three distinct types of relationship-orientated leadership behaviours, namely, transformational, transactional and passive leadership. Transformational leadership motivates subordinates by considering their individual interests, strengths and weaknesses, providing inspiring motivation (for example, through a strong vision), support for suggestions and new ideas, as well as building interpersonal trust (Bass 1985; Bass and Avolio 1997). A transactional leader motivates subordinates through reward for good or appropriate performance and gives feedback, thereby establishing interactions rooted within social exchange (Bass and Avolio 1997). Passive leadership style is often described as characterized by the absence of leadership, as well as the leader's tendency to intervene only after standards have not been met (Den Hartog et al. 1997). Studies consistently find that transformational leadership has a particularly positive role for safety outcomes at work (Barling et al. 2002; Zohar 2002; Kelloway et al. 2006). More recently, a meta-analysis (Clarke 2013) has identified that transformational as well as active forms of transactional leadership styles have their place when it comes to influencing subordinates' safety behaviours. Clarke showed that a transformational leadership style is primarily associated with safety participation (that is, willingness to engage in safety activities) and that an active transactional leadership style in particular stimulates compliance. Clarke's findings have been extended by Willis et al. (2017), who found that active transactional leadership had strong positive links with safety participation when accident likelihood was high, and that this link was negative when accident likelihood was low. Transformational leadership was less strongly related to these performance outcomes in contexts where safety was perceived as highly critical. A study by Jiang and Probst (2016) reports that transformational safety leadership enhanced the link between safety knowledge and safety participation, while weakening the link between safety motivation and safety participation.

The second theoretical perspective on interpersonal leadership is via leader–member exchange (LMX). Leader–member exchange views leadership through the dyads that leaders form with individual followers, rather than averages of leadership (Graen and Uhl-Bien 1991). Hofmann and Morgeson (1999) found that relationships between leaders and followers that are rooted in social exchange are linked with increased communication about safety issues, higher levels of safety commitment and fewer accidents.

More recent developments in the arena of interpersonal safety leadership have considered the concept of authentic leadership. This concept, consisting of self-awareness, relational transparency, moral perspective and balanced processing, has been found to be linked with safety climate and safety outcomes (Eid et al. 2012; Borgersen et al. 2014).

Decision-Based Leadership

The second group of safety leadership actions that we consider addresses the influence that leaders exert on their teams, organization and systems via the decisions and actions that they take that are not aimed at relationships with followers. We label this type of safety leadership as decision-based safety leadership. It is recognized in a number of studies and has previously been indirectly touched upon in a theoretical model by Griffin and Talati (2014), who describe decision-based leadership under the label of leader contributions to safety systems and focus on it primarily via leader influence on safety culture. We put forward a more general perspective concerning the ways in which managers shape teams, organizations and wider systems around them through actions that are not captured in leadership styles, which is directly focused on followers. This approach is rooted in traditional managerial research (for example, Mintzberg 1975; Hambrick 1989; Zaccaro and Horn 2003). In her description of different levels of safety leadership in organizations, Flin (2003) elaborates that decision-based leadership becomes more and more important for safety the further removed leaders are from the front line of business operations – that is the more senior they are. Flin (2003, p. 264) captures decision-based leadership as conveyed via 'language and actions, especially in relation to prioritization and their time allocation'.

A smaller number of studies have considered decision-based leadership in relation to safety, but no overall theoretical models that tie this stream together exist. We propose that this type of leader influence can affect safety through two pathways. First, it can directly affect safety via the immediate outcomes of the decisions that leaders make that impact safety. Second, it can also affect employee behaviour indirectly, as employees are likely to observe the decisions that their leaders make and draw conclusions from these observations. This process underlies one of the key components of safety climate and culture – perceived management commitment to safety. Notably, how leaders engage and enact safety commitment has not been extensively researched (Fruhen et al. 2014). We propose that the process of decision-based safety leadership via employee perceptions can be theorized to be rooted within the propositions put forward by the social information-processing theory (Salancik and Pfeffer 1978). The social information-processing theory stipulates that individuals make sense of their work environment by processing the social cues available to them and, in particular, consider cues from those in leadership positions (for example, Copeland 1994). Decision-based leadership has been researched, for example, in a study by Dahl and Olsen (2013), which considered leader involvement in daily work operations in relation to employee safety compliance. Their findings show that the extent to which leaders take part in the planning and preparation of work, follow up the execution of the work, and contribute to good cooperation among team members has a positive effect on safety compliance. A study by Zohar and Luria (2004) found coherence and consistency in supervisory safety scripts (that is, task-orientated actions) as critically shaping their employees' agreement and overall perception of safety climate through their clarity and consistency in prioritizing safety. Zohar and Luria's (2004) focus on priority assigned to safety relates to one of the key concepts for leader impact on safety (and safety culture and climate), namely, safety commitment. Meta-analyses show that safety commitment is one of the most commonly assessed and most influential components of safety climate (Christian et al. 2009; Beus et al. 2010; Clarke 2010). A study by Fruhen et al. (2014) identified that senior leaders can, in particular, demonstrate their safety commitment through their decision-making, personal involvement and communication with employees. Similarly, a study by Rundmo and Hale (2003) identified that time spent on activities such as motivating others and communicating about safety issues, procedures and regulations, and the senior leaders' level of engagement in training and design as actions, can convey safety commitment.

DIMENSION 2: THE CONTENT OF SAFETY LEADERSHIP

The second dimension of safety leadership that we stipulate concerns the content of safety leadership, that is the extent to which leadership in relation to safety needs to be specifically about safety, or whether general leader behaviours have an impact. Whether safety leadership needs to be thought of as safety specific or general leadership is a debated issue in the safety leadership literature (Clarke 2013). Some researchers evaluate a focus on safety-specific leadership as a distinct leadership concept as necessary (Zohar 2002, 2010), rooted in the observation that safety can conflict with other organizational issues and aspects of performance. Following this argument, a number of studies have focused on safety-specific leadership. Barling et al. (2002) report a link between safety-specific transformational leadership with injuries across two studies. They find this connection to be mediated by perceived safety climate, safety consciousness and safety-related events. Extending these findings, Kelloway et al. (2006) identify safety-specific transformational leadership and safety-specific passive leadership as having distinct effects on safety climate perceptions and safety consciousness in employees, which in turn affected safety events and injuries. More recently, Mullen (2017) explored the associations between safety-specific transformational leadership style, perceptions of employer safety obligations and employee safety performance behaviour and attitudes. Another study, by Smith et al. (2016), reports that safety-specific transformational leadership is positively linked with safety climate perceptions in firefighters.

Other research has adopted a broader, or more general, scope towards safety leadership via general dimensions of leadership as important (for example, Hofmann and Morgeson 2004). O'Dea and Flin (2001) suggest that good safety leadership can more generally be summarized under the term participative management. According to O'Dea and Flin (2001), good safety leadership is a general phenomenon rather than being limited to leader actions focused on safety. A number of studies referred to above have focused on general leadership (for example, Hofmann and Morgeson 1999; Borgersen et al. 2014; Jiang and Probst 2016). Notably, the study by Willis et al (2017) employed a measure of general leadership, but argued that the function of these behaviours may differ depending on the context in which they occur (the level of safety criticality). Their results revealed that general leadership is linked with safety outcomes and that these links are shaped by the context. Matilla et al. (1994) noted that general supervisor feedback on performance has a positive impact on safety compliance.

Some studies take both general and safety specific leadership actions into account. For example, Simard and Marchand (1997) found that cooperative supervisor–employee relationships and supervisor involvement in safety decisions are linked with safety compliance. A study by Zohar (2002) examined the joint role of supervisor general leadership and their emphasis on safety versus production. He found that both transactional and transformation leadership were linked with injury rates, and that this link was moderated by the safety priority of leaders.

It appears that there is no clear consensus in the safety leadership field as regards the safety specificity of the concepts. Studies have found both general and safety-specific conceptualizations to be useful in explaining how leadership affects safety outcomes. It is unlikely that this conceptual difference will appear relevant to leaders in the field. However, academic advancement may benefit from a closer consideration of the differences in safety versus general leadership

CONCLUSIONS

Safety culture and safety leadership are both key concepts that are relevant to understanding the social dynamics at work shaping safety outcomes. Given their central role for safety, they are also central issues that can be focused on when wanting to improve safety in workplaces. Rapid global developments such as advancement in technical knowledge and diverse organizational cultures make it more challenging to improve and maintain safety (see Noort et al. 2016). Therefore continuing research and practical changes to workplace safety via safety culture and safety leadership can drive safety in ever more complex and challenging environments. Effective approaches towards improving safety culture or safety leadership in organizations should consider both concepts, as they can support and help sustain each others' impact.

REFERENCES

Advisory Committee for the Safety of Nuclear Installations (ACSNI) (1993), *ACSNI Study Group on Human Factors: Third Report – Organising for Safety*, Sheffield: HSE Books.

Agnew, C. and Flin, R. (2014), 'Senior charge nurses' leadership behaviours in relation to hospital ward safety: a mixed method study', *International Journal of Nursing Studies*, **51** (5), 768–80.

Agnew, C., Flin, R. and Mearns, K. (2013), 'Patient safety climate and worker safety behaviours in acute hospitals in Scotland', *Journal of Safety Research*, **45** (June), 95–101.

Andriessen, J.H.T.H. (1978), 'Safe behaviour and safety motivation', *Journal of Occupational Accidents*, **1** (4), 363–76.

Barling, J., Loughlin, C. and Kelloway, E.K. (2002), 'Development and test of a model linking safety specific transformational leadership and occupational safety', *Journal of Applied Psychology*, **87** (3), 488–96.

Bass, B.M. (1985), *Leadership and Performance beyond Expectations*, New York: Free Press.

Bass, B.M. and Avolio, B.J. (1997), *Full Range Leadership Development: Manual for the Multifactor Leadership Questionnaire*, Palo Alto, CA: Mind Garden, pp. 43–4.

Battles, J.B. and Lilford, R.J. (2003), 'Organizing patient safety research to identify risks and hazards', *Quality and Safety in Healthcare*, **12** (supplement 2), ii2–ii7.

Beus, J.M., Payne, S.C., Bergman, M.E. and Arthur, W. (2010), 'Safety climate and injuries: an examination of theoretical and empirical relationships', *Journal of Applied Psychology*, **95** (4), 713–27.

Borgersen, H.C., Hystad, S.W., Larsson, G. and Eid, J. (2014), 'Authentic leadership and safety climate among seafarers', *Journal of Leadership & Organizational Studies*, **21** (4), 394–402.

Borys, D., Else, D. and Leggett, S. (2009), 'The fifth age of safety: the adaptive age', *Journal of Health Safety Research Practice*, **1** (1), 19–27.

Cavazza, N. and Serpe, N. (2009), 'Effects of safety climate on safety norm violations: exploring the mediating role of attitudinal ambivalence toward personal protective equipment', *Journal of Safety Research*, **40** (4), 277–83.

Christian, M.S., Bradley, J.S, Wallace, J.C and Burke, M.J. (2009), 'Workplace safety: a meta-analysis of the roles of person and situation factors', *Journal Applied Psychology*, **94** (5), 1103–27.

Clarke, S. (2006a), 'Contrasting perceptual, attitudinal and dispositional approaches to accident involvement in the workplace', *Safety Science*, **44** (6), 537–50.

Clarke, S. (2006b), 'Safety climate in an automobile manufacturing plant: the effects of work environment, job communication and safety attitudes on accidents and unsafe behaviour', *Personnel Review*, **35** (4), 413–30.

Clarke, S. (2006c), 'The relationship between safety climate and safety performance: a meta-analytic review', *Journal of Occupational Health Psychology*, **11** (4), 315–27.

Clarke, S. (2010), 'An integrative model of safety climate: linking psychological climate and work attitudes to individual safety outcomes using metaanalysis', *Journal of Occupational and Organizational Psychology*, **83** (3), 553–78.

Clarke, S. (2013), 'Safety leadership: a meta-analytic review of transformational and transactional leadership styles as antecedents of safety behaviours', *Journal of Occupational and Organizational Psychology*, **86** (1), 22–49.

Cooper, M.D. and Phillips, R.A. (2004), 'Exploratory analysis of the safety climate and safety behavior relationship', *Journal of Safety Research*, **35** (5), 497–512.

Cox, S.J. and Cheyne, A. (2000), 'Assessing safety culture in offshore environments', *Safety Science*, **34** (1), 111–29.

Cox, S. and Cox, T. (1996), *Safety, Systems and People*, Oxford: Butterworth-Heinemann.

Copeland, J.T. (1994), 'Prophecies of power: motivational implications of social power for behavioral confirmation', *Journal of Personality and Social Psychology*, **67** (2), 264–77.

Dahl, Ø. and Olsen, E. (2013), 'Safety compliance on offshore platforms: a multi-sample survey on the role of perceived leadership involvement and work climate', *Safety Science*, **54** (April), 17–26.

Den Hartog, D.N., Van Muijen, J.J. and Koopman, P.L. (1997), 'Transactional versus transformational leadership: an analysis of the MLQ', *Journal of Occupational and Organizational Psychology*, **70** (1), 19–34.

Denison, D.R. (1996), 'What is the difference between organizational culture and organizational climate. A native's point of view on a decade of paradigm wars', *Academy of Management Review*, **21** (3), 619–54.

Desai, V.M., Roberts, K.H. and Ciavarelli, A.P. (2006), 'The relationship between safety climate and recent accidents: behavioural learning and cognitive attributions', *Human Factors*, **48** (4), 639–50.

Eid, J., Mearns, K., Larsson, G., Laberg, J.C. and Johnsen, B.H. (2012), 'Leadership, psychological capital and safety research: conceptual issues and future research questions', *Safety Science*, **50** (1), 55–61.

Eurocontrol (2008), 'White Paper on safety culture', December, accessed 24 May 2019 at https://www.eurocontrol.int/articles/safety-culture.

Flin, R. (2003), '"Danger – men at work": management influence on safety', *Human Factors and Ergonomics in Manufacturing & Service Industries*, **13** (4), 261–8.

Flin, R. (2007), 'Measuring safety culture in healthcare: a case for accurate diagnosis', *Safety Science*, **45** (6), 653–67.

Fruhen, L.S., Mearns, K.J. Flin, R. and Kirwan, B. (2014), 'Skills, knowledge and senior managers' demonstrations of safety commitment', *Safety Science*, **69** (November), 29–36.

Gimeno, D., Felknor, S., Burau, K.D. and Delclos, G.L. (2005), 'Organisational and occupational risk factors associated with work related injuries among public hospital employees in Costa Rica', *Occupational and Environmental Medicine*, **62** (5), 337–43.

Graen, G.B. and Uhl-Bien, M. (1991), 'The transformation of professionals into self-managing and partially self-designing contributors: toward a theory of leadership-making', *Journal of Management Systems*, **3** (3), 25–39.

Graen, G.B. and Uhl-Bien, M. (1995), 'Relationship-based approach to leadership: development of leader-member exchange (LMX) theory of leadership over 25 years: applying a multi-level multi-domain perspective', *Leadership Quarterly*, **6** (2), 219–47.

Griffin, M.A. and Neal, A. (2000), 'Perceptions of safety at work: a framework for linking safety climate to safety performance, knowledge, and motivation', *Journal of Occupational Health Psychology*, **5** (3), 347–58.

Griffin, M.A. and Talati, Z. (2014), 'Safety leadership', in D. Day (ed.), *Oxford Handbook of Leadership and Organizations*, New York: Oxford University Press, pp. 638–56.

Hambrick, D.C. (1989), 'Guest Editor's introduction: putting top management back in the strategy picture', *Strategic Management Journal*, **10** (S1), 5–15.

Helmreich, R.L., Merritt, A.C. and Wilhelm, J.A. (1999), ,The evolution of crew resource management in commercial aviation', *International Journal of Aviation Psychology*, **9** (1), 19–32.

Hofmann, D.A. and Mark, A. (2006), 'An investigation of the relationship between safety climate and medication errors as well as other nurse and patient outcomes', *Personnel Psychology*, **59** (4), 847–69.

Hofmann, D.A. and Morgeson, F.P. (1999), 'Safety-related behaviour as a social exchange: the role of perceived organizational support and leader–member exchange', *Journal of Applied Psychology*, **84** (2), 286–96.

Hofmann, D.A. and Morgeson, F.P. (2004), 'The role of leadership in safety', in J. Barling and M.R. Frone (eds), *The Psychology of Workplace Safety*, Washington, DC: American Psychological Association, pp. 159–80.

Huang, Y., Ho, M., Smith, G.S. and Chen, P.Y. (2006), 'Safety climate and self-reported injury: assessing the mediating role of employee safety control', *Accident Analysis and Prevention*, **38** (3), 425–33.

Jiang, L. and Probst, T.M. (2016), 'Transformational and passive leadership as cross-level moderators of the relationships between safety knowledge, safety motivation, and safety participation', *Journal of Safety Research*, **57** (June), 27–32.

Johnson, S.E. (2007), 'The predictive validity of safety climate', *Journal of Safety Research*, **38** (5), 511–28.

Kao, L., Stewart, M. and Lee, K. (2009), 'Using structural equation modeling to predict cabin safety outcomes among Taiwanese airlines', *Transportation Research Part E*, **45** (2), 357–65.

Katz-Navon, T., Naveh, E. and Stern, Z. (2005), 'Safety climate in healthcare organizations: a multidimensional approach', *Academy of Management Journal*, **48** (6), 1075–89.

Kelloway, K.E, Mullen, J. and Francis, L. (2006), 'Divergent effects of transformational and passive leadership on employee safety', *Journal of Occupational Health Psychology*, **11** (1), 76–86.

Kennedy, I. (2001), 'Learning from Bristol: the report of the public inquiry into children's heart surgery at the Bristol Royal Infirmary 1984–1995', Command Paper CM5207, London: HMSO.

Leape, L. (1994), 'Error in medicine', *Journal of the American Medical Association* **272** (23), 1851–7.

Mann, C.J. (2003), 'Observational research methods. Research design II: cohort, cross sectional, and case-control studies', *Emergency Medicine Journal*, **20** (1), 54–60.

Mark, B., Hughes, L., Belyea, M., Chang, Y., Hofmann, D., Jones, C., et al. (2007), 'Does safety climate moderate the influence of staffing adequacy and work conditions on nurse injuries?', *Journal of Safety Research*, **38** (4), 431–46.

Matilla, M., Hyttinen, M. and Rantanen, E. (1994), 'Effective supervisory behaviour and safety at the building site', *International Journal of Industrial Ergonomics*, **13** (2), 85–93.

Mearns, K., Flin, R., Fleming, R. and Gordon, R. (1997), *Human and Organizational Factors in Offshore Safety*, HSE Report OTH 543, Norwich: HMSO.

Mearns, K., Kirwan, B., Reader, T.W., Jackson, J., Kennedy, R. and Gordon, R. (2013), 'Development of a methodology for understanding and enhancing safety culture in air traffic management', *Safety Science*, **53** (March), 123–33.

Mintzberg, H. (1975), 'The manager's role: folklore and facts', *Harvard Business Review*, **53** (4), 49–61.

Mullen, J., Kelloway, E.K. and Teed, M. (2017), 'Employer safety obligations, transformational leadership and their interactive effects on employee safety performance', *Safety Science*, **91** (January), 405–12.

Nahrgang, J.D., Morgeson, F.P. and Hofmann, D.A. (2011), 'Safety at work: a meta-analytic investigation of the link between job demands, job resources, burnout, engagement, and safety outcomes', *Journal of Applied Psychology*, **96** (1), 71–94.

Naveh, E., Katz-Navon, T. and Stern, Z. (2005), 'Treatment errors in healthcare: a safety climate approach', *Management Science*, **51** (6), 948–60.

Neal, A. and Griffin, M.A. (2006), 'A study of the lagged relationships among safety climate, safety motivation, safety behavior, and accidents at the individual and group levels', *Journal of Applied Psychology*, **91** (4), 946–53.

Neal, A., Griffin, M.A. and Hart, P.M. (2000), 'The impact of organizational climate on safety climate and individual behaviour', *Safety Science*, **34** (February), 99–109.

Noort, M.C., Reader, T., Shorrock, S. and Kirwan, B. (2016), 'The relationship between national culture and safety culture: implications for international safety culture assessments', *Journal of Occupational and Organizational Psychology*, **89** (3), 515–38.

O'Dea, A. and Flin, R. (2001), 'Site managers and safety leadership in the offshore oil and gas industry', *Safety Science*, 37 (1), 39–57.

Pilbeam, C., Doherty, N., Davidson, R. and Denyer, D. (2016), 'Safety leadership practices for organizational safety compliance: developing a research agenda from a review of the literature', *Safety Science*, **86** (July), 110–21.

Probst, T.M., Brubaker, T.L. and Barsotti, A. (2008), 'Organizational injury rate underreporting: the moderating effect of organizational climate', *Journal of Applied Psychology*, **93** (5), 1147–54.

Randell, R. (2003), 'Medicine and aviation: a review of the comparison', *Methods of Information in Medicine*, **42** (4), 433–36.

Ross, A., Plunkett, M. and Walsh, K. (2010), 'Adverse event categorisation across NHS Scotland', *Quality and Safety in Healthcare*, **19** (5), e53, doi:10.1136/qshc.2009.035279.

Rundmo, T. and A.R. Hale (2003), 'Managers' attitudes towards safety and accident prevention', *Safety Science*, **41** (7), 557–74.

Salancik, G.R. and Pfeffer, J. (1978), 'A social information processing approach to job attitudes and task design', *Administrative Science Quarterly*, **23** (2), 224–53.

Schein, E. (1990), 'Organizational culture', *American Psychologist*, **45** (2), 109–19.

Seo, D. (2005), 'An explicative model of unsafe work behaviour', *Safety Science*, **43** (3), 187–211.

Simard, M. and Marchand, A. (1997), 'Workgroups' propensity to comply with safety rules: the influence of micro-macro organisational factors', *Ergonomics*, **40** (2), 172–88.

Smith, T.D., Eldridge, F. and DeJoy, D.M. (2016), 'Safety-specific transformational and passive leadership influences on firefighter safety climate perceptions and safety behavior outcomes', *Safety Science*, **86** (July), 92–7.

Somekh, D. (2005), 'An overview of risk and patient safety across Europe', *Clinical Risk*, **11** (5), 195–8.

Toft, B. (2001), *External Inquiry into the Adverse Incident that Occurred at Queen's Medical Centre, Nottingham, 4th January 2001*, London: Department of Health.

Wachter, R.M. (2004), 'The end of the beginning: patient safety five years after "To err is human"', *Health Affairs*, **23** (July–December), 534–45, doi:10.1377/hlthaff.W4.534.

Wiegmann, D.A., Zhang, H., Thaden, T.L., Sharma, G. and Gibbons, A. (2004), 'Safety culture: an integrative review', *International Journal of Aviation*, **14** (2), 117–34.

Williams, D., Olsen, S., Crichton, W., Witte, K., Flin, R., Ingram, J., et al. (2008), 'Detection of adverse events in a Scottish hospital using a consensus-based methodology', *Scottish Medical Journal*, **53** (4), 29–33.

Willis, S., Clarke, S., and O'Connor, E. (2017), 'Contextualizing leadership: transformational leadership and management-by-exception-active in safety-critical contexts', *Journal of Occupational and Organizational Psychology*, **90** (3), 281–305.

Yukl, G. (1998), *Leadership in Organizations*, 4th edn, Upper Saddle River, NJ: Prentice Hall.

Zaccaro, S.J. and Horn, Z.N.J. (2003), 'Leadership theory and practice: fostering an effective symbiosis', *Leadership Quarterly*, **14** (6), 769–806.

Zacharatos, A., Barling, J. and Iverson, R. (2005), 'High-performance work systems and occupational safety', *Journal of Applied Psychology*, **90** (1), 77–93.

Zohar, D. (1980), 'Safety climate in industrial organizations: theoretical and applied implications', *Journal of Applied Psychology*, **65** (1), 96–102.

Zohar, D. (2000), 'A group-level model of safety climate: testing the effect of group climate on micro-accidents in manufacturing jobs', *Journal of Applied Psychology*, **85** (4), 587–96.

Zohar, D. (2002), 'The effects of leadership dimensions, safety climate, and assigned priorities on minor injuries in work groups', *Journal of Organizational Behavior*, **23** (1), 75–92.

Zohar, D (2003), 'The influence of leadership and climate on occupational health and safety', in D.A. Hofmann and L.E. Tetrick (eds), *Health and Safety in Organizations: A Multilevel Perspective*, San Francisco, CA: Jossey-Bass, pp. 201–32.

Zohar, D. (2010), 'Thirty years of safety climate research: reflections and future directions', *Accident Analysis & Prevention*, **42** (5), 1517–22.

Zohar, D. and Luria, G. (2004), 'Climate as a social-cognitive construction of supervisory safety practices: scripts as proxy of behavior patterns', *Journal of Applied Psychology*, **89** (2), 322–33.

Zohar, D. and Luria, G. (2005), 'A multilevel model of safety climate: cross-level relationships between organization and group-level climates', *Journal of Applied Psychology*, **90** (4), 616–28.

Zohar, D., Livne, Y., Tenne-Gazit, O., Admi, H. and Donchin, Y. (2007), 'Healthcare climate: a framework for measuring and improving patient safety', *Critical Care Medicine*, **35** (5), 1312–17.

17. The benefits of transformational leadership and transformational leadership training on health and safety outcomes

Tabatha Thibault, Duygu Biricik Gulseren and E. Kevin Kelloway

Researchers have studied the positive influence of effective leadership for years. These positive leadership outcomes include increased employee psychological well-being, reduced employee stress and a lower frequency of workplace accidents (Kelloway and Barling 2010). This chapter presents findings on the positive influences of leadership and leadership interventions, specifically transformational leadership, on organizational health and safety outcomes.

Transformational leadership involves behaviours that articulate a vision, provide an appropriate model, foster the acceptance of group goals, show high-performance expectations, provide individualized support and provide intellectual stimulation (Judge and Piccolo 2004). That is, it consists of leadership behaviours that go beyond the level of transactions to result in higher levels of performance (Kelloway et al. 2011). Transformational leadership consists of four dimensions: idealized influence (that is, behaving in admirable ways and acting as a role model), inspirational motivation (that is, setting high but attainable goals and challenging subordinates), intellectual stimulation (that is, encouraging creativity and soliciting ideas from subordinates) and individualized consideration (that is, attending to each subordinate's needs; Judge and Piccolo 2004; Kelloway et al. 2011; Hildenbrand et al. 2018).

Leaders influence their organization's culture (Ehrhart 2004; Tsui et al. 2006; Zohar and Tenne-Gazit 2008). Transformational leaders 'walk the walk' and 'talk the talk' (idealized influence). They spur their employees on with a sense of shared mission. They instil confidence in their followers by telling them that they can do more (inspirational motivations) and encourage them to 'think outside the box' and come up with their own solutions (intellectual stimulation). They show that their employees matter and that they care about their employee's safety and well-being (individualized consideration).

HEALTH OUTCOMES

There are various health-related constructs in industrial and organizational psychology (for example, well-being, stress, burnout and physical health). Well-being is a broad construct that can include physical and/or psychological health (Arnold et al. 2007). Physiological well-being is in effect physical health (for example, sleep quality, physical symptoms and heart rate; Liu et al. 2010). Subjective well-being can be measured through affective means (that is, feelings and arousal – Liu et al. 2010; or positive and negative affect) or through more cognitive means (for example, life or job satisfaction; Arnold et al. 2007). Psychological well-being focuses on a person's mental health and can include symptoms of depression and burnout (Arnold et al. 2007; Liu et al. 2010). Burnout is the result of chronic emotional and interpersonal stressors at work, and consists of three dimensions: exhaustion (feeling overextended), cynicism (depersonalization) and inefficacy (reduced personal accomplishment or lack of self-efficacy; Maslach and Jackson 1981). While many organizational factors can negatively impact employee well-being, stressors such as job demands can be difficult to remove or even reduce (Hildenbrand et al. 2018). Therefore, researchers and practitioners sought other methods to aid employees, one of which is having effective leadership. Meta-analytic evidence shows that leadership is associated with enhanced job well-being (for example, lower job stress, depression and anxiety; Kuoppala et al. 2008). While there are various leadership types or styles, research suggests that transformational leadership possesses the strongest relationship with employee well-being (Tafveln 2017).

Transformational leadership is positively correlated with psychological (Arnold et al. 2007; Nielsen and Munir 2009; Tafvelin et al. 2011; Zwungmann et al. 2014; Arnold and Walsh 2015) and affective (Skakon et al. 2010) well-being and dimensions of well-being (Kelloway et al. 2012), self-efficacy (Nielsen and Munir 2009; Liu et al. 2010; Hentrich et al. 2017), self-esteem (Kensbock and Boehm 2016), physical health (Zwungmann et al. 2014), job satisfaction (Nielsen et al. 2008; Liu et al. 2010; Nielsen and Daniels 2012), thriving (Hildenbrand et al. 2018) and vitality (Nielsen and Daniels 2012). It has been negatively related to job strain (Schmitt et al. 2016), negative affect (Hildenbrand et al. 2018), perceived work stress (Liu et al. 2010), stress symptoms (Liu et al. 2010), overall burnout (Hildenbrand et al. 2018), exhaustion (Kensbock and Boehm 2016; Kranabetter and Niessen 2017), cynicism (Kranabetter and Niessen 2017), sleep quality (Nielsen and Daniels 2012) and irritation (Hentrich et al. 2017).

Transformational leadership has been shown to predict various measures of well-being (Arnold 2017), including psychological well-being (Nielsen et al. 2008; Arnold and Walsh 2015), job-related affective well-being (Kelloway et al. 2012) and positive affective well-being (Arnold et al. 2007). As further evidence of the positive impact of transformational leadership, meta-analytic evidence shows that transformational leadership is linked to increased well-being and psychological functioning, and decreased affective symptoms, burnout, stress and health complaints (Montano et al. 2017).

Researchers have often asked how transformational leadership impacts well-being. Many possible mediators for the relationship between transformational leadership and well-being outcomes have been tested (Arnold 2017). These mediators include constructs such as trust in the leader (Liu et al. 2010; Kelloway et al. 2012), self-efficacy (Liu et al. 2010; Hentrich et al. 2017), meaningful work (that is, finding a purpose in work; Arnold et al. 2007; Nielsen et al. 2008), self-esteem (Kensbock and Boehm 2016), job demands (Hentrich et al. 2017), involvement (Nielsen et al. 2008) and thriving (that is, feeling alive and with a sense of learning; Hildenbrand et al. 2018).

When we have a transformational leader, we are more likely to feel optimistic, happy and enthusiastic (Bono et al. 2007). Leader support and mentoring are negatively linked to stress, burnout and depression (Sosik and Godshalk 2000; Van Dierendonck et al. 2004). Transformational leaders influence employee well-being by changing their perceptions of work (Arnold et al. 2007; Nielsen et al. 2008), by building a sense of community (McKee et al. 2011), and through recognition (Gilbert and Kelloway 2018). In their systematic review, Skakon et al. (2010) found that research consistently showed that leadership behaviours such as trust, feedback, support and integrity are positively linked to employee well-being and negatively linked to employee stress.

SAFETY OUTCOMES

With the growing awareness of occupational accidents and injuries, there is also a growing body of research focusing on the safety-related outcomes of leadership. Safety leadership can be thought of as a process where leaders interact with their employees to achieve organizational safety goals (Fernández-Muñiz et al. 2014). Safety climate refers to employees' perceptions of safety in their organization based on organizational policies, procedures and practices, and the relative priority of safety in their organization (Mullen and Kelloway 2009; Clarke 2013). Safety outcomes are measured through safety-related events such as accidents, injuries or fatalities (Christian et al. 2009). Safety behaviour or safety-related behaviour encompasses positive behaviours (for example, confronting someone when they are breaking a safety procedure) and negative behaviours (for example, taking short cuts to increase productivity at the risk of safety; Beus et al. 2015). An employee's safety performance is often broken down into safety compliance (that is, safety-related behaviours required by the organization) and safety participation (that is, voluntary behaviours that do not contribute to their personal safety but supported safety in the larger organizational context; Christian et al. 2009). A large body of research shows that transformational leadership is positively linked to perceived safety climate (for example, Barling et al. 2002; Kelloway et al. 2006; Hoffmeister et al. 2014), employee safety participation (for example, Mullen et al. 2011; Clarke 2013; Fernández-Muñiz et al. 2014) and safety compliance (for example, Mullen et al. 2011; Conchie 2013; Fernández-Muñiz et al. 2014).

Transformational leadership involves behaviours that articulate a vision, provide an appropriate model, foster the acceptance of group goals, show high-performance expectations, provides individualized support and intellectual stimulation (Clarke 2013). Similarly, safety-specific transformational leadership involves behaviours that promote shared group values around safety, and individualized support to achieve safety goals (Conchie et al. 2012). General transformational leadership is positively related to proactive risk management (that is, preventive planning, active monitoring, communication and training; Fernández-Muñiz et al. 2014). Similarly, safety-specific transformational leadership is positively related to safety consciousness (that is, an individual's personal awareness of safety issues; Barling et al. 2002), safety citizenship behaviours (specifically, whistle-blowing and safety voice; that is, seeking to improve organizational safety through identification of current issues and potential positive changes; Conchie 2013) and measures of employee trust (Conchie et al. 2012). Safety-specific transformational leadership is also linked to lower levels of safety events and injuries (for example, Kelloway et al. 2006). In summary, transformational leadership is associated with a more positive safety climate (Clarke 2013), higher levels of employee positive safety behaviours (both in role and extra role) and lower levels of occupational accidents.

While there is ample evidence to support transformational leadership's relationship to safety outcomes, inconsistently exhibiting these behaviours may not be as useful (Mullen et al. 2011). Passive leadership styles can encompass a general lack of leadership (that is, laissez-faire leadership) or only acting as a leader when a problem arises (that is, management by exception; Mullen et al., 2011). Transformational leadership and passive leadership are not opposite sides of the same coin; it is possible for a leader to exhibit both transformational behaviours and passive behaviours (Mullen et al. 2011). This inconsistency in leader behaviour can hamper the benefits of transformational leadership. Mullen and colleagues found that passive leadership moderated the relationship between transformational leadership and safety behaviour. Specifically, the link between safety behaviour (both participation and compliance) and transformational leadership was not as strong at high levels of passive leadership (Mullen et al. 2011).

In order for supervisors and managers to engage in safety leadership, the organization must have an environment that fosters and supports such behaviour. Based on qualitative interviews, Conchie et al. (2013) found that social support and autonomy help supervisors engage in safety behaviour. However, role overload, production pressure and inadequately skilled employees can hinder a supervisor's ability to engage in safety behaviour (Conchie et al. 2013). Poor safety climate and the focus of putting production ahead of safety limits the amount of power leaders have to actually engage in safety leadership.

This same reasoning can be applied to employees' ability to engage in positive safety behaviours. Employees who have a leader who prioritizes and promotes safety will have the resources and motivation necessary to work safely. For example, staff will be more motivated to go out of their way to challenge safety (for example, making constructive suggestions for change, or whistle-blowing) if their manager promotes safety. Conversely, an employee is less likely to report an accident if his or her supervisor does not enforce safety policies.

While safety leadership is linked to employees' safety performance, this link is not direct. Research shows that the relationship between safety leadership behaviours and employee safety performance is (at least partially) mediated through safety climate (for example, Barling et al. 2002; Clarke 2013; Fernández-Muñiz et al. 2014). That is, safety leadership leads to perceived safety climate, which in turn leads to employees' safety performance. The leadership–climate relationship can be explained through a social learning process in which employees observe and exchange information with their supervisors or managers in order to interpret their organizational environment (Nielsen et al. 2013). For example, when a supervisor calls attention to the importance of safety, his or her employees' perceptions of safety climate are enhanced. Furthermore, higher levels of employee safety performance (that is, participation and compliance) are linked to reduced levels of occupational injuries (Clarke 2013) and general safety outcomes (Fernández-Muñiz et al. 2014). These safety outcomes are linked to employee satisfaction (Fernández-Muñiz et al. 2014). That is, a leader's behaviour influences employee perceptions; these perceptions will serve as a guide on how to behave, and employee behaviours can then affect the risk of accidents and injuries. In addition, employees who feel safe in their working environment (owing to a low frequency of accidents) are more likely to be satisfied with their job.

Transformational leadership can help organizations foster the positive safety climate they seek to attain. Transformational leaders motivate employees to behave safely (for example, by being a role model, solving safety-related issues and developing ways to improve the organization's environment). A leader who encourages employees' feedback and suggestions (for example, for safety improvement initiatives and fixing safety hazards) allows employees to feel included in the process which will make employees more likely to support organizational safety initiatives.

TRANSFORMATIONAL LEADERSHIP INTERVENTIONS

It is better to train the leaders you currently have than to hire new ones, as recruitment and selection are often costly and more time-consuming than training (Donohoe and Kelloway 2014). Effective leadership behaviours, including transformational leadership behaviours, can be trained and improved through intervention designs (Avolio et al. 2009; Kelloway and Barling 2010; Duygulu and Kublay 2011). Transformational leadership training can have a positive influence on leaders, such as increased self-efficacy and positive affect (Mason et al. 2014). Transformational leadership interventions often involve classroom-type training, one-on-one coaching, or feedback sessions (Biggs et al. 2014; Donohoe and Kelloway 2014). All of these methods have been found to be equally effective (Kelloway et al. 2000; Biggs et al. 2014). Effective transformational leadership training involves: (1) teaching leaders about transformational leadership theory and discussing the different types of leadership styles (for example, good versus bad leaders; leadership types such as transformational vs contingent reward vs laissez-faire), (2) providing feedback on the leaders' own leadership style (by providing self-ratings and/or employee ratings), (3) discussing and/or role-playing the newly learned behaviours; and (4) goal-setting (that is, setting three to five specific challenging but attainable goals; Donohoe and Kelloway 2014).

It is important to note that the primary motivation of changing leaders' behaviours is not to help the leaders themselves, but to help their subordinates (Kelloway and Barling 2010). Just as good leadership is associated with good employee outcomes; improved leadership is associated with improved employee outcomes (for example, McKee and Kelloway 2009; Mullen and Kelloway 2009). Leadership development can be seen as a primary intervention (that is, improving workplace conditions to influence employee health and safety; Kelloway and Barling 2010).

LEADERSHIP INTERVENTIONS AND HEALTH OUTCOMES

There are three levels of health interventions: primary (focused on reducing or eliminating the sources of stress), secondary (focused on changing people's appraisal and/or reaction to said sources of stress) and tertiary (treating the consequences that resulted from stress; Kelloway and Barling 2010). Primary interventions, or really trying to fix the problem before it arises, are said to be the best type of health intervention (Kelloway and Day 2005). Leadership interventions such as transformational leadership training can be effective types of primary interventions (Kelloway and Barling, 2010) as changing leaders can lead to changes in work conditions.

Previous studies have found that transformational leadership interventions lead to positive employee and leader health outcomes. For example, Barling et al. (1996) showed that employees whose managers had participated in a transformational leadership intervention had higher levels of affective organizational commitment three months after the intervention. Furthermore, increased transformational leadership as the result of a leadership intervention has been related to increased employee mental health (McKee et al. 2018). In an attempt to increase employees' resources at work, Biggs et al. (2014) implemented a transformational leadership development program that included both feedback and training. Subordinates' work engagement and job satisfaction significantly improved seven months after the intervention (Biggs et al. 2014). Similarly, both Brown and May (2012) and MacKie (2014) found that not only did employee-rated transformational leadership increase, but satisfaction with their leader also increased, each using different methods of delivering transformational leadership training.

Transformational leadership interventions not only increase the health outcomes of the subordinates, but also the health of the leaders (McKee et al. 2018). For example, O'Connor and Cavanagh (2013) found that the 20 middle- and senior-level managers from an academic organizational network who received a coaching intervention showed an increase in their levels of transformational leadership as well as psychological well-being following the intervention. Similarly, Rivers et al. (2011) found that nurses who worked on their transformational leadership skills with a coach during a 20-week coaching programme reported increased life satisfaction and decreased burnout. All of these interventions show that developing transformational leaders is not only beneficial to employee health, but also to the health of the leaders.

LEADERSHIP INTERVENTIONS AND SAFETY OUTCOMES

Transformational leadership training has been found to improve (employee-rated) safety climate (for example, von Thiele Schwarz et al. 2016). Various studies have examined the impact of safety-specific transformational leadership (that is, applying these leadership behaviours to a safety context) to safety-related outcomes (for example, Conchie et al. 2012). Mullen and Kelloway (2009) found that leaders in a half-day safety-specific transformational leadership training workshop reported enhanced safety attitudes, self-efficacy and intent to promote safety. In addition, it was found that the safety-specific leadership training significantly improved employees' safety climate perceptions and safety participation, and reduced safety-related events and safety injuries three months after training compared with control groups where managers either did not receive any training or received general (non-safety specific) transformational leadership training (Mullen and Kelloway 2009).

It is recommended that supervisors and managers explain the safety intervention process (and any leadership training) to their staff so that employees will continue to (or begin to) trust their leaders and modify their own behaviours to meet the new safety policies and/or practices (Nielsen 2017). Conchie et al. (2012) found that the influence of safety-specific transformational leadership on safety voice behaviour was mediated by affect-based trust (that is, belief that a leader will act unselfishly and care about their employees' welfare). This finding suggests that safety citizenship behaviours (that is, going above and beyond what is occupationally required) such as safety voice will probably be higher when supervisors recognize the importance of developing relationships and making their concerns for safety visible. Similarly, Conchie (2013) found that the link between safety-specific transformational leadership and employee safety motivation was positive at high levels of trust (in leader) but not at low levels of trust. When engaging in a safety leadership intervention, employees may be hesitant when their manager begins to exhibit different safety behaviours. During the intervention process and post-intervention, it is important that the leader appears genuine in his or her actions, especially if there is a significant change.

Safety leadership training improves leader behaviours and organizational safety. However, it is important to note that in order to maintain intervention effects, leaders need to work continuously to ensure they are engaging in safety leadership and make these new supervisory practices habitual. By making these positive behaviours habitual, it becomes easier to engage in such behaviours and creates new organizational norms (Zohar 2002).

The focus of leadership interventions is on the indirect effects of these interventions on employees who are not directly participating (that is, subordinates of the leaders; Kelloway and Barling 2010). Changing or improving managers' or supervisors' behaviours influences their employees' perceptions and behaviours in turn. Safety leadership training as a safety intervention is more efficient, and even cost-effective, compared with simply training or retraining employees (Zohar 2002). The effects of leadership interventions can be seen as a top-down approach. Training leaders changes their behaviours; these behavioural changes are seen by their subordinates; these subordinates, in turn, start engaging in similar behaviours, thereby increasing employee safety behaviours and then decreasing safety outcomes such as accidents and injuries. Given that safety leadership is linked to safety climate and employee safety performance, simply retraining employees without changing anything about leader behaviour may not have a strong effect on safety outcomes. If employees still see that their leader is unconcerned about workplace safety, they can easily ignore the training they were given in order to keep up with their leader's production demands.

LEADERS' ROLE IN HEALTH AND SAFETY INTERVENTIONS

Simply having a transformational leader can improve outcomes of occupational health interventions (Lundmark et al. 2017; Nielsen 2017). Lundmark et al. (2017) found that transformational leadership was indirectly associated with increased self-rated employee health following an occupational health intervention. The relationship between transformational leadership and health was mediated by managers' attitudes and actions (Lundmark et al. 2017). Transformational leaders can act as role models and express their vision of a workplace where the intervention succeeds (Lundmark et al. 2017). Leaders who inspire their followers during a health-related intervention can instil positive change and spur their employees to participate in the intervention (Lundmark et al. 2017).

Leaders' behaviours and attitudes can make or break occupational health initiatives (Nielsen 2017). For an intervention to succeed, the intended targets need to want to change. Employees' readiness to change is linked to their leader's readiness to change (Nielsen and Randall 2011). Giving leaders a role in, or having them assume responsibility for, an intervention can help this intervention succeed (Kompier et al. 2000; Nielsen 2017). For instance, leaders should promote the intervention and ensure employees are aware of the intervention taking place. Leaders should also facilitate change during an intervention through job crafting (that is, changing aspects of work, such as tasks or interactions). Improved work conditions following an intervention are associated with an increase in employee affective well-being and job satisfaction (Nielsen and Randall 2011). Finally, leaders can play a role in designing the intervention itself (Nielsen 2017). Having leaders get their say in the design of the intervention may aid in targeting specific areas in need of change. Also, including leaders in the intervention design increases leader responsibility and buy-in to the intervention process.

CONCLUSION

Good leadership goes a long way. Having transformational leaders in an organization can and does improve employee health and safety. Targeting leader behaviour through education and training can help to make organizations more positive environments for employees. Transformational leadership interventions are cost-effective, can empower leaders, and can influence employee well-being and their behaviours and attitudes towards safety (Kelloway and Barling 2010; Donohoe and Kelloway 2014).

Having a transformational leader can help counteract the negative consequences of job stressors and act as a resource for employees. Leadership development can improve employee outcomes such as psychological well-being (McKee and Kelloway 2009). Occupational safety can be enhanced through more than just ergonomic design or regulator approaches. Transformational and safety-specific leadership interventions can increase safety climate, increase employees' safety performance, and subsequently reduce occupational accidents and injuries. These safety leadership interventions can be one of many safety initiatives to be implemented to improve an organization's occupational safety.

Increasing occupational health and safety in workplaces

REFERENCES

Arnold, K.A. (2017), 'Transformational leadership and employee psychological well-being: a review and directions for future research', *Journal of Occupational Health Psychology*, **22** (3), 381–93, doi:10.1037/ocp0000062.

Arnold, K.A. and Walsh, M.M. (2015), 'Customer incivility and employee well-being: testing the moderating effects of meaning, perspective taking and transformational leadership', *Work & Stress*, **29** (4), 362–78, doi:10.1080/02678 373.2015.1075234.

Arnold, K.A., Turner, N., Barling, J., Kelloway, E.K. and McKee, M.C. (2007), 'Transformational leadership and psychological well-being: the mediating role of meaningful work', *Journal of Occupational Health Psychology*, **12** (3), 193–203, doi:10.1037/1076-8998.12.3.193.

Avolio, B.J., Reichard, R.J., Hannah, S.T., Walumba, F.O. and Chan, A. (2009), 'A meta-analytic review of leadership impact research: experimental and quasi-experimental studies', *Leadership Quarterly*, **20** (5), 764–84.

Barling, J., Loughlin, C. and Kelloway, E.K. (2002), 'Development and test of a model linking safety-specific transformational leadership and occupational safety', *Journal of Applied Psychology*, **87** (3), 488–96, doi:10.1037/0021-9010.87.3.488.

Barling, J., Weber, T. and Kelloway, E.K. (1996), 'Effects of transformational leadership training on attitudinal and financial outcomes: a field experiment', *Journal of Applied Psychology*, **81** (6), 827–32.

Beus, J.M., Dhanani, L.Y. and McCord, M.A. (2015), 'A meta-analysis of personality and workplace safety: addressing unanswered questions', *Journal of Applied Psychology*, **100** (2), 481–98, doi:10.1037/a0037916.

Biggs, A., Brough, P. and Barbour, J.P. (2014), 'Enhancing work-related attitudes and work engagement: a quasi-experimental study of the impact of an organizational intervention', *International Journal of Stress Management*, **21** (1), 43–68.

Bono, J.E., Foldes, H J., Vinson, G. and Muros, J.P. (2007), 'Workplace emotions: the role of supervision and leadership', *Journal of Applied Psychology*, **92** (5), 1357–67.

Brown, W. and May, D. (2012), 'Organizational change and development: the efficacy of transformational leadership training', *Journal of Management Development*, **31** (6), 520–36.

Christian, M.S., Bradley, J.C., Wallace, J.C. and Burke, M.J. (2009), 'Workplace safety: a meta-analysis of the roles of person and situation factors', *Journal of Applied Psychology*, **94** (5), 1103–27, doi:10.1037/a0016172.

Clarke, S. (2013), 'Safety leadership: a meta-analytic review of transformational and transactional leadership styles as antecedents of safety behaviours', *Journal of Occupational and Organizational Psychology*, **86** (1), 22–49, doi:10.1111/j.2044-8325.2012.02064.x.

Conchie, S.M. (2013), 'Transformational leadership, intrinsic motivation, and trust: a moderated-mediated model of workplace safety', *Journal of Occupational Health Psychology*, **18** (2), 198–210, doi:10.1037/a0031805.

Conchie, S.M., Moon, S. and Duncan, M. (2013), 'Supervisors' engagement in safety leadership: factors that help and hinder', *Safety Science*, **51** (1), 109–17, doi:10.1016/j.ssci.2012.05.020.

Conchie, S.M., Taylor, P.J. and Donald, I.J. (2012), 'Promoting safety voice with safety-specific transformational leadership: the mediating role of two dimensions of trust', *Journal of Occupational Health Psychology*, **17** (1), 105–15, doi:10.1037/a0025101.

Donohoe, M. and Kelloway, E.K. (2014), 'Transformational leadership training for managers: effects on employee well-being', in C. Biron and R.J. Burke (eds), *Creating Healthy Workplaces*, Abingdon and New York: Routledge, pp. 205–21.

Duygulu, S. and Kublay, G. (2011), 'Transformational leadership training programme for charge nurses', *Journal of Advanced Nursing*, **67** (3), 633–42, doi:10.1111/j.1365-2648.2010.05507.x.

Ehrhart, M.G. (2004), 'Leadership and procedural justice climate as antecedents of unit-level organizational citizenship behavior', *Personnel Psychology*, **57** (1), 61–94, doi:10.1111/j.1744-6570.2004.tb02484.x.

Fernández-Muñiz, B., Montes-Peón, J.M. and Vázquez-Ordás, C.J. (2014), 'Safety leadership, risk management and safety performance in Spanish firms', *Safety Science*, **70** (December), 295–307, doi:10.1016/j.ssci.2014.07.010.

Gilbert, S. and Kelloway, E.K. (2018), 'Self-determined leader motivation and follower perceptions of leadership', *Leadership & Organization Development Journal*, **39** (5), 608–19, doi:10.1108/LODJ-09-2017-0262

Hentrich, S., Zimber, A., Garbade, S.F., Gregersen, S. and Nienhaus, A. (2017), 'Relationships between transformational leadership and health: the mediating role of perceived job demands and occupational self-efficacy', *International Journal of Stress Management*, **24** (1), 34–61, doi:10.1037/str0000027.

Hildenbrand, K., Sacramento, C.A. and Binnewies, C. (2018), 'Transformational leadership and burnout: the role of thriving and followers' openness to experience', *Journal of Occupational Health Psychology*, **23** (1), 31–43, doi:10.1037/ocp0000051.

Hoffmeister, K., Gibbons, A.M., Johnson, S.K., Cigularov, K.P., Chen, P.Y. and Rosecrance, J.C. (2014), 'The differential effects of transformational leadership facets on employee safety', *Safety Science*, **62** (February), 68–78, doi:10.1016/j.ssci.2013.07.004.

Judge, T.A. and Piccolo, R.F. (2004), 'Transformational and transactional leadership: a meta-analytic test of their relative validity', *Journal of Applied Psychology*, **89** (5), 755–68.

Kelloway, E.K. and Barling, J. (2010), 'Leadership development as an intervention in occupational health psychology', *Work & Stress*, **24** (3), 260–79, doi:10.1080/02678373.2010.518441.

Kelloway, E.K. and Day, A. (2005), 'Building healthy workplaces. What we know so far', *Canadian Journal of Behavioral Science*, **34** (7), 223–35.

Kelloway, E.K., Barling, J. and Helleur, J. (2000), 'Enhancing transformational leadership: the roles of training and feedback', *Leadership and Organizational Development Journal*, **21** (3), 145–9, doi:10.1108/01437730010325022.

Kelloway, E.K., Catano, V.M. and Day, A.L. (2011), *People and Work in Canada*, Toronto: Nelson.

Kelloway, E.K., Mullen, J. and Francis, L. (2006), 'Divergent effects of transformational and passive leadership on employee safety', *Journal of Occupational Health Psychology*, **11** (1), 76–86, doi:10.1037/1076-8998.11.1.76.

Kelloway, E.K., Turner, N., Barling, J. and Loughlin, C. (2012), 'Transformational leadership and employee psychological well-being: the mediating role of employee trust in leadership', *Work & Stress*, **26** (1), 39–55, doi:10.1080/02678373.2012.660774.

Kensbock, J.M. and Boehm, S.A. (2016), 'The role of transformational leadership in the mental health and job performance of employees with disabilities', *International Journal of Human Resource Management*, **27** (14), 1580–609, doi:1 0.1080/09585192.2015.1079231.

Kompier, M., Cooper, C. and Geurts, S. (2000), 'A multiple case study approach to work stress prevention in Europe', *European Journal of Work and Organizational Psychology*, **9** (3), 371–400, doi:10.1080/135943200417975.

Kranabetter, C. and Niessen, C. (2017), 'Managers as role models for health: moderators of the relationship of transformational leadership with employee exhaustion and cynicism', *Journal of Occupational Health Psychology*, **22** (4), 492–502, doi:10.1037/ocp0000044.

Kuoppala, J., Lamminpaa, A., Liira, J. and Vainio, H. (2008), 'Leadership, job well-being, and health effects: a systematic review and meta-analysis', *Journal of Occupational and Environmental Medicine*, **60** (8), 904–15.

Liu, J., Siu, O. and Shi, K. (2010), 'Transformational leadership and employee well-being: the mediating role of trust in the leader and self-efficacy', *Applied Psychology: An International Review*, **59** (3), 454–79, doi:10.1111/j.1464-0597.2009.00407.x.

Lundmark, R., Hasson, H., von Thiele Schwarz, U., Hasson, D. and Tafvelin, S. (2017), 'Leading for change: line managers' influence on the outcomes of an occupational health intervention', *Work & Stress*, **31** (3), 276–96, doi:10.1080/02 678373.2017.1308446.

MacKie, D. (2014), 'The effectiveness of strength-based executive coaching in enhancing full range leadership development: a controlled study', *Consulting Psychology Journal: Practice and Research*, **66** (2), 118–37.

Maslach, C. and Jackson, S.E. (1981), 'The measurement of experienced burnout', *Journal of Organizational Behavior*, **2** (2), 99–113, doi:10.1002/job.4030020205.

Mason, C., Griffin, M. and Parker, S. (2014), 'Transformational leadership development: connecting psychological and behavioral change', *Leadership & Organization Development Journal*, **35** (3), 174–94, doi:10.1108/LODJ-05-2012-0063.

McKee, M. and Kelloway, E.K. (2009), 'Leading to wellbeing', paper presented at the annual meeting of the European Academy of Work and Organizational Psychology, Santiago de Compostella, Spain, May.

McKee, M.C., Driscoll, C., Kelloway, E.K. and Kelley, E. (2011), 'Exploring linkages among transformational leadership, workplace spirituality and well-being in health care workers', *Journal of Management, Spirituality & Religion*, **8** (3), 233–55, doi:10.1080/14766086.2011.59914.

McKee, M., Driscoll, C., Kelloway, E.K. and Kelley, E. (2018), 'Lifting leaders and their employees to higher levels of well-being: results from a transformational leadership field development program', manuscript under review.

Montano, D., Reeske, A., Franke, F. and Hüffmeier, J. (2017), 'Leadership, followers' mental health and job performance in organizations: a comprehensive meta-analysis from an occupational health perspective', *Journal of Organizational Behavior*, **38** (3), 327–50, doi:10.1002/job.2124.

Mullen, J.E. and Kelloway, E.K. (2009), 'Safety leadership: a longitudinal study of the effects of transformational leadership on safety outcomes', *Journal of Occupational and Organizational Psychology*, **82** (2), 253–72, doi:10.1348/096317908X325313.

Mullen, J., Kelloway, E.K. and Teed, M. (2011), 'Inconsistent style of leadership as a predictor of safety behaviour', *Work & Stress*, **25** (1), 41–54, doi:10.1080/0 2678373.2011.569200.

Nielsen, K. (2017), 'Leaders can make or break an intervention – but are they the villains of the piece?', in E.K. Kelloway, K. Nielsen and J.K. Dimoff (eds), *Leading to Occupational Health and Safety: How Leadership Behaviours Impact Organizational Safety and Well-Being*, Chichester: John Wiley & Sons, pp. 197–210.

Nielsen, K. and Daniels, K. (2012), 'Does shared and differentiated transformational leadership predict followers' working conditions and well-being?', *Leadership Quarterly*, **23** (3), 383–97, doi:10.1016/j.leaqua.2011.09.001.

Nielsen, K. and Munir, F. (2009), 'How do transformational leaders influence followers' affective well-being? Exploring the mediating role of self-efficacy', *Work & Stress*, **23** (4), 313–29, doi:10.1080/02678370903385106.

Nielsen, K. and Randall, R. (2011), 'The importance of middle manager support for change: a case study from the financial sector in Denmark', in P.-A. Lapointe, J. Pelletier and F. Vaudreuil (eds), *Different Perspective on Work Changes*, Quebec: Université Laval, pp. 95–102.

Nielsen, K., Yarker J., Brenner S.O., Randall, R. and Borg, V. (2008), 'The importance of transformational leadership style for the well-being of employees working with older people', *Journal of Advanced Nursing*, **63** (5), 465–75, doi:10.1111/j.1365-2648 .2008.04701.x.

Nielsen, M.B., Eid, J., Mearns, K. and Larsson, G. (2013), 'Authentic leadership and its relationship with risk perception and safety climate', *Leadership & Organization Development Journal*, **34** (4), 308–25, doi:10.1108/LODJ-07-2011-0065.

O'Connor, S. and Cavanagh, M. (2013), 'The coaching ripple effect: the effects of developmental coaching on wellbeing across organisational networks', *Psychology of Well-Being: Theory, Research and Practice*, **3** (2), doi:10.1186/2211-1522-3-2.

Rivers, R., Pesata, V., Beasley, M. and Dietrich, M. (2011), 'Transformational leadership: creating a prosperity-planning coaching model for RN retention', *Nurse Leader*, **9** (5), 48–51.

Schmitt, A., Den Hartog, D.N. and Belschak, F.D. (2016), 'Transformational leadership and proactive work behaviour: a moderated mediation model including work engagement and job strain', *Journal of Occupational and Organizational Psychology*, **89** (3), 588–610, doi:10.1111/joop.12143.

Skakon, J., Nielsen, K., Borg, V. and Guzman, J. (2010), 'Are leaders' well-being, behaviours and style associated with the affective well-being of their employees? A systematic review of three decades of research', *Work & Stress*, **24** (2), 107–39, doi:10.1080/02678373.2010.495262.

Sosik, J. and Godshalk, V. (2000), 'Leadership styles, mentoring functions received, and job-related stress: a conceptual model and preliminary study', *Journal of Organizational Behavior*, **21** (4), 365–90, doi:10.1002/(SICI)1099-1379 (200006)21:4<365::AID-JOB14>3.0.CO;2-H.

Tafvelin, S. (2017), 'Antecedents and consequences of transformational leadership on occupational health and safety', in E.K. Kelloway, K. Nielsen and J.K. Dimoff (eds), *Leading to Occupational Health and Safety: How Leadership Behaviours Impact Organizational Safety and Well-Being*, Chichester: John Wiley & Sons, pp. 69–92.

Tafvelin, S., Armelius, K. and Westerberg, K. (2011), 'Toward understanding the direct and indirect effects of transformational leadership on well-being: a longitudinal study', *Journal of Leadership & Organizational Studies*, **18** (4), 480–92, doi:10.1177/1548051811418342.

Tsui, A.S., Zhang, Z., Wang, H., Xin, K.R. and Wu, J.B. (2006), 'Unpacking the relationship between CEO leadership behavior and organizational culture', *Leadership Quarterly*, **17** (2), 113–37, doi:10.1016/j.leaqua.2005.12.001.

Van Dierendonck, D., Haynes, C., Borrill, C. and Stride, C. (2004), 'Leadership behavior and subordinate well-being', *Journal of Occupational Health Psychology*, **9** (2), 165–75.

Von Thiele Schwarz, U., Hasson, H. and Tafvelin, S. (2016), 'Leadership training as an occupational health intervention: improved safety and sustained productivity', *Safety Science*, **81** (January), 35–45, doi:10.1016/j.ssci.2015.07.020.

Zohar, D. (2002), 'Modifying supervisory practices to improve subunit safety: a leadership-based intervention model', *Journal of Applied Psychology*, **87** (1), 156–63, doi:10.1037/0021-9010.87.1.156.

Zohar, D. and Tenne-Gazit, O. (2008), 'Transformational leadership and group interaction as climate antecedents: a social network analysis', *Journal of Applied Psychology*, **93** (4), 744–57, doi:10.1037/0021-9010.93.4.744.

Zwungmann, I., Wegge, J., Wolf, S., Rudolf, M., Schmidt, M. and Richter, P. (2014), 'Is transformational leadership healthy for employees? A multilevel analysis in 16 nations', *German Journal of Research in Human Resource Management*, **28** (1–2), 24–51, doi:10.1688/ZfP-2014-01-Zwingmann.

18. Crew resource management (CRM) and non-technical skills

Rhona Flin, Amy Irwin and Oliver Hamlet

1. INTRODUCTION

The concept of CRM has now permeated many work settings since its inception in civil aviation in the late 1970s (Helmreich and Foushee 2010). This training method was originally prompted by a series of aircraft accidents with no primary technical cause, forcing investigators to broaden their attention to consider the behaviours of the humans operating, maintaining and controlling aviation operations. Psychological research examining pilots' attitudes, communications and actions in flight-deck simulators and the analysis of cockpit voice recordings confirmed the need to consider the behavioural, as well as the technical, aspects of flight safety. Crew resource management in aviation can be defined as the effective utilisation of all available resources (for example, crew members, aircraft systems, supporting facilities and persons) to achieve safe and efficient operations. In aviation, its objective is to enhance the flight management skills of crew members with an emphasis placed on the non-technical knowledge, skills and attitudes of crew performance. The term non-technical skills has been used by the European aviation regulator for many years as a synonym for CRM skills. Over the past 40 years, aviation CRM has evolved (Kanki et al. 2019) and it is now widely adopted in other settings, especially in safety-critical work domains. More generally, non-technical skills can be defined as the cognitive, social and personal resource skills that complement technical skills, to achieve safe and efficient task performance (Flin et al. 2008).

This chapter describes non-technical (CRM) skills training and assessment, with particular reference to the current regulations for airline pilots in Europe. It then considers applications of this approach in other domains, particularly healthcare, the oil and gas industry and agriculture.

2. AVIATION

The aviation sector continues to place strong reliance on CRM training and assessment. In Europe, the aviation regulator (European Aviation Safety Agency, EASA) recently reviewed current practice and future requirements for CRM (Boettcher 2016). Questions had been raised as to whether CRM training was meeting current demands, given changes in operations (for example, evolving technologies, larger aircraft and new generations of pilots) and emerging threats (for example, increasing traffic and loss of control identified in several accidents). The changes modernised the existing CRM training scheme and the requirements came into force in 2016 (see Flin 2019). National regulators in Europe may issue their own guidance on CRM (for example, CAA 2016) as well as related advice (for example, CAA 2014b) and on an international level, industry reports from IATA (International Aviation Training Association) and ICAO (International Civil Aviation Organisation) can also have an influence on CRM training.

Aviation CRM Training Content

The main non-technical (CRM) skills topics taught in aviation are listed in Box 18.1.

The threat and error management (TEM) model (Helmreich 1997) has been a core component of aviation CRM training since the 1990s, as it emphasises that threats can be present in every operation and that errors are normal and have to be managed using non-technical skills. The first ten

---

BOX 18.1   MAIN AVIATION CRM TRAINING COURSE TOPICS

- Application of the principles of CRM and threat and error management
- Situation awareness
- Problem-solving and decision-making
- Workload management/task sharing
- Communication/assertiveness
- Leadership and teamwork
- Stress and stress management
- Fatigue and vigilance
- Automation
- Monitoring and intervention
- Resilience development
  - Surprise and startle effects
- Operator's safety culture and company culture.

topics in Box 18.1 have been described well elsewhere (for example, Kanki et al. 2010, 2019) and are therefore not explained here. The last three topics are more recent introductions and therefore are explained below.

Influenced by renewed attention to loss of control events (for example, the Air France 447 crash into the Atlantic Ocean on a flight between Rio de Janeiro and Paris; BEA 2012), two related subjects which now have to be incorporated into CRM training in Europe are the development of resilience and the skills to cope with unexpected events that could produce surprise, or worse, a startle reaction in the pilots. Resilience has been examined at both an organisational (Rankin et al. 2014) and an individual level (Drath 2017), and can be defined as 'the ability of a system to adjust its functioning prior to, during, or following changes and disturbances, so that it can sustain required operations under both expected and unexpected conditions' (EASA 2017a, p. 14). Crew resource management training addresses resilience development in terms of pilots' mental flexibility and performance adaptation, that is, mitigating frozen behaviours, overreactions and inappropriate hesitation, and adjusting actions to current conditions. The main aspects of resilience development are the ability to: learn (knowing what has happened), monitor (knowing what to look for), anticipate (finding out and knowing what to expect) and respond (knowing what to do and being capable of doing it). When an unexpected event occurs, such as an alarm for an equipment malfunction, sudden extreme turbulence, loud noise or very strong vibration, then a normal human reaction is to experience a sense of surprise, which may have physiological aspects. If this event is perceived as very threatening, then a stronger emotional and physical reaction may be experienced, known as startle. These phenomena have been studied in laboratories for many years but research with airline pilots is only just emerging (Landsman et al. 2017; Martin 2019).

Another new topic that CRM training should now include is the operator's safety culture, its company culture and multicultural crews. National, organisational and professional cultural differences can have a significant influence on the behaviour of flight crew members (Helmreich and Merritt 1998) and large airlines can have more than 50 nationalities working on their flight decks.

Aviation CRM Training Requirements

Pilots in Europe receive initial in-depth CRM training (minimum 18 training hours with a minimum of 12 training hours in the classroom) from their operating company, plus annual refresher training. The airlines have to update their CRM recurrent training programme over a period not exceeding three years, to cover all the topics, taking into account information (for example, hazards and risks) from the safety management system, including the identified results of the pilots' non-technical skills assessments. Operators must provide combined training for flight crew, cabin crew and technical crew during the recurrent CRM training. In the UK, guidance from the regulator (CAA 2014b, p. 166) advises that 'experienced crew-members should leave the training feeling motivated to continue or improve their use of non-technical skills and behaviours in line operations. To be effective, recurrent training material should be more carefully researched than other CRM training course material'.

Crew resource management training is typically classroom based, sometimes with online components, as well as sessions in simulators and in aircraft. Moreover, the CRM principles have to be mentioned during flight crew (technical) training and operations, including the use of checklists, briefings and non-routine procedures. To teach the topics in the classroom, lectures are used, as well as problem-solving exercises, group discussions, scenario analysis, role-plays, task simulations and videos. It is advised that case studies are included, and these can cover specific aircraft events, based on in-house information, for example, accident and incident reviews that can show the role of non-technical skills in increasing flight safety or risk.

Crew resource management trainers have to be approved (for example, CAA 2016) and the basic training for a flight crew CRM ground school trainer is 40 training hours. The trainers responsible for classroom CRM training need to have an adequate knowledge of: (1) the relevant flight operations and (2) human performance and limitations. They should also (3) have completed flight crew initial operator's CRM training, (4) have received training in group facilitation skills, (5) have received additional training in the fields of group management, group dynamics and personal awareness, and (6) have demonstrated the knowledge, skills and credibility required to teach the CRM training elements in the non-operational environment. The training of new CRM trainers is carried out by flight crew CRM trainers with a minimum of three years' experience and then, once qualified, the CRM trainer will have regular assessments (see CAA 2016). Moreover, all trainers of flight crew (not just CRM trainers) have to be able to integrate TEM and CRM during any instruction, and the importance of facilitation skills is increasingly recognised (Kearns et al. 2015). The new CAA (2017) CRM *Handy Guide* provides useful guidance, for example on 'easy to spot markers' of effective CRM training and trainers.

Evidence-Based Training (EBT) in Aviation

In recent years, an EBT approach has been developed in order to orientate pilots' training more directly towards the current situations and conditions encountered in their day-to-day operations. Evidence-based training is intended to enhance the confidence and capability of flight crews to operate the aircraft in all flight regimes and to be able to recognise and manage unexpected situations, and this is now included in the regulations (EASA 2015). The extensive database to support EBT was derived from sources such as opinion surveys of pilots, accident investigations, airlines' flight data monitoring archives, line-orientated safety audits (LOSAs) and other reporting systems. The resulting analyses produced a set of core competencies each with a set of six to 13 positive behavioural indicators (IATA 2013; ICAO 2013). The European version has nine competencies, of which five are non-technical: situation awareness, problem-solving and decision-making, leadership and teamwork, communication, and workload management (EASA 2015). Thus technical and non-technical skills are now listed as one set of competencies, and are interdependent.

Within Europe, airlines such as British Airways, Air France and KLM are beginning to implement EBT. Iberia is now moving to a mixed implementation of EBT (UBF 2017) and the first Lufthansa Group airlines were expecting to begin implementing EBT in 2018 (Read 2017). (For more information on EBT see www.ebt-foundation.com.)

Non-Technical (CRM) Skills Assessment for Pilots

The assessment of a pilot's skills (technical and non-technical) takes place on an aircraft or in a flight simulator as part of the regular licence and operator proficiency checks, and this includes a debriefing. The non-technical skills assessment is based on observable behaviours that reflect skills that result in enhanced safety, as well as those that result in an unacceptable reduction in safety margin. Operators have various procedures, including additional training, to be applied if a pilot does not achieve or maintain the required CRM standards. The assessment practices either follow (1) a competency-based approach which combines technical and non-technical skills into one set of competencies (usually nine), or (2) the earlier approach which has mandated events for the technical skills and a set of non-technical skills categories evaluated with an approved rating method (NOTECHS). These approaches are outlined in Flin (2019) and a description of the NOTECHS method is given below.

NOTECHS (see Van Avermaete and Kruijsen 1988; Flin et al. 2003) is recognised by the regulator (EASA 2017b, GM6 ORO.FC.115) as one of the validated tools for assessing an individual pilot's CRM skills. It consists of four categories: (1) cooperation, (2) leadership and managerial skills, (3) situation awareness and (4) decision-making. Each category is subdivided into elements, and each of these has a set of behavioural markers illustrating good and poor performance. The categories or elements are assessed by a rating scale established by the operator. The assessor must be able to justify the ratings using a standardised vocabulary, and the underlying judgements should not be based on a vague general impression or on an isolated behaviour or action. Repetition of the behaviour during the flight is usually required to explicitly identify the nature of a problem. All the non-technical skills assessment methods are designed provide guidance to look beyond failure during recurrent checks or training, and to help identify possible underlying deficiencies in CRM competence in relation to task performance.

The recommended training period for the use of NOTECHS is two full days or longer (depending on the level of the experience of rating pilots' non-technical skills) (Flin 2019). The assessor should have a good knowledge of CRM concepts and of human performance limitations. Training courses should focus on understanding the NOTECHS method, how to use the rating form, calibration of judgements and debriefing skills. Most European airlines will have their own customised, non-technical skills assessment systems, whereas some smaller operators use NOTECHS or versions of it. The original NOTECHS rating scale had five points, but airlines now use a range of scales in their systems, with four- and five-point scales being commonly adopted. In some companies, there is a requirement for an associated technical failure to accompany a failure based on non-technical skills but in other operators, it is now possible for a pilot's check to be failed on non-technical grounds alone, although this is likely to be an unusual situation.

The above material refers to both fixed-wing and helicopter pilots within civil aviation, but there are some more unusual types of helicopter operations which may require a slightly different set of non-technical skills, as discussed below.

3. HELICOPTER PILOTS' NON-TECHNICAL SKILLS

Helicopters and fixed-wing aircraft undertake different roles based upon their respective capabilities. While fixed-wing aircraft can fly at greater speeds over longer distances, helicopters can operate with minimal ground infrastructure, fly at low altitudes and access locations that would not be possible in a fixed-wing aircraft (de Voogt et al. 2009; Morowsky and Funk 2016). Owing to the flexibility provided by using a helicopter, it can be used for various roles such as transport, emergency medical response, aerial application and search and rescue. Given the environments in which helicopters operate, there are specific risks factors that pilots face, such as obstructions within the environment on land (for example, power lines, trees and tall structures). Offshore helicopter operations come under an increased risk of adverse weather, high winds and obstructions related to offshore structures (HSE 2001). Of all US helicopter accidents from 1982 to 2006, those which occurred close to the ground (such as rollovers) had a lower risk of fatalities, whereas flights that operated within restricted visual conditions, such as at night or in bad weather, were noted to result in a higher rate of fatal accidents (de Voogt and Van Doorn 2007). Accident rates vary based on their operating environment; in a comparison of offshore and onshore helicopter operations in Norway, for example, it was found that differences in pilot training and the effects of employment conditions on decision-making and fatigue management may impact the likelihood of unsafe acts in the cockpit (Bye et al. 2018).

Helicopter pilots experience various pressures in association with their roles. Offshore transport pilots, for example, have been noted to experience commercial pressure to continue to fly despite inclement circumstances, such as technical problems (Gomes et al. 2009). In contrast, emergency response operations may experience time-related pressures relating to a quick response to incidents. The UK search and rescue (SAR) contract holder, Bristow, maintains a 15-minute daytime and 45-minute night-time air readiness position (Bristowgroup 2013). In Norway, air ambulance services are expected to reach the majority of the population within 40 minutes (Zakariassen et al. 2015). For UK SAR, aircraft are expected to arrive on the scene within an hour (Bristowgroup 2014). Within emergency response operations, differences in crew sizes, mission range and helicopter equipment have been noted as having an impact on pilots' decisions and communications on a task. Search and rescue flights typically operate in more remote locations and with more members on board than helicopter air ambulances. The addition of a hoist to SAR aircraft allow the crew to undertake extraction-based rescues. With these unique mission types comes the prospect of diverse errors, suggesting generic countermeasures will be less effective (Morowsky and Funk 2016).

Mitigating the risk of human-error through use of non-technical skills is emphasised within rotorcraft aviation as it is within fixed-wing aviation, with recent incidents highlighting the need for such training in commercial operators. One incident, in August 2013, left four dead after a Super Puma helicopter carrying 16 passengers and two crew members crashed into the sea near Sumburgh airport. The accident report identified lapses in situation awareness and decision-making as primary contributing factors; flight instruments were ineffectively monitored and the chosen approach type was determined suboptimal (AAIB 2016). In response, the CAA in partnership with EASA and the Norwegian Civil Aviation Authority conducted a safety review to examine the risks associated with offshore helicopter operation and compare safety differences. This identified several actions to reduce the fatality of offshore helicopter accidents, such as prohibiting transport flights in the most severe conditions, placing restrictions on flights in serious sea conditions relative to the aircraft's certifications, and placing passengers next to push-out windows unless the helicopter is fitted with side-floats or passengers have emergency breathing apparatus (CAA 2014a).

Helicopter operators within the UK are subject to the same CRM regulations as fixed-wing aircraft and are therefore mandated to undertake modulated CRM training tailored to the basis of their operations manual, part D (EASA 2017b, GM1 ORO.GEN.110(c)). Research has found CRM training within military helicopter crews to be largely effective. Notable improvements in teamwork behaviours during pre-flight and high-workload sections in flight, in addition to an increase in pilots' knowledge of teamwork principles, have been noted when comparing helicopter pilots who have been trained in CRM with a control group (Salas et al. 1999). Military helicopter pilots who underwent standardised team-based training over the course of one week displayed improvements in team communication patterns and resource management, and less team-based errors (Leedom and Simon 1995). Similarly, helicopter pilots who received a one-day training course on concepts related to teamwork, communication and assertiveness demonstrated improved knowledge structure and more desirable teamwork behaviours when assessed in a flight simulation scenario (Stout et al. 1997). Recent research has begun to explore the differences in the utilisation of non-technical skills based upon helicopter flight mission parameters. In a comparison of offshore transport and SAR pilots, differences in the utilisation of core non-technical skills were noted between pilot groups. An additional skill category of cognitive readiness was identified for SAR pilots in order to encompass a range of dynamic behaviours relating to quick response, problem-solving and flexibility (Hamlet et al. 2018).

Some helicopter operators (for example, CHC) are extending their CRM training to all staff currently within UK operations, with the intention of introducing it globally, owing to a belief that everyone is involved in aviation safety and would benefit from the training. However, evaluating the effectiveness of CRM on an organisational level is difficult as there are a variety of factors which may influence safety improvements within the organisation, such as changes to organisational structure or aircraft type (O'Connor et al. 2002a).

4. OFFSHORE OIL AND GAS INDUSTRY

Following the Piper Alpha disaster in 1988 in which 167 men died following explosions and fire on an offshore production platform (Cullen 1990), a number of safety research projects were initiated, including one which examined the possible application of CRM for this sector. Initial work sought to identify the competencies required for safe and efficient offshore operations and produced a framework that presented six key competencies, as illustrated in Figure 18.1.

A two-day CRM course was designed and delivered to groups of offshore crews and was then assessed using self-reported reactions, attitudes and knowledge development. The results indicated that participants reacted positively to the course and reported a positive attitude change towards decision-making and personal limitations, although there did not appear to be a significant change in participant knowledge (O'Connor and Flin 2003). This preliminary evaluation suggested that this type of training could have benefits in the oil and gas industry. However at this time, five years after the Piper Alpha accident, there appeared to be very little interest from the companies and their professional bodies in identifying and training non-technical skills in the offshore industry, until the investigation of a subsequent major accident once again revealed human factors in the pattern of failures that had contributed to that event.

In 2010, the *Deepwater Horizon*, a large drilling rig operating in the Gulf of Mexico, suffered a blowout which killed 11 workers and resulted in an environmental disaster when 5 million barrels of crude oil were released into the sea. In the aftermath, multiple analyses were undertaken to understand what had gone wrong, and the importance of non-technical skill failures in the lead-up to the incident was highlighted (CSB 2018). Specifically, failures in situation awareness, risk perception, decision-making, communication, teamwork and leadership were all identified. Although the regulations for offshore installation safety do not currently stipulate CRM training, advisory bodies, including the Energy Institute (EI 2014) and the International Association of Oil and Gas Producers (IOGP 2014), have recommended CRM training for energy sector workers, especially for offshore crews engaged in well control.

A key component for well control is situation awareness, that is, an individual's awareness and understanding of their environment, together with the ability to anticipate future problems (Endsley 1995). In drilling operations, workers must monitor equipment continuously, be alert to any changes and be able to comprehend early warning signs (Roberts et al. 2015). Team situation awareness, where workers develop a shared knowledge and understanding in order to coordinate and engage in shared

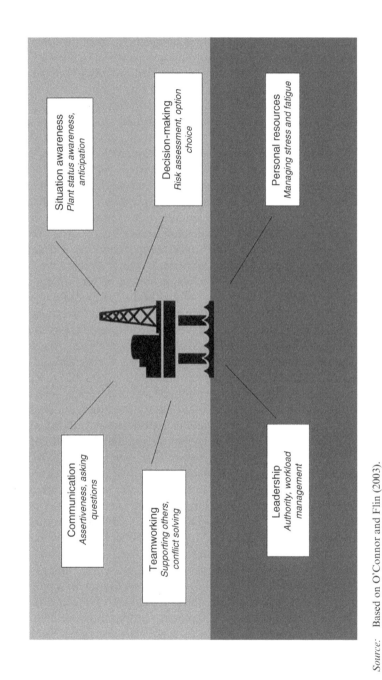

Source: Based on O'Connor and Flin (2003).

Figure 18.1 Six key CRM competencies for offshore crews

decision-making, is required in addition to individual situation awareness. However, concerns have been raised regarding offshore worker situation awareness levels, such as problems owing to lapses in attention (O'Dea and Flin 1998) which have been linked to several factors, including stress, fatigue, lack of experience, poor communication and conflict within a team (Sneddon et al. 2006, 2013). Researchers suggested that management of fatigue and stress should be a key element of any CRM training programme for the offshore sector, as it is for other industries.

More recently, a team-based drilling specific behavioural marker system has been designed using observation data from a high-fidelity drilling simulator which has two 'cyber chairs' for the driller and assistant driller, and also enables participants to engage in control room actions. The set of behavioural markers was developed on the basis of 25 simulator exercises and included four key elements: team situation awareness, team decision-making, teamwork and communication, and team workload and stress management (Crichton et al. 2017). The behavioural ratings were used as the basis for feedback sessions with drill crews, and was reported to lead to enhanced CRM behaviours over a period of five exercises. The authors suggest that introducing CRM principles to an entire drilling team, rather than on an individual basis, should enable crew to support team learning, with some members observed coaching others during exercises. However, crews do not always have the same composition during a project, so there are also benefits in assuring that every individual crew member has competence in non-technical skills.

The International Well Control Forum (IWCF 2018) have recently launched an online training course in CRM specially designed for drilling and other well-control personnel. There have also been developments in using innovative training methods to enhance and assess non-technical skills, such as tactical decision games where participants engage in a detailed decision-making scenario (Crichton et al. 2000) and computer-based monitoring tasks to examine cognitive skills, such as situation awareness (Roberts et al. 2019).

5.   AGRICULTURE

Agriculture is a high-risk industry, with new figures released by the Health and Safety Executive (HSE) reporting 27 fatalities in the UK during 2017–18, which is a mortality rate of 8.44 per 100000 workers (HSE 2018). This fatal injury rate is 16 times higher than the general industry average (HSE 2018), a rate that has been relatively static across the past ten years. On a more global scale, farming is associated with approximately 170000 farm worker fatalities per annum (Douphrate et al. 2013). The main hazards that farmers have to deal with on a daily basis include working at height (HSE 2013), managing livestock (Mitloehner and Calvo 2008) and operating heavy machinery such as tractors and combine harvesters (Rautiainen and Reynolds 2002). In addition to these risks, farmers commonly work alone, sometimes in isolated locations, where help can be slow to arrive if an incident occurs (Olsen and Schellenberg 1986).

Farmers in the UK operate under a regulatory framework laid out by the HSE, this includes guidelines on safe operation of machinery, legal requirements (such as wearing a helmet while operating a quad bike) and recommended protocols (such as performing a 'safe stop' each time a farmer exits a vehicle). Currently this framework does not include distinct training or regulation regarding CRM skills, although non-technical skills are mentioned within some protocols (such as the benefits of planning ahead before conducting a task; HSE 2005). It is within this environment that new research is being conducted to identify relevant non-technical skills, with the long-term aim of developing agriculture-specific behavioural markers, and CRM to enhance farmer safety and help to reduce the accident and fatality rate.

Initial interview-based research sought to identify the skills most relevant and important to farmers (Irwin and Poots 2015). The result was a list of five key skills for team-based farming and a subset of three non-technical skills identified as important when working alone, as illustrated in Figure 18.2.

These skill categories match the key competencies identified within aviation and illustrate the importance of these skills across high-risk industries, even when the required technical knowledge and environment are very different.

The Irwin and Poots (2015) study highlighted the importance of risk assessment and management as an aspect of decision-making, with farmers reporting their consideration of weather conditions, the availability of helpers and the equipment available during risk assessment. A follow-up study examined risk assessment and decision-making within tractor-based scenarios (Irwin and Poots 2018). Fatalities associated with heavy machinery account for 35 per cent of farm-related fatalities in the UK, predominantly caused by being struck by a moving or overturned vehicle (HSE 2015). Despite guidance on safe tractor operation from the HSE, farmers have been reported engaging in unsafe activities such as operating tractors without the appropriate safety equipment (Sorensen et al. 2017).

The aim of the follow-up study was to assess farmer attitudes towards risk management across vignette scenarios that included hazards such as

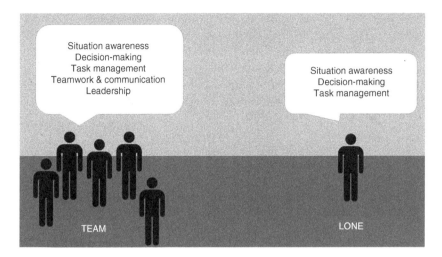

Source: Based on Irwin and Poots (2015).

Figure 18.2   Non-technical skills identified as important when working in a team, and alone, in agriculture

farmer tiredness, faulty equipment and lack of safety equipment (Irwin and Poots 2018). The results identified several areas for further study; farmers did not consider stress as a factor that could influence performance, with many reporting that stress was just part of the job. This is at odds with other research which has identified stress as a causal factor in farm accidents (Kogler et al. 2014). The study also highlighted that farmers were impacted by work-based and financial pressure, with several farmers reporting they would go ahead in each scenario because farm tasks are time critical, there is no one else to complete the task and they need to complete the work to produce the end product.

As research in this sector progresses, CRM training for farmers will eventually become a concern. Devising training methods for farmers may be complicated as one of the main avenues for sharing knowledge and expertise is generational, with each generation educating the next in preparation for succession within the family business (Schreiner et al. 2018). As a consequence, the route into farming in terms of qualifications is diverse, with some farmers primarily learning on the job, others engaging in practical internships and a proportion following an educational route via agricultural colleges (Schreiner et al. 2018). With this in mind, within the US state of California, general farm safety training is delivered through a coalition of farming associations and academic institutions, and comprises multiple short (three hours or less) modules that include interactive activities (Lee et al. 2010). A similar scheme has been launched in the UK, coordinated by the Farm Safety Foundation, with short half-day safety sessions provided for agricultural colleges and young farmer associations. Based on these formats, future CRM training for this population should seek to be flexible, relatively short in duration and delivered on location to ensure participation from farmers.

6. HEALTHCARE

As part of the international drive to improve patient safety, after it was discovered that the rates of adverse events experienced by hospital patients could be as high as 10 per cent of admissions (Vincent 2010), healthcare organisations were encouraged to adopt safety management practices favoured by the high-reliability industries. Helmreich (1997) who had provided key psychological research findings to support the introduction of aviation CRM, realised that this type of approach could also be applied in healthcare. His collaboration with doctors from Switzerland showed that the same CRM skills were required in medical environments, such as operating theatres. From observations of surgical operations, they identified instances of error relating to inadequate teamwork, failures in preparation, briefings, communication and workload distribution. From this research, tools were designed to rate operating team behaviours (Schaefer et al. 1995). Similarly, anaesthesiologists, who were starting to use simulation facilities, began to run CRM-type courses on anaesthesia crisis resource management (ACRM) (Howard et al. 1992).

Studies to identify the non-technical skills required by specific medical professions have resulted in the formulation of skill sets for their CRM training, together with behavioural rating tools, for example, for anaesthetists (anaesthetists' non-technical skills, ANTS; Flin et al. 2010), surgeons (non-technical skills for surgeons, NOTSS; Yule et al. 2008; Flin et al. 2015), scrub practitioners (scrub practitioners' list of intraoperative non-technical skills, SPLINTS; Mitchell et al. 2012) and anaesthetic nurses (Rutherford et al. 2015), and emergency physicians (Flowerdew 2012). Now CRM-type courses are run in many hospital settings, with an evidence base beginning to build on their effectiveness (Haerkens et al. 2015; Schwartz et al. 2018). In the UK, the level of regulatory guidance, independent scrutiny and assessment of CRM in healthcare is currently at a much lower level compared with that in aviation, although the skill sets are being introduced into training. As healthcare students and practitioners learn the value of non-technical skills for clinical efficiency and patient safety, then more formal recognition of their importance in competence frameworks is likely to be introduced.

7. CRM EFFECTIVENESS

The degree to which CRM training is effective, and how that effectiveness should be measured, is still the subject of debate. This is a key concern for organisations that invest in CRM, with training evaluation necessary to support continued financial investment, as a method of continuing to improve training methods, and as a feedback mechanism to allow any issues or areas for development to be identified (Salas et al. 2001). The accepted framework for evaluating training is the Kirkpatrick (1976) typology where reactions, learning, behaviours and organisational effectiveness should all be considered. This allows the consideration of multiple learning outcomes, providing a holistic evaluation picture.

Reactions

Reactions to CRM training in aviation are generally positive, with participants reporting that they like the programme and find it interesting (Salas et al. 2001). There is some variation in response to the methods of training used, with participants more positive towards interactive activities such as role-play, rather than passive learning provided by a lecture format (Schiewe 1995). Similarly, a meta-analysis of research evaluating CRM training in healthcare settings indicated that reactions to the training were positive, with participants reporting that the training was relevant and likely to improve patient safety (O'Dea et al. 2014).

Learning

Within aviation, learning has been assessed via the Cockpit Management Attitudes Questionnaire (CMAQ; Helmreich 1984), with studies indicating a positive attitude change after training (Salas et al. 2001). However, although the CMAQ continues to be used as a form of assessment, it does not cover the full range of CRM competencies, most notably excluding cognitive aspects such as situation awareness (O'Connor et al. 2002b). Researchers generally report positive learning outcomes for aviation CRM, such as increased confidence and heightened levels of awareness (Simpson and Wiggins 1995). Knowledge gains and attitude change have also been reported within the fire service (Griffith and Roberts 2015) and healthcare (O'Dea et al. 2014), although the reported effect sizes were small within healthcare. One study of aviation operators indicated that fewer than 25 per cent of the respondents conducted an attitude evaluation, this number dropped even lower for small operators (15 per cent). Companies reported they were not always sure of how to approach measurement, indicating a need for a rigorous and usable method of evaluation (O'Connor et al. 2002b). Salas et al. (2001) also note that the long-term effects of training should be considered in more detail, with some studies (for example, Helmreich 1991) indicating that the positive effects recorded immediately after training can decline over time.

Behaviour

Behavioural evidence collected via observations of simulation exercises or online tasks indicates a positive change in behaviour following CRM training in aviation (Salas et al. 2001). Behaviour change is usually assessed via behavioural checklists, such as NOTECHS, with researchers reporting a range of positive changes, including improved decision-making, teamwork and leadership behaviours (Salas et al. 2001). A similar pattern has also been reported within healthcare, although the authors note that there is a great deal of variation in the evaluation methods used (O'Dea et al. 2014).

Organisational Effectiveness

This level of evaluation is the most complex and difficult to determine and so it is perhaps not surprising that very few research studies have sought to address this element. In their meta-analysis, Salas et al. (2001) report that data in this area are predominantly based on anecdotal reports, with only a few researchers attempting to collect stronger evidence. There is some suggestion that CRM training may reduce discrepancy reports (Byrnes and Black 1993), although that data may be subject to environmental confounds. Similar issues are reported within the healthcare literature, with methods of evaluating the impact of training on clinical outcomes highly varied and with low, or no, effects reported (O'Dea et al. 2014). Some headway is being made however; for example, one study used a three-year prospective design to examine paediatric clinical outcomes across a baseline (no training), implementation (CRM training provided) and assessment year. The results indicated an association between CRM training and a reduction in serious complications and mortality rate (Haerkens et al. 2015).

In order to evaluate the effectiveness of CRM training, it is important that organisations are encouraged to utilise appropriate evaluation methods. There appears to be a need for a toolkit of rigorous, valid, methods of evaluation, together with appropriate guidance, that can be used to work towards solving the issues of variation across companies in the type, and usefulness, of data collected. It is also apparent that training in the use of such methods is required to ensure the collected data are valid. Finally, the key outcome that the majority of users are concerned with is the impact of CRM on safety and, as that is the aspect currently most under-researched, it is important that future research focuses on this issue and explores novel methods of evaluation.

REFERENCES

Air Accidents Investigation Branch (AAIB) (2016), 'Report on the accident to AS332 L2 Super Puma helicopter, G-WNSB on approach to Sumburgh Airport on 23 August 2013', accessed 1 October 2018 at https://assets.publishing.service. gov.uk/media/56e7eaeaed915d0379000023/AAR_1-2016_G-WNSB.pdf.

Boettcher, J. (2016), 'Regulatory update', paper presented at the EASA Workshop 'CRM in Practice', Cologne, 8 November, accessed 16 August 2018 at https:// www.easa.europa.eu/newsroom-and-events/events/easa-crm-workshop-%E2%80 %9Ccrm-practice%E2%80%9D.

Bristow Group (2013), 'Bristow helicopter to provide UK search and rescue', accessed 20 July 2018 at http://bristowgroup.com/_assets/filer/2013/09/04/bristo wworld-25-lowres.pdf.

Bristow Group (2014), 'UK search and rescue partners – frequently asked questions', accessed 20 July 2018 from: http://www.scottishmountainrescue.org/ wp-content/uploads/2014/05/Bristows-SAR-H-FAQ.pdf.

Bureau d'Enquêtes et d'Analyses pour la sécurité de l'aviation civile (BEA) (2012), *Final Report. Flight AF 447 on 1st June 2009 A330-203*, Paris: BEA.

Bye, R.J., Johnsen, S.O. and Lillehammer, G. (2018), 'Addressing differences in safety influencing factors – a comparison of offshore and onshore helicopter operations', *Safety*, **4** (1), 4.

Byrnes, R.E. and Black, R. (1993), 'Developing and implementing CRM programs: the Delta experience', in E.L. Wiener, B.G. Kanki, and R.L. Helmreich (eds), *Cockpit Resource Management*, San Diego, CA: Academic Press, pp. 421–43.

Civil Aviation Authority (CAA) (2014a), *Safety Review of Offshore Public Transport Helicopter Operations in Support of the Exploitation of Oil and Gas*, CAP1145, February version, London: Civil Aviation Authority.

Civil Aviation Authority (CAA) (2014b), *Flight-Crew Human Factors Handbook*, CAP737, December 2016 version, London: Civil Aviation Authority.

Civil Aviation Authority (CAA) (2016) *Guidance on the Requirements that Pertain to Flightcrew for the Training and Testing of Human Factors under EASA Part-ORO and EASA Part-FCL*, Standards Document No. 29, version 7, London: Civil Aviation Authority.

Civil Aviation Authority (CAA) (2017) *Practical Crew Resource Management (CRM) Standards: The Handy Guide*, CAP 1607, London: Civil Aviation Authority.

Crichton, M., Flin, R. and Rattray, W. (2000), 'Training decision makers – tactical decision games', *Journal of Contingencies and Crisis Management*, **8** (4), 208–17.

Crichton, M., Moffat, S. and Crichton, L. (2017), 'Developing a team behavioural marker framework using observations of simulator-based exercises to improve team effectiveness: a drilling team case study', *Simulation and Gaming*, **48** (3), 299–313.

Chemical Safety and Hazard Investigation Board (CSB) (2018), *Drilling Rig Explosion and Fire at the Macondo Well. Investigation Report 3*, (Organizational and human factors), No. 2010-10-I-OS 4/17/2016, Washington, DC: US Chemical Safety and Hazard Investigation Board.

Cullen, D. (1990), *The Public Inquiry into the Piper Alpha Disaster*, (CM1310), London: HMSO.

De Voogt, A.J. and van Doorn, R.A. (2007), 'Helicopter accident analysis: datamining the NTSB database', in Curran Associates, Inc., *Proceedings of the European Rotorcraft Forum, Kazan, Russia, September 11–13*, Kazan: European Rotorcraft Forum 33 (ERF 33), pp. 2517–23.

De Voogt, A.J., Uitdewilligen, S. and Eremenko, N. (2009), 'Safety in high-risk helicopter operations: the role of additional crew in accident prevention', *Safety Science*, **47** (5), 717–21.

Douphrate, D.I., Stallones, L., Lostrup, C.L., Nonnenmann, M.W., Pinzke, S., Hagevoort, G.R., et al. (2013), 'Work-related injuries and fatalities on dairy farm operations – a global perspective', *Journal of Agromedicine*, **18** (3), 256–64.

Drath, K. (2017), *Resilient Leaders*, Oxford: Routledge.

Endsley, M.R. (1995), 'Toward a theory of situation awareness in dynamic systems', *Human Factors*, **37** (1), 32–64.

Energy Institute (EI) (2014), *Guidance on Crew Resource Management (CRM) and Non-Technical Skills Training Programmes*, London: Energy Institute.

European Union Aviation Safety Agency (EASA) (2015), *ED Decision 2015/027/R Implementation of Evidence-Based Training (EBT) within the European Regulatory Framework*, accessed 16 February 2018 at https://www.easa.europa.eu/document-library/agency-decisions/ed-decision-2015027r.

European Union Aviation Safety Agency (EASA) (2017a), *Crew Resource Management in Practice. Version 1, December 2017*, accessed 9 February 2018 at https://www.easa.europa.eu/document-library/general-publications/crm-training-implementation#group-easa-downloads.

European Union Aviation Safety Agency (EASA) (2017b), *Acceptable Means of Compliance (AMC) and Guidance Material (GM) to annex III Organisation Requirements for Air Operations (Part-ORO)*, accessed 29 September 2018 at https://www.easa.europa.eu/sites/default/files/dfu/Consolidated%20unofficial%20 AMC-GM_Annex%20III%20Part-ORO%20Dec%202017.pdf.

Flin, R. (2019), 'CRM training and non-technical skills assessment: a European perspective', in B. Kanki, J. Anca and T. Chidester (eds), *Crew Resource Management*, 3rd edn, San Diego, CA: Elsevier.

Flin, R., Martin, L., Goeters, K.-M., Hörmann, H.-J., Amalberti, R., Valot, C., et al. (2003), 'Development of the NOTECHS (Non-Technical Skills) system for assessing pilots' CRM skills', *Human Factors and Aerospace Safety*, **3** (2), 97–119.

Flin, R., O'Connor, P. and Crichton, M. (2008), *Safety at the Sharp End. A Guide to Non-Technical Skills*, Aldershot: Ashgate.

Flin, R., Patey, R., Glavin, R. and Maran, N. (2010), 'Anaesthetists' non-technical skills', *British Journal of Anaesthesia*, **105** (1), 38–44.

Flin, R., Youngson, G. and Yule, S. (2015), *Enhancing Surgical Performance. A Primer in Non-Technical Skills*, London: CRC Press.

Flowerdew, L., Brown, R., Vincent, C. and Woloshynowych, M. (2012), 'Development and validation of a tool to assess emergency physicians' nontechnical skills', *Annals of Emergency Medicine*, **59** (5), 376–85.

Gomes, J.O., Woods, D.D., Carvalho, P.V., Huber, G.J. and Borges, M.R. (2009), 'Resilience and brittleness in the offshore helicopter transportation system: the identification of constraints and sacrifice decisions in pilots' work', *Reliability Engineering and System Safety*, **94** (2), 311–19.

Griffith, J.C. and Roberts, D.L. (2015), 'A meta-analysis of crew resource management/incident command systems implementation studies in the fire and emergency services', *Journal of Aviation/Aerospace Education and Research*, **25** (1), 1–25.

Haerkens, M., Kox, M., Lemson, J., Houterman, S., Van Der Hoeven, J. and Pickkers, P. (2015), 'Crew resource management in the intensive care unit: a prospective 3-year cohort study', *Acta Anaesthesiologica Scandinavica*, **59** (10), 1319–29.

Hamlet, O., Irwin, A. and McGregor, M. (2018), 'Is it all about the mission? Comparing non-technical skills across offshore transport and search and rescue helicopter pilots', accessed 6 September 2018 at https://doi.org/10.31234/osf.io/qpj2f.

Health and Safety Executive (HSE) (2001), *Helicopter Safety Offshore*, London: HSE.

Health and Safety Executive (HSE) (2005), *Farmers, Farm Workers and Work-Related Stress, 2005*, London: HSE.

Health and Safety Executive (HSE) (2013), *Health and Safety in Agriculture in Great Britain, 2013*, London: HSE.

Health and Safety Executive (HSE) (2015), *Health and Safety in Agriculture, Forestry and Fishing in Great Britain, 2014/15*, London: HSE.

Health and Safety Executive (HSE) (2018), *Workplace Fatal Injuries in Great Britain 2018*, London: HSE.

Helmreich, R. (1997), 'Managing human error in aviation', *Scientific American*, May, 62–7.

Helmreich, R. and Foushee, C. (2010), 'Why CRM? Empirical and theoretical bases of human factors training', in B. Kanki, R. Hemreich and J. Anca (eds), *Crew Resource Management*, 2nd edn, San Diego, CA: Academic Press, pp. 3–45.

Helmreich, R. and Merritt, A. (1998), *Culture at Work in Aviation and Medicine. National, Organizational and Professional Influences*, Aldershot: Ashgate.

Helmreich, R.L. (1984), 'Cockpit management attitudes', *Human Factors*, **26** (5), 583–89.

Helmreich, R.L. (1991), 'Strategies for the study of flight crew behavior', in R. Jensen (ed.), *Proceedings of the 6th International Symposium on Aviation Psychology*, Columbus, OH: Ohio State University, pp. 338–43.

Howard, S., Gaba, D., Fish, K., Yang, G. and Sarnquist, F. (1992), 'Anesthesia crisis resource management training: teaching anesthesiologists to handle critical incidents', *Aviation, Space, and Environmental Medicine*, **63** (9), 763–70.

International Air Transport Association (IATA) (2013), *Evidence-Based Training Implementation Guide*, Montreal: IATA.

International Civil Aviation Organization (ICAO) (2013), *Manual of Evidence-Based Training. DOC 9995*, Montreal: ICAO.

International Oil and Gas Producers (IOGP) (2014), *Crew Resource Management for Well Operations Teams*, Report 501, London: IOGP, accessed 17 August

2018 at https://www.iogp.org/bookstore/product/crew-resource-management
-for-well-operations-teams/.
International Well Control Forum (IWCF) (2018), 'IWCF launches global crew
resource management digital initiative', accessed 17 August 2018 at http://
www.iwcf.org/news/news/142-iwcf-launches-global-crew-resource-management-
digital-initiative.
Irwin, A. and Poots, J. (2018), 'Investigation of UK farmer go/no-go decisions in
response to tractor-based risk scenarios', *Journal of Agromedicine*, **23** (2), 154–65.
Irwin, A. and Poots, J. (2015), 'The human factor in agriculture: an interview study
to identify farmers' non-technical skills', *Safety Science*, **74** (April), 114–21.
Kanki, B., Anca, J. and Chidester, T. (eds) (2019), *Crew Resource Management*, 3rd
edn, San Diego, CA: Elsevier.
Kanki, B., Hemreich, R. and Anca. J. (eds) (2010), *Crew Resource Management*,
2nd edn, San Diego, CA: Academic Press.
Kearns, S., Mavin, T. and Hodge, S. (2015), *Competency-Based Education in
Aviation: Exploring Alternate Training Pathways*, London: Routledge.
Kirkpatrick, D.L. (1976), 'Evaluation of training', in R. Craig (ed.), *Training and
Development Handbook: A Guide to Human Resources Development*, New York:
McGraw-Hill. pp. 18.1–18.27.
Kogler, R., Quendler, E. and Boxberger, J. (2014), 'Accident at work with fertilizer
distributors in Austrian agriculture', *Agricultural Engineering International:
CIGR Journal*, **16** (3), 157–65.
Landsman, A., Groen, E., van Paassen, M., Bronkhorst, A. and Mulder, M.
(2017), 'Dealing with unexpected events on the flight deck: a conceptual model
of startle and surprise', *Human Factors*, **59** (8), 1161–72.
Lee, B.C., Wolfe, A. and Meyers, J.M. (2010), 'Agricultural safety training:
California style', *Journal of Agromedicine*, **15** (3), 300–306.
Leedom, D.K. and Simon, R. (1995), 'Improving team coordination: a case for
behavior-based training', *Military Psychology*, **7** (2), 109–22.
Martin, W. (2019), 'CRM and individual resilience', in B. Kanki, J. Anca and
T. Chidester (eds), *Crew Resource Management*, 3rd edn, San Diego, CA:
Elsevier, ch. 7
Mitchell, L., Flin, R., Yule, S., Mitchell, J., Coutts, K. and Youngson, G. (2012),
'Evaluation of the scrub practitioners' list of intraoperative non-technical skills
(SPLINTS) system', *International Journal of Nursing Studies*, **49** (2), 201–11.
Mitloehner, F.M. and Calvo, M.S. (2008), 'Worker health and safety in concentrated
feeding operations', *Journal of Agricultural Safety and Health*, **14** (2), 163–87.
Morowsky, K. and Funk, K.H. (2016), 'Understanding differences in helicopter
mission sets prior to human error analysis', *Proceedings of the Human Factors
and Ergonomics Society Annual Meeting*, **60** (1), 1439–43.
O'Connor, P. and Flin, R. (2003), 'Crew resource management training for offshore
production platform teams', *Safety Science*, **41** (7), 591–609.
O'Connor, P., Flin R., Fletcher, G. (2002a), 'Methods used to evaluate the effec-
tiveness of CRM training: a literature review', *Journal of Human Factors and
Aerospace Safety*, **2** (3), 217–33.
O'Connor, P., Flin, R., Fletcher, G. and Hemsley, P. (2002b), 'Methods used to
evaluate the effectiveness of CRM training: a survey of UK aviation operators',
Journal of Human Factors and Aerospace Safety, **2** (3), 235–56.
O'Dea, A. and Flin, R. (1998), 'Site managers and safety leadership in the offshore
oil and gas industry', *Safety Science*, **37** (1), 39–57.

O'Dea, A., O'Connor, P. and Keogh, I. (2014), 'A meta-analysis of the effectiveness of crew resource management training in acute care domains', *Postgraduate Medical Journal*, **90** (1070), 699–708.

Olsen, K.R. and Schellenberg, R.P. (1986), 'Farm stressors', *American Journal of Community Psychology*, **14** (5), 555–69.

Rankin, A., Lundberg, J., Woltjer, R., Rollenhagen, C. and Hollnagel, E. (2014), 'Resilience in everyday operations: a framework for analyzing adaptations in high-risk work', *Journal of Cognitive Engineering and Decision Making*, **8** (1), 78–97.

Rautiainen, R.H. and Reynolds, S.J. (2002), 'Mortality and morbidity in agriculture in the United States', *Journal of Agricultural Safety and Health*, **8** (3), 259–76.

Read, B. (2017), 'Training for the new millennium', *Aerospace*, December, 28–31.

Roberts, R., Flin, R. and Cleland, J. (2015), 'Staying in the zone: offshore drillers' situation awareness', *Human Factors*, **57** (4), 573–90.

Roberts, R., Flin, R., Cleland, J. and Urquhart, J. (2019), 'Drillers' cognitive skills monitoring task', *Ergonomics in Design*, **27** (2), 13–20.

Rutherford, J., Flin, R., Irwin, A. and McFadyen, A. (2015), 'Evaluation of the prototype Anaesthetic Non-technical Skills for Anaesthetic Practitioners (ANTS-AP) system: a behavioural rating system to assess the non-technical skills used by staff assisting the anaesthetist', *Anaesthesia*, **70** (8), 907–14.

Salas, E., Burke, C.S., Bowers, C.A. and Wilson, K.A. (2001), 'Team training in the skies: does crew resource management (CRM) training work?', *Human Factors*, **43** (4), 641–74.

Salas, E., Fowlkes, J.E., Stout, R.J., Milanovich, D.M. and Prince, C. (1999), 'Does CRM training improve teamwork skills in the cockpit? Two evaluation studies', *Human Factors*, **41** (2), 326–43.

Schaefer, H., Helmreich, R. and Scheidegger, D. (1995), 'Safety in the operating theatre – part 1: interpersonal relationships and team performance', *Current Anaesthesia and Critical Care*, **6**, 48–53.

Schiewe, A. (1995), 'On the acceptance of CRM-methods by pilots: results of a cluster analysis', in R.S. Jensen and L.A. Rakovan (eds), *Proceedings of the 8th International Symposium on Aviation Psychology*, Columbus, OH: Ohio State University, pp. 540–45.

Schreiner, L., Levkoe, C.Z. and Schumilas, T. (2018), 'Categorizing practical training programs for new farmers', *Journal of Agriculture, Food Systems, and Community Development*, **8** (2), 1–9.

Schwartz, M., Welsh, D., Paull, D., Knowles, R., DeLeeuw, L., Hemphill, R. et al. (2018), 'The effects of crew resource management on teamwork and safety climate at Veterans Health Administration facilities', *Journal of Healthcare Risk Management*, **38** (1), 17–37.

Simpson, P. and Wiggins, M. (1995), 'Human factor attitudes', in B.J. Hayward and A.R. Lowe (eds), *Applied Aviation Psychology: Achievement, Change, and Challenge. Proceedings of the Third Australian Aviation Psychology Symposium*, Aldershot: Avebury Aviation, pp. 185–92.

Sneddon, A., Mearns, K. and Flin, R. (2006), 'Situation awareness and safety in offshore drill crews', *Cognition, Technology and Work*, **8** (4), 255–67.

Sneddon, A., Mearns, K. and Flin, R. (2013), 'Stress, fatigue, situation awareness and safety in offshore drilling crews', *Safety Science*, **56** (July), 80–88.

Sorensen, J.A., Tinc, P.J., Weil, R. and Droullard, D. (2017), 'Symbolic interactionism: a framework for understanding risk-taking behaviors in farm communities', *Journal of Agromedicine*, **22** (1), 26–35.

Stout, R.J., Salas, E. and Kraiger, K. (1997), 'The role of trainee knowledge structures in aviation team environments', *International Journal of Aviation Psychology*, **7** (3), 235–50.

Use Before Flight (UBF) (2017), EBT blog, accessed 16 December 2018 at https://evidencebased.training/blog/2017/03/31/news-and-events/.

Van Avermaete, J. and Kruijsen, E. (eds) (1998), 'NOTECHS: non-technical skill evaluation in JAR-FCL', NLR-TP-98518, National Aerospace Laboratory (NLR), Amsterdam.

Vincent, C. (2010) *Patient Safety*, Chichester: Wiley Blackwell.

Yule, S., Flin, R., Maran, N., Rowley, D., Youngson, G. and Paterson-Brown, S. (2008), 'Surgeons' non-technical skills in the operating room: reliability testing of the NOTSS behaviour rating system', *World Journal of Surgery*, **32** (4), 548–56.

Zakariassen, E., Uleberg, O. and Røislien, J. (2015), 'Helicopter emergency medical services response times in Norway: do they matter?', *Air Medical Journal*, **34** (2), 98–103.

19. Health protection and health promotion in small business

Natalie V. Schwatka, Liliana Tenney and Lee S. Newman

INTRODUCTION

Owing to the large number of workers throughout the world who work for small enterprises (United States Census Bureau 2016; Papadopoulos et al. 2018), strategies are needed for improving how such businesses address safety, health and well-being. This chapter emphasizes some of the challenges and approaches being used to promote safer and healthier conditions in these smaller work settings.

Emerging lines of evidence suggest that a more holistic approach is needed to optimize worker health, safety and well-being in businesses of all sizes and across all industrial sectors. Job injuries and fatalities persist at unacceptably high rates (Concha-Barrientos et al. 2005). The workforce, and the rest of society, suffers from chronic health conditions that are both related and unrelated to work, and at increasing rates (Sorensen et al. 2011). Research shows that job hazards impact personal health. Inversely, personal health impacts job safety and performance (Schulte et al. 2012). In addition, personal health conditions and health behaviors have the potential to improve when organizations engage in health promotion (Goetzel et al. 2012). Worker well-being is gaining traction as a unifying central objective in the occupational safety and health field that can be achieved by enhancing workplace physical environment and safety climate, workplace policies and culture, worker health status, and contributors to the nature of work and the work experience (Schill and Chosewood 2013; Chari et al. 2018). This chapter focuses on the application of an integrated approach, referred to by the US Centers for Disease Control's National Institute for Occupational Safety and Health (NIOSH) as Total Worker Health® (TWH). The goal is to address safety first, while also contributing to health promotion, so that employees end their work shift intact and at least as healthy, if not healthier, than when they started (Schulte et al. 2012; NIOSH 2016).

In this chapter, we approach the topic by first providing an overview of TWH and our mechanistic model by which TWH policies, programs, and practices impact worker health, safety and well-being outcomes. Second, within this framework, we discuss why small businesses require special consideration and provide an overview of how TWH concepts can be applied. Through case studies, we illustrate the application of TWH principles in practice. The chapter concludes with a discussion of research needs, based on identified gaps in published research literature and stemming from the questions raised in our case studies.

OVERVIEW OF TOTAL WORKER HEALTH

Total Worker Health is a growing field that is defined as policies, programs and practices that integrate protection from work-related safety and health hazards with promotion of injury and illness prevention efforts to advance worker well-being (NIOSH 2018). The NIOSH developed TWH as a holistic approach to keeping workers safe as the foundation, and integrating a broad spectrum of workplace interventions to also improve worker health and well-being. As one of NIOSH's six national TWH Centers of Excellence, we consider occupational health and safety as a platform to not only prevent work-related injuries, illnesses and fatalities, but also to treat the workplace as a social determinant of health that can enhance health and well-being (Schill and Chosewood 2013). Safety comes first. However, when employers also steward health, they benefit the workforce, communities and economies at large (Robert Wood Johnson Foundation 2016).

Total Worker Health researchers identified key organizational characteristics that are linked to positive business and worker outcomes as well as with high levels of health and safety integration (Sorensen et al. 2016; McLellan et al. 2017). Organization leadership and support for a developing a climate of safety and health are pivotally important. Total Worker Health leadership reflects practices that demonstrate business values, mission and strategy to achieve a high level of health, safety, and well-being and a climate of both health and safety. Total Worker Health climate combines elements of both safety climate and health climate and reflects whether employees perceive that their organization supports these practices to protect and enhance their safety, health and well-being (Schwatka et al. 2018c). Ultimately, both leadership and climate contribute to the organization's TWH culture (Denison 1996).

Internationally, groups are calling upon business to create environments that keep workers safe and promote health and well-being. In 2006, the International Labour Organization convened world leaders to discuss workers' health and a global plan of action to tackle 'the enormous humanitarian and economic cost of work-related accidents and diseases globally'. The last assembly resulted in an international instrument linking occupational health and public health to consider not only work-related hazards, but also social and individual factors and access to health services (International Labour Organization 2006). In 2015, the United Nations called upon the private sector to prioritize healthy business practices, including establishing tobacco-free workplaces and safe and healthy workplaces (Sustainable Development Goals Fund, Harvard Kennedy School: Corporate Social Responsibility Initiative, & Business Fights Poverty 2015). In 2018, the International Social Security Administration launched the Vision Zero campaign to raise global awareness around three core values – safety, health and wellbeing at work – in an effort to create a healthy and motivated workforce (International Social Security Association 2018). These agencies stress the importance of worker health as a prerequisite for productivity and economic development. As a core part of these efforts, they also communicate that in order to have real impact, steps must be taken to provide full coverage of health and safety for all workers, 'including those in the informal economy, small- and medium-sized enterprises' (WHO 2007, p. 2).

To address these challenges, occupational health and safety researchers and practitioners must think not only about how to find the evidence of effectiveness, but also conceptualize what leadership practices and climate perceptions lend themselves to be of practical use to the highly heterogeneous small business sector. Our model, the Colorado TWH model, stems from the fundamental view that achieving TWH begins by engineering organizational change through leadership and climate (Schwatka et al. 2018c). Our model presents the levers that improve attitudes, knowledge and behavior at the community, workplace and worker levels.

From Theory to Practice: The Colorado Total Worker Health Model

To facilitate TWH research and practice, we propose a mechanism model displayed in Figure 19.1 by which community and organizational TWH interventions improve employee and employer outcomes. Since we believe that the creation of a culture of safety and a culture of health and wellbeing requires both organizational behavior change and individual behavior change, we have merged core concepts from two existing theoretical models: Burke and Signal's (2010) multi-level model of safety and Burke and Litwin's (1992) model of organizational change.

Occupational safety and health professionals are familiar with models of safety, but may be less familiar with causal models of organizational

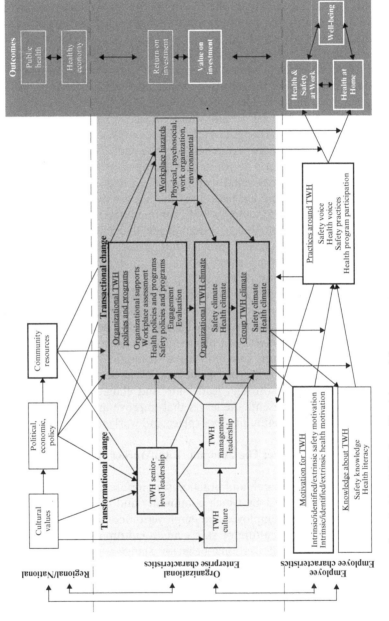

Figure 19.1 Colorado Total Worker Health model

performance and change. In 1992, Burke and Litwin combined a number of theories into a composite view of organizational structure as it relates to change. The so-called Burke–Litwin model categorizes organization behaviors into two categories: transformational and transactional. Transformational elements are deeply embedded in an organization, and include leadership, organizational culture and mission or strategy. They require more from top leadership to effect change that transforms the rest of the enterprise. Transactional elements address more of the day-to-day operations of the business and depend more on performance and rewards that are usually controlled at the managerial level. It is at the transactional level that management practices result in the execution of an organization's strategy, including the activities that directly impact performance, such as systems (policies and procedures), work unit climate, motivation or incentives, matching of work tasks and abilities, and meeting individual workers' needs and values with that of the organization.

For each element in Figure 19.1, it is possible to identify opportunities for practitioners and employers to improve workplace practices and for researchers to develop and test hypotheses to address health promotion and protection at work. The model applies to businesses of any size, although the emphasis for intervention design will depend, in part, on the enterprise, including its size. In this chapter, we discuss the model especially as it relates to small enterprises. This model is predicated on the notion that alignment of both organizational-level goals and outcomes and individual-level goals and outcomes contribute to success in meeting the business's mission (for example, sustainability, profitability, and quality of goods and services), a healthier economy, and a healthier, safer and more productive workforce.

WHY SMALL BUSINESS?

There are many ways to define small business. The US Small Business Administration, for example, defines a small business as having less than 100 to 1500 employees and typically makes a maximum of $750 000 to $38.5 million in annual revenue, depending on industry (US Small Business Administration 2018). Internationally, small and medium-sized enterprises (SMEs) have been studied extensively owing to their dominance in employment and the key constraints they face, including financing and sustainability, and the barriers they face in relation to occupational health and safety (International Labour Organization 2015).

However, to date there have been few studies that describe what TWH leadership and TWH climate look like in practice (Schwatka et al. 2018c).

Specific to small business, there is even less known about how these firms are currently implementing TWH and what approaches are effective and sustainable in meeting the goals of changing organizational performance and improving worker health, safety and well-being (Anger et al. 2015). We lack insight into the perspective of small business owners about health and safety in general.

In order to begin to understand small business leadership views on TWH, with the goal of identifying better ways of engaging and effecting organizational change, we recently conducted semi-structured interviews with a diverse group of 18 small business leaders. A qualitative coding approach was used to determine themes regarding leaders' practices for health, safety and well-being (Thompson et al. 2018). Common themes that emerged were that leaders mainly focus on business health, secondarily focus on their own health and to a much lesser extent consider the health of their employees. Our interviews demonstrate that when small business leaders express interest in the health of their employees, it is largely because of the perceived value to their business. Also, our interviews confirm that these small business leaders often lack the knowledge and skills needed to successfully implement TWH, or any occupational safety and health agenda (Thompson et al. 2018).

There are data showing that SMEs in both developed and developing countries require different considerations when we design TWH interventions, compared with large, well-resourced organizations. Also, other factors must be considered in addition to the numerical size of organizations (International Labor Organization 2015). Some research has enumerated the contextual factors that should be taken into account when designing and implementing TWH in small businesses, including organizational structure (Cunningham et al. 2014). Globally, there is strong support that the small business sector cannot be considered a homogeneous group (Cunningham et al. 2014), adding to the potential complexity of developing effective solutions that can be readily adopted, implemented and scaled across the range of SMEs.

Small business owners report that there are many service providers approaching them, ranging from insurance brokers, to vendors, to public health professionals – all seeking the attention of owners and administrators who face competing demands of running a business. These businesses are focused on the bottom line (that is, staying in business, being profitable and growing). After meeting the payroll, most small businesses spend their money on technology, marketing, insurance and other operating costs (Storey 1994). The pressure is great. While many small businesses are striving to grow, one in five will close within the first two years, and the majority of those that are small today are likely to remain small-scale operations

(Storey 1994). In this environment, we have a lot to learn about ways to make the case for how TWH can benefit workers and organizations, and why it is worthwhile to invest in TWH interventions (Anger et al. 2015).

The good news is that workplace health is emerging as a topic of interest internationally. In 2010, the World Health Organization (WHO) published the 'WHO healthy workplace framework and model: background and supporting literature and practices' (Burton 2010). They stated in the publication that, at the time, if you searched the term 'healthy workplace' using the Google search engine, it resulted in about 2 million results. About a decade later, in 2018, the search yielded over 207 million results. An informal reading of this sea of information on the Internet reveals how many different definitions, interpretations, models, tools and types of business services have surfaced. This can be potentially beneficial, but also creates potential confusion. Employers, especially those working in resource-poor, time-pressured small enterprises, will find it challenging to identify trusted sources of evidence-based approaches. Part of our objective in this chapter is to help identify the role that occupational health and safety professionals can play in helping businesses cut through the sea of information. To begin with, we must find ways to effectively reach, educate and connect with small business leaders and influencers to change organizational behaviors that effect TWH outcomes. In the next section of this chapter, we describe one approach to use with small businesses to target both transformational and transactional change in support of better health outcomes.

A MODEL OF TWH IN SMALL BUSINESSES

In this section, we provide examples of the ways in which small businesses differ from larger businesses in the context of our TWH model in Figure 19.1. Building on this, we describe our approach to helping small businesses apply emerging best practices.

In our experience in working with hundreds of small enterprises, we find that small business owners and managers are quick to recognize the value of investing (VOI) in TWH and are less likely to expect to receive a return on investment (ROI) from health, safety, or well-being initiatives (Schwatka et al. 2018c). In the context of TWH, ROI measures the amount of money saved after investing money into a TWH initiative. Some research indicates that there is an ROI for safety programs (OSHA 2018) and worksite wellness programs (Baxter et al. 2014; Goetzel et al. 2014). However, most of these studies reflect ROIs calculated with large companies. Smaller businesses have a harder time calculating ROI. Work-related injuries are infrequent, which keeps both the direct (for example, workers' compensation) and indirect (for example, absence) costs low. Small businesses in countries that do not have national healthcare coverage, such as the US, may not provide health insurance benefits for their employees, and thus do not have a large financial risk tied to employee healthcare expenses. The VOI, on the other hand, measures other benefits of TWH programs. Value-of-investing measures include health and safety indicators, such as health risks and work-related injuries, as well as productivity and employee engagement (National Business Group on Health 2014). The VOI can produce 'soft' ROI in the form of lower absenteeism, more success in hiring in a competitive market and in retention of employees. In our work with small businesses, we have learned that small businesses are looking to attract and retain top talent, improve employee engagement and boost productivity.

Businesses of any size may benefit when leaders and managers implement TWH principles and practices (NIOSH 2016). However, the importance of owners and other senior-level managers is more pronounced among smaller businesses than it is among larger businesses (Schwatka et al. 2018c). These leaders are in a more powerful position to influence transactional and transformational organizational change (Burke and Litwin 1992). Small business owners and senior managers play major roles in setting TWH values and bear much of the direct responsibility for ensuring that business practices align with these values. Therefore, senior small business leaders who build their TWH leadership skills will facilitate a more positive TWH culture that supports the stability, depth, breadth and integration of TWH into their business (Schein 2010). When small business leaders work on their TWH climate, they are able to communicate a shared understanding of how to behave and think in ways that facilitate good health. These shared understandings then become frames of reference that all employees can draw upon when making business and personal decisions that affect employee health (Clarke 2013; Schwatka et al. 2018c).

There are several reasons why owners and senior managers are particularly influential in small businesses. First, by nature, small businesses have fewer units, department, locations, and so on. This narrows the workplace social network and increases the direct influence that they exert over business practices and employee perceptions of these practices (Gray et al. 2003). Second, there may be lower turnover among small business leadership teams, depending on the industry, age and structure of the business, especially in family-owned businesses. Lower management-level turnover rates can help small businesses maintain continuity and consistent messaging around business, as well as safety and health, objectives. Finally, owners and managers have closer employee contact, a greater influence and more responsibility in their organizations. Many small businesses are family owned or employee owned, making them environments where all employees are invested in the health of the business. The small business owner also has many jobs and responsibilities. Successful small business owners are shrewd in learning and practicing news skills from a variety of specialties.

Smaller businesses present occupational safety and health professionals with a challenging, yet significant, opportunity for prevention. They do not typically have as many TWH policies and programs in place compared with larger businesses. Several intervention studies demonstrate that employers who adopt health protection (Robson et al. 2007; Breslin et al. 2009) and health promotion programs (Kaspin et al. 2013) observe positive changes in health, safety and well-being outcomes. The evidence for integrated TWH approaches is more limited (Anger et al. 2015; Feltner et al. 2016). Of the studies conducted in a small business context, there is promising evidence that these programs help improve workforce health and safety (Breslin et al. 2009). However, small businesses often do not have as robust safety and health programs as large business. For example, Barbeau et al. (2004) found several opportunities for improving small business safety programs including, implementation of a systematic approach, management commitment, employee involvement, and assistance with overcoming barriers such as expertise and time. We found similar results in a review of small business health promotion programs (McCoy et al. 2014).

Total Worker Health policies and programs include a focus on communication strategies that convey to employees that health, safety and well-being are valued and important to the business's mission. In practice, lines of communication tend to be more direct and impactful in small work units, such as small businesses. In the context of TWH, communication efforts focus on physical, psychosocial, work organization, and job tasks and demands hazards (Schill and Chosewood 2013; NIOSH 2016; Sorensen et al. 2016). However, the quality of the policies and programs are often not perceived as adequate or supported in practice, which indicates that employees perceive poor climates for both safety and health (Zohar 2011). Safety climate and health climate reflect employee perceptions of whether safety and health are valued and supported by their organization, and these perceptions act as prompts for how to speak and act at work (Denison 1996; Zohar 2010, 2011). Thus, climate perceptions are important moderators in the relationship between TWH policies and programs, and employee engagement in them. Ultimately, this can result in a poor culture of health and safety (Schneider et al. 2017).

In the small business context, it is also important for there to be worker involvement in TWH efforts. Participatory approaches to TWH recognize the value of employee engagement and leadership in the process of adopting and implementing TWH policies and programs. As beneficiaries of business TWH efforts, workers who have the knowledge and values, and who demonstrate buy-in, can facilitate the transformational and transactional change process (Punnett et al. 2013). Implicit in the participatory process is the inclusion of workers at all stages of change implementation, including during evaluation of change efforts. For example, in small businesses where employees often wear multiple hats, management can benefit from workers leading and voicing concerns and participating in the process to develop, implement and maintain solutions. Thus, while the model in Figure 19.1 emphasizes the influence of many facets of the organization on worker-level outcomes, the influence that small businesses workers can have should not be understated.

The model shown in Figure 19.1 is not intended to imply an exclusively top-down approach to health, safety and well-being. It is more appropriate to consider the significant interdependence and influence exerted by society (for example, customers and government), organizations and workers on one another. First, workers' voice in creating a healthy and safe work environment is necessary (Punnett et al. 2013), and may be especially impactful in small businesses. Second, the organization can influence regional or national values, policy and resources via corporate social responsibility strategies. Research indicates that small business leaders typically serve their communities in leadership positions, by making financial and technical contributions, and contributing to the 'social and political development of the community' (Fitzgerald et al. 2010, p. 545).

Example Small Business TWH Intervention

Recently, we developed a comprehensive small business TWH intervention named Small+Safe+Well (SSWell) that is described in more detail in Schwatka et al. (2018c). SSWell addresses major elements displayed in Figure 19.1, including transactional and transformational change activities to improve TWH leadership, TWH policies and programs, safety and health climates, and, ultimately, worker motivation, behavior, health and injury. The core of our intervention is a business-facing TWH initiative developed and operated by our Center for Health, Work & Environment called Health Links™ (Tenney et al. 2019). Health Links began in 2013 with the aim of giving any enterprise, but especially small businesses, access to online assessment tools, recognition, and advising. As part of the SSWell study, we developed a TWH leadership program to complement existing Health Links services. Our framework for engaging and helping small businesses using Health Links and the complementary TWH leadership program is illustrated in Figure 19.2.

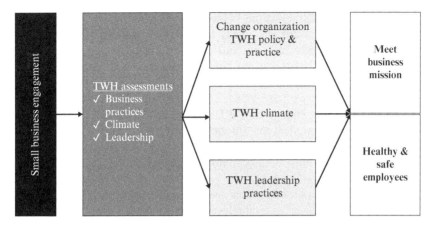

Figure 19.2   SS Well intervention small business engagement strategy

We first engage small businesses through an assessment process. The first survey they complete is the Health Links business assessment, which evaluates their TWH policies and programs in six benchmark areas: organizational supports, workplace assessment, safety policies and programs, health policies and programs, engagement, and evaluation. Second, all employees are invited to complete an anonymous health and safety culture survey. The survey contains validated questions about health and safety climates as well as other indicators of company support for employees and employee engagement in the TWH program. Third, we ask the small business owner or other senior manager to take an online TWH leadership assessment, which we developed based on key leadership theories.

We then provide the small businesses with assessment reports and assistance with how to take action to improve their scores. To facilitate transactional change, we provide short mentoring sessions to help create an action plan for improving their Health Links Healthy Workplace Assessment™ results. The advising session is used to set goals and recommendations based on the results of the assessments. To facilitate transformational change, we provide longer-term TWH leadership training for the small business owner or other senior leader as well as the individual who has managerial or implementation responsibility for TWH in their business. They are invited to participate in a four-month program composed of assessments, in-person training, goal setting, coaching and peer engagement.

This longitudinal study is ongoing. However, there are already some key lessons to take from this study:

- First, and most importantly for researchers, it has proven feasible to design a small business study that tests the potential contributions of both transformational and transactional elements shown in our model as well as target outcomes of relevance at both the organizational and individual levels (Figure 19.1).
- Qualitatively, we have learned that although business owners may be willing to engage in TWH programming, small businesses are under a great deal of financial and time pressure to complete mission-critical work. As a result, any health, safety and well-being intervention must be easy to implement, communicate and translate into action.
- Finally, although businesses are all unique, there are six core benchmarks that have universal relevance to small businesses, regardless of their industry sector, location or operating principles.

CASE STUDIES

The following two case studies offer qualitative examples of how the elements described in Figure 19.1 can be applied to small business TWH research and occupational safety and health practice. The first case study offers a way in which data can be integrated to understand the connections between TWH business policies and programs and employee health. The second case study examines how data can be integrated to understand the connections between TWH leadership, TWH business policies and programs, and employee perceptions of TWH leadership. Both case studies point to the need to integrate data from multiple sources and from multiple levels of the organization in order to address and improve TWH outcomes for businesses and their employees.

Case Study 1: Transactional Intervention Linking Business Practices to Health Outcomes

The question addressed in this case study is, 'Do healthier small businesses have healthier employees?' Over the past five years, we have worked with several hundred small businesses in collaboration with a Colorado state-based workers' compensation insurance carrier. The objective was to determine if addition of a worksite wellness program to existing safety services for small businesses improved health and safety outcomes. A description of this project and study findings can be found in previously published studies (Goetzel et al. 2014; Newman et al. 2015; Jinnett et al. 2017; Schwatka et al. 2017, 2018a, 2018b). Some of the businesses in this program also participated in Health Links. Through Health Links and the previously mentioned workers' compensation initiatives, we have been able to link and compare the organizational- and worker-level data on TWH policies and programs with employee health status of workers in the small businesses that participated in both.

Data from two small Colorado businesses are presented below to illustrate the relationship between the extent of TWH programming and employees' personal health risk factors.

Company A was in the public administration industry. It had 311 employees. Company B was in the healthcare and social assistance industry. It had 212 employees. We collected information on their organizational level TWH policies and programs through the Health Links Business Assessment. We also collected information on their employees' self-report health status via a Wellsource health risk assessment completed by employees (Wellsource 2015) and the Work Performance and Health Questionnaire (Kessler et al. 2003; Goetzel et al. 2014; Newman et al. 2015; Jinnett et al., 2017; Schwatka et al. 2017, 2018a, 2018b) also completed by employees. The central hypothesis of this case study was that small businesses that offer more comprehensive TWH programming have healthier employees than do small businesses that offer less TWH programming.

Key findings

- Organizational data: while both companies had similar levels of organizational supports, company A had more health policies and programs and safety policies and programs. They also did more to engage their workers in these policies and programs. The only benchmark where company B outperformed company A was workplace assessments (see Figure 19.3).

- Employee health data: company A's employees, on average, reported a better health culture and lower work stress levels. They also reported better overall health, fewer chronic health conditions, more exercise days per week and fewer days of absence owing to illness. Company B employees exceeded employees in company A in self-reported job performance (see Figure 19.4).

- Data integration: while we cannot infer causality in this case study, we observe that employees who work for a company with more organizational TWH policies and programs report better health than employees who work for a company with fewer TWH policies and programs.

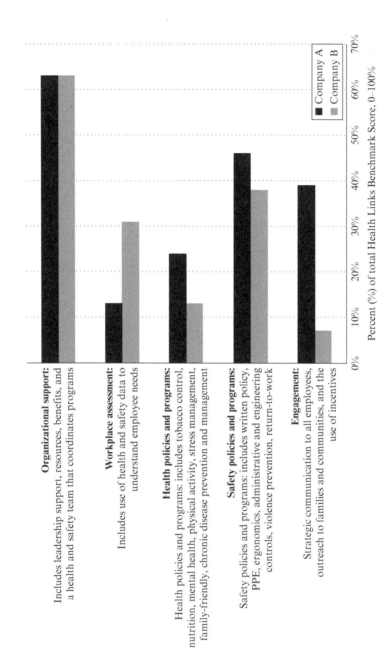

Organizational support:

Includes leadership support, resources, benefits, and a health and safety team that coordinates programs

Workplace assessment:

Includes use of health and safety data to understand employee needs

Health policies and programs:

Health policies and programs: includes tobacco control, nutrition, mental health, physical activity, stress management, family-friendly, chronic disease prevention and management

Safety policies and programs:

Safety policies and programs: includes written policy, PPE, ergonomics, administrative and engineering controls, violence prevention, return-to-work

Engagement:

Strategic communication to all employees, outreach to families and communities, and the use of incentives

0%   10%   20%   30%   40%   50%   60%   70%

Percent (%) of total Health Links Benchmark Score, 0–100%

■ Company A
■ Company B

Figure 19.3   Case study 1: comparison of Health Links Business Assessment results by company

Work-related health indicators

	Health culture (Best culture = 4.0)	Stress at work (Low stress = 4.0)
Company A	2.5	2.0
Company B	2.1	1.4

Health indicators

	Overall health (Best health = 4.0)	Number of chronic conditions	Number of > 30 minute exercise days per week
Company A	1.7	1.8	2.9
Company B	1.2	3.3	2.5

Productivity indicators

	Job performance (Best performance = 10)	Number of sick hours in the past month
Company A	8.1	2.7
Company B	8.4	3.2

Figure 19.4   Case study 1: comparison of health risk assessments results by company

Case Study 2: Transformational Intervention Linking TWH Leadership to Business Practices and Health Outcomes

The questions addressed in this case study are (1) do small businesses owners who report engaging in more TWH leadership practices adopt more TWH policies and programs? (2) Do those who work for small businesses owners who engage in more TWH leadership practices report better perceptions of leadership commitment to TWH policies and programs? The four businesses in this case study participated in leadership training designed to help small business owners develop their TWH leadership skills to ensure the effectiveness of their TWH policies and programs.

Company A was in construction with 248 employees. Company B was in technology with 30 employees. Company C was in healthcare with 219 employees. Company D was in education with 75 employees.

In this case study, we were able to link the Health Links business benchmarks data to specific company leaders and to the companies' worker-level assessment data. The leaders in the training completed a self-assessment of their TWH leadership practices (for example, leading by example). All of their employees were invited to participate in a health and safety culture survey. For the purposes of this case study, only one facet of that survey is presented: perceptions of leadership commitment to TWH. This was measured by adapting questions from the Health Links organizational supports benchmark to the employee level.

Key findings

- Small businesses that scored well in one area (for example, company C scored a 93 percent in organizational supports benchmark) did not necessarily score well in another area (for example, company C scored a 42 percent in workplace assessment benchmark). This suggests a need to understand small business strengths and weaknesses and then provide specific, customized assistance.
- Across all three assessment areas – Health Links business benchmarks, leader self-rating of TWH leadership, and employee perceptions of leadership commitment to TWH – company C generally had the best scores and company D had the worst. This suggests some degree of correlation between all three assessment areas.
- When comparing the leader self-rating of TWH leadership and the Health Links benchmarks, we observed a positive correlation whereby leaders who rated their TWH leadership skills more positively had more TWH policies and programs in place. This suggests that small business owners who possess positive TWH leadership attributes have small businesses that do more to protect and promote their employees' health and safety, as reflected by policies and programs.
- There is a positive correlation between leaders' self-rating of TWH leadership and employees' perceptions of leadership commitment to TWH. This suggests that employees experience the effects of a company's TWH supportive behaviors and demonstrations of commitment.

In addition to the survey assessments described above, we also collected immediate post-training evaluation surveys. The leaders in the training appreciated the awareness that the assessments brought around how their business goes about TWH (see Box 19.1). However, leaders indicated that there may be several barriers to implementing TWH leadership practices. All leaders indicated that time pressures would be a barrier and 80 percent of leaders indicated that stress and too much work would be barriers as well. Some (40 percent) even indicated that fatigue would be a barrier. Thus, in practice, small business leaders need realistic strategies to incorporate TWH leadership practices into their daily work flow.

BOX 19.1   QUOTES FROM LEADERS IN THE TWH
LEADERSHIP TRAINING

'As a CEO who truly cares about my staff, I thought I had a pretty good awareness
of how to support them through our health and safety efforts. However, what I
learned through this training is how to deeply root wellness and safety in our culture
by listening to staff and empowering them to help shape their work environment.'
(CEO of a small healthcare company)

'[This] training helped us understand that there are many factors that affect worker
productivity ... I personally found it interesting and helpful to recognize that the
techniques that work for improving workplace safety can be used to improve
mental health or personal wellness and vice versa.' (Co-owner of a small construction company)

CONCLUSIONS AND RESEARCH NEEDS

From a research standpoint, the field of TWH dissemination and implementation science is still in its infancy, especially with regard to the needs
of workers in small enterprises (Anger et al. 2015; Bradley et al. 2016).
Even those few TWH interventions that have demonstrated efficacy in
controlled research settings have rarely been evaluated for effectiveness
when taken into field application. Interventions and intervention study
designs are needed that take into consideration the very pragmatic issues
of an intervention's reach, implementation, adoption and maintenance.
Theoretical frameworks should be considered, applied and assessed.

Small businesses, in particular, are in need of practical interventions
that can promote safety, health and well-being of employees. They face
some unique challenges, but also present occupational safety and health
practitioners with opportunities to have major impact. By nature, TWH
interventions attend to the multilevel influences on workforce health,
safety and well-being. Through these types of TWH interventions, there
are opportunities to help small business company and leadership support
as well as workforce participation, through helping them connect to community resources. Furthermore, small business TWH interventions that
focus on specific transactional activities for wellness or safety are less likely
to produce sustained benefits compared with interventions that seek to
effect transformational system change. To demonstrate benefits, evaluation
metrics should be employed, regardless of business size.

Given the importance of safety and health climate and leadership,
small business research is needed to more fully characterize climate

perception and leadership differences between small and large businesses (Cunningham et al. 2014). To our knowledge, there is no research that explicitly compares safety climate or health climate perceptions by size of business. When considering climate differences by size of business it is important to consider differences in both strength (that is, employee consensus) and level (that is, quality) (Beus et al. 2010). Research is needed to better understand how factors such as a company's network size, its relationship with community, the level of employee interaction and development of shared perceptions affect TWH adoption and effectiveness. Research is also needed to characterize differences between small and large businesses in how consistently policies are implemented and how implementation, including communication, affect health and safety outcomes. While promising data are emerging regarding the role and outsized impact of small business leaders on employee health and safety, further research is also needed to understand how TWH leadership at top leadership, mid-level management and supervisor levels affect outcomes. Top leadership, in particular, may be especially influential in a small business context. Ultimately, there is a need for pragmatic research examining whether the creation of TWH policies and programs result in improved health and safety of workers. Rather than extrapolating from the experience of large businesses, researchers need to test fundamental premises about workplace safety, health and well-being at the small enterprise level.

REFERENCES

Anger, W.K., Elliot, D.L., Bodner, T., Olson, R., Rholman, D.S., Truxillo, D.M., et al. (2015), 'Effectiveness of Total Worker Health interventions', *Journal of Occupational Health Psychology*, **20** (2), 226–47, doi:10.1037/a0038340.supp.

Barbeau, E., Roelofs, C., Youngstrom, R., Sorensen, G., Stoddard, A. and LaMontagne, A.D. (2004), 'Assessment of occupational safety and health programs in small businesses', *American Journal of Industrial Medicine*, **45** (4), 371–9, doi:10.1002/ajim.10336.

Baxter, S., Sanderson, K., Venn, A.J., Blizzard, C.L. and Palmer, A.J. (2014), 'The relationship between return on investment and quality of study methodology in workplace health promotion programs', *American Journal of Health Promotion*, **28** (6), 347–63, doi:10.4278/ajhp.130731-LIT-395.

Beus, J.M., Bergman, M.E. and Payne, S.C. (2010), 'The influence of organizational tenure on safety climate strength: a first look', *Accident Analysis and Prevention*, **42** (5), 1431–7, doi:10.1016/j.aap.2009.06.002.

Bradley, C.J., Grossman, D.C., Hubbard, R.A., Ortega, A.N, and Curry, S.J. (2016), 'Integrated interventions for improving total worker health: a panel report from the National Institutes of Health Pathways to Prevention workshop: Total Worker Health – what's work got to do with it? *Annals of Internal Medicine*, **165** (4), 279–83, doi:10.7326/M16-0740.

Breslin, F.C., Kyle, N., Bigelow, P., Irvin, E., Morassaei, S., MacEachen, E., et al. (2009), 'Effectiveness of health and safety in small enterprises: a systematic review of quantitative evaluations of interventions', *Journal of Occupational Rehabilitation*, **20** (2), 163–79, doi:10.1007/s10926-009-9212-1.

Burke, M.J. and Signal, S.M. (2010), 'Workplace safety: a multilevel, interdisciplinary perspective', in H. Liao, J. Martocchio and A. Joshi (eds), *Research in Personnel and Human Resources Management*, vol. 29, Bingley: Emerald Group, pp. 1–47.

Burke, W. and Litwin, G. (1992), 'A causal model of organizational performance', *Journal of Management*, **18** (3), 532–45.

Burton, J. (2010), 'WHO healthy workplace framework and model: background and supporting literature and practices', February, WHO, Geneva, accessed 24 May 2019 at https://www.who.int/occupational_health/healthy_workplace_frame work.pdf.

Chari, R., Chang, C.C., Sauter, S.L., Petrun Sayers, E.L., Cerully, J.L., Schulte, P., et al. (2018), 'Expanding the paradigm of occupational safety and health: a new framework for worker well-being', *Journal of Occupational and Environmental Medicine*, **60** (7), 589–93, doi:10.1097/JOM.0000000000001330.

Clarke, S. (2013), 'Safety leadership: a meta-analytic review of transformational and transactional leadership styles as antecedents of safety behaviours', *Journal of Occupational and Organizational Psychology*, **86** (1), 22–49, doi:10.1111/j.2044-8325.2012.02064.x.

Concha-Barrientos, M., Nelson, D.I., Fingerhut, M., Driscoll, T. and Leigh, J. (2005), 'The global burden due to occupational injury', *American Journal of Industrial Medicine*, **48** (6), 470–81.

Cunningham, T.R., Sinclair, R. and Schulte, P. (2014), 'Better understanding the small business construct to advance research on delivering workplace health and safety', *Small Enterprise Research*, **21** (2), 148–60, doi:10.1080/13215906.2014.1 1082084.

Denison, D.R. (1996), 'What is the difference between organizational culture and organizational climate? A native's point of view on a decade of paradigm wars', *Academy of Management Review*, **21** (3), 619–54, doi:10.2307/258997.

Feltner, C., Peterson, K., Weber, R.P., Cluff, L., Coker-Schwimmer, E., Viswanathan, M., et al. (2016), 'The effectiveness of Total Worker Health interventions: a systematic review for a National Institutes of Health Pathways to Prevention workshop', *Annals of Internal Medicine*, **165** (4), 262–9, doi:10.7326/M16-0626.

Fitzgerald, M.A., Haynes, G.W., Schrank, H.L. and Danes, S.M. (2010), 'Socially responsible processes of small family business owners: exploratory evidence from the National Family Business Survey', *Journal of Small Business Management*, **48** (4), 524–51.

Goetzel, R.Z., Pei, X., Tabrizi, M.J., Henke, R.M., Kowlessar, N.M., Nelson, C.F., et al. (2012), 'Ten modifiable health risk factors are linked to more than one-fifth of employer-employee health care spending', *Health Affairs (Millwood)*, **31** (11), 2474–84, doi:10.1377/hlthaff.2011.0819.

Goetzel, R.Z., Tabrizi, M., Henke, R.M., Benevent, R., Brockbank, C.V., Stinson, K., et al. (2014), 'Estimating the return on investment from a health risk management program offered to small Colorado-based employers', *Journal of Occupational and Environmental Medicine*, **56** (5), 554–60, doi:10.1097/ JOM.0000000000000152.

Gray, J.H., Densten, I.L. and Sarros, J.C. (2003), 'Size matters: organizational culture in small, medium, and large Australian organizations', *Journal of Small Business & Entrepreneurship*, **17** (1), 31–46.

International Labour Organization (2006), International Labour Conference, 95th Session, 31 May–16 June, accessed 15 December 2018 at https://www.ilo.org/public/english/standards/relm/ilc/ilc95/index.htm.

International Labour Organization (2015), 'Small and medium-sized enterprises and decent and productive employment creation', conference paper, 1 April, accessed 15 December 2018 at https://www.ilo.org/ilc/ILCSessions/104/reports/reports-to-the-conference/WCMS_358294/lang--en/index.htm.

International Social Security Association (2018), 'Vision Zero – ISSA launches global campaign for zero accidents, diseases and harm at work', accessed 15 December 2018 at https://www.issa.int/en/-/vision-zero-issa-launches-global-campaign-for-zero-accidents-diseases-and-harm-at-work.

Jinnett, K., Schwatka, N., Tenney, L., Brockbank, C.V. and Newman, L.S. (2017), 'Chronic conditions, workplace safety, and job demands contribute to absenteeism and job performance', *Health Affairs (Millwood)*, **36** (2), 237–44, doi:10.1377/hlthaff.2016.1151.

Kaspin, L.C., Gorman, K.M. and Miller, R.M. (2013), 'Systematic review of employer-sponsored wellness strategies and their economic and health-related outcomes', *Population Health Management*, **16** (1), 14–21, doi:10.1089/pop.2012.0006.

Kessler, R.C., Barber, C., Beck, A., Berglund, P., Cleary, P.D., McKenas, D., et al. (2003), 'The World Health Organization Health and Work Performance Questionnaire (HPQ)', *Journal of Occupational and Environmental Medicine*, **45** (2), 156–74.

McCoy, K., Stinson, K., Scott, K., Tenney, L. and Newman, L.S. (2014), 'Health promotion in small business: a systematic review of factors influencing adoption and effectiveness of worksite wellness programs', *Journal of Occupational and Environmental Medicine*, **56** (6), 579–87, doi:10.1097/JOM.0000000000000171.

McLellan, D.L., Williams, J.A., Katz, J.N., Pronk, N.P., Wagner, G.R., Caban-Martinez, A.J., et al. (2017), 'Key organizational characteristics for integrated approaches to protect and promote worker health in smaller enterprises', *Journal of Occupational and Environmental Medicine*, **59** (3), 289–94, doi:10.1097/JOM.0000000000000949.

National Business Group on Health (2014), 'Value of investment in employee health, productivity, and well-being', accessed 15 December 2018 at https://www.businessgrouphealth.org/tools-resources/toolkits/value-of-investment/.

Newman, L.S., Stinson, K.E., Metcalf, D., Fang, H., Brockbank, C.V., Jinnett, K., et al. (2015), 'Implementation of a worksite wellness program targeting small businesses: the Pinnacol Assurance Health Risk Management Study', *Journal of Occupational and Environmental Medicine*, **57** (1), 14–21, doi:10.1097/JOM.0000000000000279.

National Institute for Occupational Safety and Health (NIOSH) (2016), 'Fundamentals of *Total Worker Health*® approaches: essential elements for advancing worker safety, health, and well-being', DHHS (NIOSH) Publication No. 2017-112, accessed 21 May 2019 at https://www.cdc.gov/niosh/docs/2017-112/pdfs/2017_112.pdf.

National Institute for Occupational Safety and Health (NIOSH) (2018), 'What is *Total Worker Health*?', accessed 15 December 2018 at https://www.cdc.gov/niosh/twh/default.html.

Occupational Safety and Health Administration (OSHA) (2018), 'Business case for safety and health', accessed 15 December 2018 at https://www.osha.gov/dcsp/products/topics/businesscase/benefits.html.

Papadopoulos, G., Riama, S., Alajaasko, P., Salah-Eddine, Z., Airaksinen, A. and Luomaranta, H. (2018), 'Statistics on small and medium-sized enterprises', Eurostat, accessed 15 December 2018 at https://ec.europa.eu/eurostat/statistics-explained/index.php/Statistics_on_small_and_medium-sized_enterprises.

Punnett, L., Warren, N., Henning, R., Nobrega, S., Cherniack, M. and CPH-NEW Research Team (2013), 'Participatory ergonomics as a model for integrated programs to prevent chronic disease', *Journal of Occupational and Environmental Medicine*, **55** (12) supplement, S19–S24. doi:10.1097/JOM.0000000000000040.

Robert Wood Johnson Foundation (2016), 'Why healthy communities matter to business', 4 May, accessed 15 December 2018 at https://www.rwjf.org/en/library/research/2016/05/why-healthy-communities-matter-to-business.html.

Robson, L.S., Clarke, J.A., Cullen, K., Bielecky, A., Sèverin, C., Bigelow, P.L., et al. (2007), 'The effectiveness of occupational health and safety management system interventions: a systematic review', *Safety Science*, **45** (3), 329–53. doi:10.1016/j.ssci.2006.07.003.

Schein, E.H. (2010), *Organizational Culture and Leadership*. San Francisco, CA: Jossey-Bass.

Schill, A.L. and Chosewood, L.C. (2013), 'The NIOSH Total Worker Health™ program', *Journal of Occupational and Environmental Medicine*, **55** (12) supplement, S8–S11, doi:10.1097/JOM.0000000000000037.

Schneider, B., Gonzalez-Roma, V., Ostroff, C. and West, M.A. (2017), 'Organizational climate and culture: reflections on the history of the constructs in the Journal of Applied Psychology', *Journal of Applied Psychology*, **102** (3), 468–82, doi:10.1037/apl0000090.

Schulte, P.A., Pandalai, S., Wulsin, V. and Chun, H. (2012), 'Interaction of occupational risk factors in workforce health and safety', *American Journal of Public Health*, **102** (3), 434–48, doi:10.2105/AJPH.2011.

Schwatka, N., Atherly, A., Dally, M.J., Fang, H., Brockbank, C.V., Tenney, L., et al. (2017), 'Health risk factors as predictors of workers' compensation claim occurrence and cost', *Occupational and Environmental Medicine*, **74** (1), 14–23, doi:10.1136/oemed-2015-103334.

Schwatka, N., Shore, E., Atherly, A., Weitzenkamp, D., Dally, M., Brockbank, C.V., et al. (2018a), 'Reoccurring injury, chronic health conditions, and behavioral health: gender differences in the causes of workers' compensation claims', *Journal of Occupational and Environmental Medicine*, **60** (8), 710–16, doi:10.1097/JOM.0000000000001301.

Schwatka, N., Smith, D., Weitzenkamp, D., Atherly, A., Dally, M.J., Brockbank, C.V, et al. (2018b), 'The impact of worksite wellness programs by size of business: a three-year longitudinal study of participation, health benefits, absenteeism, and presenteeism', *Annals of Work Exposures and Health*, **62** (1) supplement, S42–S54.

Schwatka, N., Tenney, L., Dally, M., Scott, J., Brown, C., Weitzenkamp, D., et al. (2018c), 'Small business Total Worker Health: a conceptual and methodological approach to facilitating organizational change', *Occupational Health Science*, **2** (1), 25–41.

Sorensen, G., Landsbergis, P.A., Hammer, L., Amick, B.C., Linnan, L., Yancey, A., et al. (2011), 'Preventing chronic disease in the workplace: a workshop report and recommendations', *American Journal of Public Health*, **101** (1) supplement, S196–S207.

Sorensen, G., McLellan, D., Sabbath, E., Dennerlein, J., Nagler, E., Hurtado, D., et al. (2016), 'Integrating worksite health protection and health promotion: a conceptual model for intervention and research', *Preventative Medicine*, **91** (October), 188–96, doi:10.1016/j.ypmed.2016.08.005.

Storey, D.J. (1994), *Understanding the Small Business Sector*, London and New York: Routledge.

Sustainable Development Goals Fund, Harvard Kennedy School: Corporate Social Responsibility Initiative, & Business Fights Poverty (2015), 'Business and the United Nations: working together towards the sustainable development goals: a framework for action', accessed 15 December 2018 at https://www.sdgfund.org/sites/default/files/business-and-un/SDGF_BFP_HKSCSRI_Business_and_SDGs-Web_Version.pdf.

Tenney, L., Fan, W., Dally, M., Scott, J., Haan, M., Rivera, K., Newman, M. and Newman, L. (2019), 'Health Links™ assessment of Total Worker Health® practices as indicators of organizational behavior in small business', *Journal of Occupational & Environmental Medicine*, doi:10.1097/JOM.0000000000001623.

Thompson, J., Schwatka, N., Tenney, L. and Newman, L. (2018), 'Total Worker Health: a small business leader perspective', *International Journal of Environmental Research and Public Health*, **15** (11), 2416, doi:10.3390/ijerph15112416.

US Small Business Administration (2018), 'Size standards: the SBA's size standards determine whether or not your business qualifies as small', accessed 15 December 2018 at https://www.sba.gov/federal-contracting/contracting-guide/size-standards.

United States Census Bureau (2016), 'Statistics of U.S. businesses main', accessed 15 December 2018 at http://www.census.gov/data/tables/2014/econ/susb/2014-susb-annual.html.

Wellsource (2015), 'What is a health risk assessment?', accessed 15 December 2018 at http://www.wellsource.com/health-risk-assessments.html.

World Health Organization (WHO) (2007), 'Workers' health: global plan of action', Sixtieth World Health Assembly, ref. no. WHA60.26, accessed 24 May 2019 at https://www.who.int/occupational_health/publications/global_plan/en/.

Zohar, D. (2010), 'Thirty years of safety climate research: reflections and future directions', *Accident Analysis and Prevention*, **42** (5), 1517–22, doi:10.1016/j.aap.2009.12.019.

Zohar, D. (2011), 'Safety climate: conceptual and measurement issues' in J.C. Quick and L.E. Tretrick (eds), *Handbook of Occupational Health Psychology*, 2nd edn, Washington, DC: APA, pp. 141–64.

Index

Note: OH&S is an abbreviation for occupational health and safety.